Mathematical Logic for Computer Science

Mordechai Ben-Ari

Mathematical
Logic for
Computer Science

Third Edition

 Springer

Prof. Mordechai (Moti) Ben-Ari
Department of Science Teaching
Weizmann Institute of Science
Rehovot, Israel

ISBN 978-1-4471-4128-0   ISBN 978-1-4471-4129-7 (eBook)
DOI 10.1007/978-1-4471-4129-7
Springer London Heidelberg New York Dordrecht

Library of Congress Control Number: 2012941863

1st edition: © Prentice Hall International Ltd. 1993
© Springer-Verlag London 2009, 2012

Printed on acid-free paper

Springer is part of Springer Science+Business Media (www.springer.com)

For Anita

Preface

Students of science and engineering are required to study mathematics during their first years at a university. Traditionally, they concentrate on calculus, linear algebra and differential equations, but in computer science and engineering, logic, combinatorics and discrete mathematics are more appropriate. Logic is particularly important because it is the mathematical basis of software: it is used to formalize the semantics of programming languages and the specification of programs, and to verify the correctness of programs.

Mathematical Logic for Computer Science is a *mathematics* textbook, just as a first-year calculus text is a mathematics textbook. A scientist or engineer needs more than just a facility for manipulating formulas and a firm foundation in mathematics is an excellent defense against technological obsolescence. Tempering this requirement for mathematical competence is the realization that applications use only a fraction of the theoretical results. Just as the theory of calculus can be taught to students of engineering without the full generality of measure theory, students of computer science need not be taught the full generality of uncountable structures. Fortunately (as shown by Raymond M. Smullyan), tableaux provide an elegant way to teach mathematical logic that is both theoretically sound and yet sufficiently elementary for the undergraduate.

Audience

The book is intended for undergraduate computer science students. No specific mathematical knowledge is assumed aside from informal set theory which is summarized in an appendix, but elementary knowledge of concepts from computer science (graphs, languages, programs) is used.

Organization

The book can be divided into four parts. Within each part the chapters should be read sequentially; the prerequisites between the parts are described here.

Propositional Logic: Chapter 2 is on the syntax and semantics of propositional logic. It introduces the method of semantic tableaux as a decision procedure for the logic. This chapter is an essential prerequisite for reading the rest of the book. Chapter 3 introduces deductive systems (axiomatic proof systems). The next three chapters present techniques that are used in practice for tasks such as automatic theorem proving and program verification: Chap. 4 on resolution, Chap. 5 on binary decision diagrams and Chap. 6 on SAT solvers.

First-Order Logic: The same progression is followed for first-order logic. There are two chapters on the basic theory of the logic: Chap. 7 on syntax, semantics and semantic tableaux, followed by Chap. 8 on deductive systems. Important application of first-order logic are automatic theorem proving using resolution (Chap. 10) and logic programming (Chap. 11). These are preceded by Chap. 9 which introduces an essential extension of the logic to terms and functions. Chapter 12 surveys fundamental theoretical results in first-order logic. The chapters on first-order logic assume as prerequisites the corresponding chapters on propositional logic; for example, you should read Chap. 4 on resolution in the propositional logic before the corresponding Chap. 10 in first-order logic.

Temporal Logic: Again, the same progression is followed: Chap. 13 on syntax, semantics and semantic tableaux, followed by Chap. 14 on deductive systems. The prerequisites are the corresponding chapters on propositional logic since first-order temporal logic is not discussed.

Program Verification: One of the most important applications of mathematical logic in computer science is in the field of program verification. Chapter 15 presents a deductive system for the verification of sequential programs; the reader should have mastered Chap. 3 on deductive systems in propositional logic before reading this chapter. Chapter 16 is highly dependent on earlier chapters: it includes deductive proofs, the use of temporal logic, and implementations using binary decision diagrams and satisfiability solvers.

Supplementary Materials

Slides of the diagrams and tables in the book (in both PDF and LaTeX) can be downloaded from http://www.springer.com/978-1-4471-4128-0, which also contains instructions for obtaining the answers to the exercises (qualified instructors only). The source code and documentation of Prolog programs for most of the algorithms in the book can be downloaded from http://code.google.com/p/mlcs/.

Third Edition

The third edition has been totally rewritten for clarity and accuracy. In addition, the following major changes have been made to the content:

- The discussion of logic programming has been shortened somewhat and the Prolog programs and their documentation have been removed to a freely available archive.
- The chapter on the $\mathscr{Z}$ notation has been removed because it was difficult to do justice to this important topic in a single chapter.
- The discussion of model checking in Chap. 16 has been significantly expanded since model checking has become a widely used technique for program verification.
- Chapter 6 has been added to reflect the growing importance of SAT solvers in all areas of computer science.

Notation

If and only if is abbreviated *iff*. Definitions by convention use *iff* to emphasize that the definition is restrictive. For example: A natural number is even iff it can be expressed as $2k$ for some natural number k. In the definition, *iff* means that numbers expressed as $2k$ are even and these are the only even numbers.

Definitions, theorems and examples are consecutively numbered within each chapter to make them easy to locate. The end of a definition, example or proof is denoted by ■.

Advanced topics and exercises, as well as topics outside the mainstream of the book, are marked with an asterisk.

Acknowledgments

I am indebted to Jørgen Villadsen for his extensive comments on the second edition which materially improved the text. I would like to thank Joost-Pieter Katoen and Doron Peled for reviewing parts of the manuscript. I would also like to thank Helen Desmond, Ben Bishop and Beverley Ford of Springer for facilitating the publication of the book.

Rehovot, Israel   Mordechai (Moti) Ben-Ari

Contents

Chapter 1
Introduction

1.1 The Origins of Mathematical Logic

Logic formalizes valid methods of reasoning. The study of logic was begun by the ancient Greeks whose educational system stressed competence in reasoning and in the use of language. Along with rhetoric and grammar, logic formed part of the *trivium*, the first subjects taught to young people. Rules of logic were classified and named. The most widely known set of rules are the *syllogisms*; here is an example of one form of syllogism:

Premise All rabbits have fur.
Premise Some pets are rabbits.
Conclusion Some pets have fur.

If both premises are true, the rules ensure that the conclusion is true.

Logic must be formalized because reasoning expressed in informal natural language can be flawed. A clever example is the following 'syllogism' given by Smullyan (1978, p. 183):

Premise Some cars rattle.
Premise My car is some car.
Conclusion My car rattles.

The formalization of logic began in the nineteenth century as mathematicians attempted to clarify the foundations of mathematics. One trigger was the discovery of non-Euclidean geometries: replacing Euclid's parallel axiom with another axiom resulted in a different theory of geometry that was just as consistent as that of Euclid. Logical systems—axioms and rules of inference—were developed with the understanding that different sets of axioms would lead to different theorems. The questions investigated included:

Consistency A logical system is consistent if it is impossible to prove both a formula and its negation.
Independence The axioms of a logical system are independent if no axiom can be proved from the others.

M. Ben-Ari, *Mathematical Logic for Computer Science*,
DOI 10.1007/978-1-4471-4129-7_1, © Springer-Verlag London 2012

Soundness All theorems that can be proved in the logical system are true.
Completeness All true statements can be proved in the logical system.

Clearly, these questions will only make sense once we have formally defined the
central concepts of *truth* and *proof*.

During the first half of the twentieth century, logic became a full-fledged topic
of modern mathematics. The framework for research into the foundations of mathematics was called *Hilbert's program*, (named after the great mathematician David
Hilbert). His central goal was to prove that mathematics, starting with arithmetic,
could be axiomatized in a system that was both consistent and complete. In 1931,
Kurt Gödel showed that this goal cannot be achieved: any consistent axiomatic system for arithmetic is incomplete since it contains true statements that cannot be
proved within the system.

In the second half of the twentieth century, mathematical logic was applied in
computer science and has become one of its most important theoretical foundations.
Problems in computer science have led to the development of many new systems
of logic that did not exist before or that existed only at the margins of the classical
systems. In the remainder of this chapter, we will give an overview of systems of
logic relevant to computer science and sketch their applications.

1.2 Propositional Logic

Our first task is to formalize the concept of the *truth* of a statement. Every statement
is assigned one of two values, conventionally called *true* and *false* or *T* and *F*.
These should be considered as arbitrary symbols that could easily be replaced by
any other pair of symbols like 1 and 0 or even ♣ and ♠.

Our study of logic commences with the study of *propositional logic* (also called
the *propositional calculus*). The *formulas* of the logic are built from *atomic propositions*, which are statements that have no internal structure. Formulas can be combined using *Boolean operators*. These operators have conventional names derived
from natural language (*and, or, implies*), but they are given a formal meaning in the
logic. For example, the Boolean operator *and* is defined as the operator that gives
the value *true* if and only if applied to two formulas whose values are *true*.

Example 1.1 The statements 'one plus one equals two' and 'Earth is farther from the
sun than Venus' are both true statements; therefore, by definition, so is the following
statement:

'one plus one equals two' *and* 'Earth is farther from the sun than Venus'.

Since 'Earth is farther from the sun than Mars' is a false statement, so is:

'one plus one equals two' *and* 'Earth is farther from the sun than Mars'. ∎

Rules of *syntax* define the legal structure of formulas in propositional logic. The
semantics—the meaning of formulas—is defined by *interpretations*, which assign

one of the *(truth) values T* or *F* to every atomic proposition. For every legal way that a formula can be constructed, a semantical rule specifies the truth value of the formula based upon the values of its constituents.

Proof is another syntactical concept. A proof is a deduction of a formula from a set of formulas called *axioms* using *rules of inference*. The central theoretical result that we prove is the soundness and completeness of the axiom system: the set of provable formulas is the same as the set of formulas which are always true.

Propositional logic is central to the design of computer hardware because hardware is usually designed with components having two voltage levels that are arbitrarily assigned the symbols 0 and 1. Circuits are described by idealized elements called *logic gates*; for example, an and-gate produces the voltage level associated with 1 if and only if both its input terminals are held at this same voltage level.

Example 1.2 Here is a *half-adder* constructed from and, or- and not-gates.

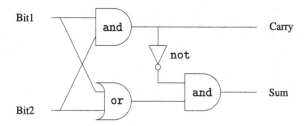

The half-adder adds two one-bit binary numbers and by joining several half-adders we can add binary numbers composed of many bits. ∎

Propositional logic is widely used in software, too. The reason is that any program is a finite entity. Mathematicians may consider the natural numbers to be infinite $(0, 1, 2, \ldots)$, but a word of a computer's memory can only store numbers in a finite range. By using an atomic proposition for each bit of a program's state, the meaning of a computation can be expressed as a (very large) formula. Algorithms have been developed to study properties of computations by evaluating properties of formulas in propositional logic.

1.3 First-Order Logic

Propositional logic is not sufficiently expressive for formalizing mathematical theories such as arithmetic. An arithmetic expression such as $x + 2 > y - 1$ is neither true nor false: (a) its truth depends on the values of the *variables x* and *y*; (b) we need to formalize the meaning of the operators $+$ and $-$ as *functions* that map a pair of numbers to a number; (c) *relational* operators like $>$ must be formalized as mapping pairs of numbers into truth values. The system of logic that can be interpreted by values, functions and relations is called *first-order logic* (also called *predicate logic* or the *predicate calculus*).

The study of the foundations of mathematics emphasized first-order logic, but it has also found applications in computer science, in particular, in the fields of automated theorem proving and logic programming. Can a computer carry out the work of a mathematician? That is, given a set of axioms for, say, number theory, can we write software that will find proofs of known theorems, as well as statements and proofs of new ones? With luck, the computer might even discover a proof of Goldbach's Conjecture, which states that every even number greater than two is the sum of two prime numbers:

$$4 = 2 + 2, \quad 6 = 3 + 3, \quad \ldots,$$

$$100 = 3 + 97, \quad 102 = 5 + 97, \quad 104 = 3 + 101, \quad \ldots.$$

Goldbach's Conjecture has not been proved, though no counterexample has been found even with an extensive computerized search.

Research into automated theorem proving led to a new and efficient method of proving formulas in first-order logic called *resolution*. Certain restrictions of resolution have proved to be so efficient they are the basis of a new type of programming language. Suppose that a theorem prover is capable of proving the following formula:

Let A be an array of integers. Then there exists an array A' such that the elements of A' are a permutation of those of A, and such that A' is ordered: $A'(i) \leq A'(j)$ for $i < j$.

Suppose, further, that given any specific array A, the theorem prover constructs the array A' which the required properties. Then the formula is a *program* for sorting, and the proof of the formula generates the *result*. The use of theorem provers for computation is called *logic programming*. Logic programming is attractive because it is *declarative*—you just write what you *want* from the computation—as opposed to classical programming languages, where you have to specify in detail *how* the computation is to be carried out.

1.4 Modal and Temporal Logics

A statement need not be absolutely true or false. The statement 'it is raining' is sometimes true and sometimes false. *Modal logics* are used to formalize statements where finer distinctions need to be made than just 'true' or 'false'. Classically, modal logic distinguished between statements that are *necessarily* true and those that are *possibly* true. For example, $1 + 1 = 2$, as a statement about the natural numbers, is necessarily true because of the way the concepts are defined. But any historical statement like 'Napoleon lost the battle of Waterloo' is only possibly true; if circumstances had been different, the outcome of Waterloo might have been different.

Modal logics have turned out to be extremely useful in computer science. We will study a form of modal logic called *temporal logic*, where 'necessarily' is interpreted as *always* and 'possibly' is interpreted as *eventually*. Temporal logic has turned out to be the preferred logic for program verification as described in the following section.

1.5 Program Verification

One of the major applications of logic to computer science is in *program verification*. Software now controls our most critical systems in transportation, medicine, communications and finance, so that it is hard to think of an area in which we are not dependent on the correct functioning of a computerized system. Testing a program can be an ineffective method of verifying the correctness of a program because we test the scenarios that we think will happen and not those that arise unexpectedly. Since a computer program is simply a formal description of a calculation, it can be verified in the same way that a mathematical theorem can be verified using logic.

First, we need to express a *correctness specification* as a formal statement in logic. Temporal logic is widely used for this purpose because it can express the dynamic behavior of program, especially of *reactive* programs like operating systems and real-time systems, which do not compute an result but instead are intended to run indefinitely.

Example 1.3 The property '*always* not deadlocked' is an important correctness specification for operating systems, as is 'if you request to print a document, *eventually* the document will be printed'.   ∎

Next, we need to formalize the semantics (the meaning) of a program, and, finally, we need a formal system for deducing that the program fulfills a correctness specification. An axiomatic system for temporal logic can be used to prove concurrent programs correct.

For sequential programs, verification is performed using an axiomatic system called *Hoare logic* after its inventor C.A.R. Hoare. Hoare logic assumes that we know the truth of statements of the program's domain like arithmetic; for example, $-(1-x) = (x-1)$ is considered to be an axiom of the logic. There are axioms and rules of inference that concern the structure of the program: assignment statements, loops, and so on. These are used to create a proof that a program fulfills a correctness specification.

Rather than deductively prove the correctness of a program relative to a specification, a *model checker* verifies the truth of a correctness specification in every possible state that can appear during the computation of a program. On a physical computer, there are only a finite number of different states, so this is always possible. The challenge is to make model checking feasible by developing methods and algorithms to deal with the very large number of possible states. Ingenious algorithms and data structures, together with the increasing CPU power and memory of modern computers, have made model checkers into viable tools for program verification.

1.6 Summary

Mathematical logic formalizes reasoning. There are many different systems of logic: propositional logic, first-order logic and modal logic are really families of logic with many variants. Although systems of logic are very different, we approach each logic in a similar manner: We start with their syntax (what constitutes a formula in the logic) and their semantics (how truth values are attributed to a formula). Then we describe the method of *semantic tableaux* for deciding the validity of a formula. This is followed by the description of an axiomatic system for the logic. Along the way, we will look at the applications of the various logics in computer science with emphasis on theorem proving and program verification.

1.7 Further Reading

This book was originally inspired by Raymond M. Smullyan's presentation of logic using semantic tableaux. It is still worthwhile studying Smullyan (1968). A more advanced logic textbook for computer science students is Nerode and Shore (1997); its approach to propositional and first-order logic is similar to ours but it includes chapters on modal and intuitionistic logics and on set theory. It has a useful appendix that provides an overview of the history of logic as well as a comprehensive bibliography. Mendelson (2009) is a classic textbook that is more mathematical in its approach.

Smullyan's books such as Smullyan (1978) will exercise your abilities to think logically! The final section of that book contains an informal presentation of Gödel's incompleteness theorem.

1.8 Exercise

1.1 What is wrong with Smullyan's 'syllogism'?

References

E. Mendelson. *Introduction to Mathematical Logic* (*Fifth Edition*). Chapman & Hall/CRC, 2009.
A. Nerode and R.A. Shore. *Logic for Applications* (*Second Edition*). Springer, 1997.
R.M. Smullyan. *First-Order Logic*. Springer-Verlag, 1968. Reprinted by Dover, 1995.
R.M. Smullyan. *What Is the Name of This Book?—The Riddle of Dracula and Other Logical Puzzles*. Prentice-Hall, 1978.

Chapter 2
Propositional Logic: Formulas, Models, Tableaux

Propositional logic is a simple logical system that is the basis for all others. Propositions are claims like 'one plus one equals two' and 'one plus two equals two' that cannot be further decomposed and that can be assigned a truth value of *true* or *false*. From these *atomic propositions*, we will build complex *formulas* using *Boolean operators*:

> 'one plus one equals two' *and* 'Earth is farther from the sun than Venus'.

Logical systems formalize reasoning and are similar to programming languages that formalize computations. In both cases, we need to define the syntax and the semantics. The *syntax* defines what strings of symbols constitute legal formulas (legal programs, in the case of languages), while the *semantics* defines what legal formulas mean (what legal programs compute). Once the syntax and semantics of propositional logic have been defined, we will show how to construct *semantic tableaux*, which provide an efficient *decision procedure* for checking when a formula is true.

2.1 Propositional Formulas

In computer science, an *expression* denoted the computation of a value from other values; for example, $2 * 9 + 5$. In propositional logic, the term *formula* is used instead. The formal definition will be in terms of trees, because our the main proof technique called structural induction is easy to understand when applied to trees. Optional subsections will expand on different approaches to syntax.

M. Ben-Ari, *Mathematical Logic for Computer Science*,
DOI 10.1007/978-1-4471-4129-7_2, © Springer-Verlag London 2012

2.1.1 Formulas as Trees

Definition 2.1 The symbols used to construct formulas in propositional logic are:

- An unbounded set of symbols $\mathscr{P}$ called *atomic propositions* (often shortened to *atoms*). Atoms will be denoted by lower case letters in the set $\{p, q, r, \ldots\}$, possibly with subscripts.
- *Boolean operators.* Their names and the symbols used to denote them are:

negation	$\neg$
disjunction	$\vee$
conjunction	$\wedge$
implication	$\rightarrow$
equivalence	$\leftrightarrow$
exclusive or	$\oplus$
nor	$\downarrow$
nand	$\uparrow$

The negation operator is a *unary operator* that takes one operand, while the other operators are *binary operators* taking two operands. ■

Definition 2.2 A *formula* in propositional logic is a tree defined recursively:

- A formula is a leaf labeled by an atomic proposition.
- A formula is a node labeled by $\neg$ with a single child that is a formula.
- A formula is a node labeled by one of the binary operators with two children both of which are formulas. ■

Example 2.3 Figure 2.1 shows two formulas. ■

2.1.2 Formulas as Strings

Just as we write expressions as strings (linear sequences of symbols), we can write formulas as strings. The string associated with a formula is obtained by an *inorder traversal* of the tree:

Algorithm 2.4 (Represent a formula by a string)
Input: A formula A of propositional logic.
Output: A string representation of A.

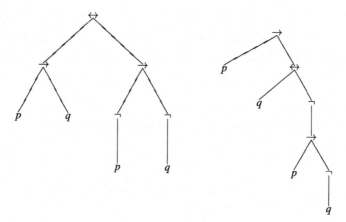

Fig. 2.1 Two formulas

Call the recursive procedure `Inorder(A)`:

```
Inorder(F)
   if F is a leaf
      write its label
      return
   let F1 and F2 be the left and right subtrees of F
   Inorder(F1)
   write the label of the root of F
   Inorder(F2)
```

If the root of F is labeled by negation, the left subtree is considered to be empty and the step `Inorder(F1)` is skipped. ∎

Definition 2.5 The term *formula* will also be used for the string with the understanding that it refers to the underlying tree. ∎

Example 2.6 Consider the left formula in Fig. 2.1. The inorder traversal gives: write the leftmost leaf labeled p, followed by its root labeled $\rightarrow$, followed by the right leaf of the implication labeled q, followed by the root of the tree labeled $\leftrightarrow$, and so on. The result is the string:

$$p \rightarrow q \leftrightarrow \neg p \rightarrow \neg q.$$

Consider now the right formula in Fig. 2.1. Performing the traversal results in the string:

$$p \rightarrow q \leftrightarrow \neg p \rightarrow \neg q,$$

which is precisely the same as that associated with the left formula. ∎

Although the formulas are *not* ambiguous—the trees have entirely different structures—their representations as strings are ambiguous. Since we prefer to deal with strings, we need some way to resolve such ambiguities. There are three ways of doing this.

2.1.3 Resolving Ambiguity in the String Representation

Parentheses

The simplest way to avoid ambiguity is to use parentheses to maintain the structure of the tree when the string is constructed.

Algorithm 2.7 (Represent a formula by a string with parentheses)
Input: A formula A of propositional logic.
Output: A string representation of A.
Call the recursive procedure `Inorder(A)`:

```
Inorder(F)
  if F is a leaf
    write its label
    return
  let F1 and F2 be the left and right subtrees of F
  write a left parenthesis '('
  Inorder(F1)
  write the label of the root of F
  Inorder(F2)
  write a right parenthesis ')'
```

If the root of `F` is labeled by negation, the left subtree is considered to be empty and the step `Inorder(F1)` is skipped.   ∎

The two formulas in Fig. 2.1 are now associated with two different strings and there is no ambiguity:

$$((p \to q) \leftrightarrow ((\neg q) \to (\neg p))),$$

$$(p \to (q \leftrightarrow (\neg (p \to (\neg q)))))).$$

The problem with parentheses is that they make formulas verbose and hard to read and write.

Precedence

The second way of resolving ambiguous formulas is to define *precedence* and *associativity* conventions among the operators as is done in arithmetic, so that we immediately recognize $a * b * c + d * e$ as $(((a * b) * c) + (d * e))$. For formulas the order of precedence from high to low is as follows:

$$\neg$$
$$\wedge, \uparrow$$
$$\vee, \downarrow$$
$$\rightarrow$$
$$\leftrightarrow, \oplus$$

Operators are assumed to associate to the right, that is, $a \vee b \vee c$ means $(a \vee (b \vee c))$.

Parentheses are used only if needed to indicate an order different from that imposed by the rules for precedence and associativity, as in arithmetic where $a * (b + c)$ needs parentheses to denote that the addition is done before the multiplication. With minimal use of parentheses, the formulas above can be written:

$$p \rightarrow q \leftrightarrow \neg q \rightarrow \neg p,$$

$$p \rightarrow (q \leftrightarrow \neg (p \rightarrow \neg q)).$$

Additional parentheses may always be used to clarify a formula: $(p \vee q) \wedge (q \vee r)$.

The Boolean operators $\wedge, \vee, \leftrightarrow, \oplus$ are associative so we will often omit parentheses in formulas that have repeated occurrences of these operators: $p \vee q \vee r \vee s$. Note that $\rightarrow, \downarrow, \uparrow$ are *not* associative, so parentheses must be used to avoid confusion. Although the implication operator is assumed to be right associative, so that $p \rightarrow q \rightarrow r$ unambiguously means $p \rightarrow (q \rightarrow r)$, we will write the formula with parentheses to avoid confusion with $(p \rightarrow q) \rightarrow r$.

Polish Notation *

There will be no ambiguity if the string representing a formula is created by a *preorder traversal* of the tree:

Algorithm 2.8 (Represent a formula by a string in Polish notation)
Input: A formula A of propositional logic.
Output: A string representation of A.
    Call the recursive procedure `Preorder(A)`:

```
Preorder(F)
   write the label of the root of F
   if F is a leaf
      return
   let F1 and F2 be the left and right subtrees of F
   Preorder(F1)
   Preorder(F2)
```

If the root of F is labeled by negation, the left subtree is considered to be empty and the step `Preorder(F1)` is skipped.   ∎

Example 2.9 The strings associated with the two formulas in Fig. 2.1 are:

$$\leftrightarrow \rightarrow p\, q \rightarrow \neg\, p \neg q \, ,$$

$$\rightarrow p \leftrightarrow q \neg \rightarrow p \neg q$$

and there is no longer any ambiguity.   ■

The formulas are said to be in *Polish notation*, named after a group of Polish logicians led by Jan Łukasiewicz.

We find infix notation easier to read because it is familiar from arithmetic, so Polish notation is normally used only in the internal representation of arithmetic and logical expressions in a computer. The advantage of Polish notation is that the expression can be evaluated in the linear order that the symbols appear using a stack. If we rewrite the first formula backwards (*reverse Polish notation*):

$$q \neg p \neg \rightarrow q p \rightarrow \leftrightarrow \, ,$$

it can be directly compiled to the following sequence of instructions of an assembly language:

```
Push q
Negate
Push p
Negate
Imply
Push q
Push p
Imply
Equiv
```

The operators are applied to the top operands on the stack which are then popped and the result pushed.

2.1.4 Structural Induction

Given an arithmetic expression like $a * b + b * c$, it is immediately clear that the expression is composed of two terms that are added together. In turn, each term is composed of two factors that are multiplied together. In the same way, any propositional formula can be classified by its top-level operator.

Definition 2.10 Let $A \in \mathscr{F}$. If A is not an atom, the operator labeling the root of the formula A is the *principal operator* of the A.   ■

Example 2.11 The principal operator of the left formula in Fig. 2.1 is $\leftrightarrow$, while the principal operator of the right formulas is $\rightarrow$.   ■

Structural induction is used to prove that a property holds for *all* formulas. This form of induction is similar to the familiar numerical induction that is used to prove that a property holds for all natural numbers (Appendix A.6). In numerical induction, the *base case* is to prove the property for 0 and then to prove the *inductive step*: assume that the property holds for arbitrary n and then show that it holds for $n + 1$. By Definition 2.10, a formula is either a leaf labeled by an atom or it is a tree with a principal operator and one or two subtrees. The base case of structural induction is to prove the property for a leaf and the inductive step is to prove the property for the formula obtained by applying the principal operator to the subtrees, assuming that the property holds for the subtrees.

Theorem 2.12 (Structural induction) *To show that a property holds for all formulas* $A \in \mathscr{F}$:

1. *Prove that the property holds all atoms* p.
2. *Assume that the property holds for a formula A and prove that the property holds for* $\neg A$.
3. *Assume that the property holds for formulas A_1 and A_2 and prove that the property holds for A_1 op A_2, for each of the binary operators.*

Proof Let A be an arbitrary formula and suppose that (1), (2), (3) have been shown for some property. We show that the property holds for A by numerical induction on n, the height of the tree for A. For $n = 0$, the tree is a leaf and A is an atom p, so the property holds by (1). Let $n > 0$. The subtrees A are of height $n - 1$, so by numerical induction, the property holds for these formulas. The principal operator of A is either negation or one of the binary operators, so by (2) or (3), the property holds for A. ∎

We will later show that all the binary operators can be defined in terms negation and either disjunction or conjunction, so a proof that a property holds for all formulas can be done using structural induction with the base case and only two inductive steps.

2.1.5 Notation

Unfortunately, books on mathematical logic use widely varying notation for the Boolean operators; furthermore, the operators appear in programming languages with a different notation from that used in mathematics textbooks. The following table shows some of these alternate notations.

Operator	Alternates	Java language
$\neg$	$\sim$	!
$\wedge$	&	&, &&
$\vee$		$\mid$, $\mid\mid$
$\rightarrow$	$\supset$, $\Rightarrow$	
$\leftrightarrow$	$\equiv$, $\Leftrightarrow$	
$\oplus$	$\not\equiv$	^
$\uparrow$	$\mid$	

2.1.6 A Formal Grammar for Formulas *

This subsection assumes familiarity with formal grammars.

Instead of defining formulas as trees, they can be defined as strings generated by a context-free formal grammar.

Definition 2.13 Formula in propositional logic are derived from the context-free grammar whose terminals are:

- An unbounded set of symbols $\mathscr{P}$ called *atomic propositions.*
- The *Boolean operators* given in Definition 2.1.

The productions of the grammar are:

$$
\begin{aligned}
\textit{fml} \;\; &::= \;\; p && \text{for any } p \in \mathscr{P} \\
\textit{fml} \;\; &::= \;\; \neg\textit{fml} \\
\textit{fml} \;\; &::= \;\; \textit{fml op fml} \\
\textit{op} \;\; &::= \;\; \vee \mid \wedge \mid \rightarrow \mid \leftrightarrow \mid \oplus \mid \uparrow \mid \downarrow
\end{aligned}
$$

A formula is a word that can be derived from the nonterminal *fml*. The set of all formulas that can be derived from the grammar is denoted $\mathscr{F}$. ■

Derivations of strings (words) in a formal grammar can be represented as trees (Hopcroft et al., 2006, Sect. 4.3). The word generated by a derivation can be read off the leaves from left to right.

Example 2.14 Here is a derivation of the formula $p \rightarrow q \leftrightarrow \neg p \rightarrow \neg q$ in propositional logic; the tree representing its derivation is shown in Fig. 2.2.

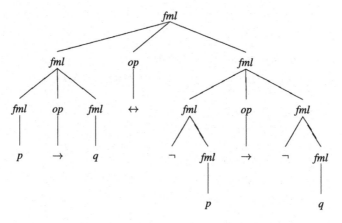

Fig. 2.2 Derivation tree for $p \to q \leftrightarrow \neg\, p \to \neg\, q$

   1. *fml*
   2. *fml op fml*
   3. *fml ↔ fml*
   4. *fml op fml ↔ fml*
   5. *fml → fml ↔ fml*
   6. $p \to$ *fml ↔ fml*
   7. $p \to q \leftrightarrow$ *fml*
   8. $p \to q \leftrightarrow$ *fml op fml*
   9. $p \to q \leftrightarrow$ *fml → fml*
10. $p \to q \leftrightarrow \neg$ *fml → fml*
11. $p \to q \leftrightarrow \neg\, p \to$ *fml*
12. $p \to q \leftrightarrow \neg\, p \to \neg$ *fml*
13. $p \to q \leftrightarrow \neg\, p \to \neg\, q$

                                                                     ■

The methods discussed in Sect. 2.1.2 can be used to resolve ambiguity. We can change the grammar to introduce parentheses:

$$fml \; ::= \; (\neg fml)$$
$$fml \; ::= \; (fml \; op \; fml)$$

and then use precedence to reduce their number.

$v_{\mathscr{I}}(A) = \mathscr{I}_A(A)$	if A is an atom
$v_{\mathscr{I}}(\neg A) = T$	if $v_{\mathscr{I}}(A) = F$
$v_{\mathscr{I}}(\neg A) = F$	if $v_{\mathscr{I}}(A) = T$
$v_{\mathscr{I}}(A_1 \vee A_2) = F$	if $v_{\mathscr{I}}(A_1) = F$ and $v_{\mathscr{I}}(A_2) = F$
$v_{\mathscr{I}}(A_1 \vee A_2) = T$	otherwise
$v_{\mathscr{I}}(A_1 \wedge A_2) = T$	if $v_{\mathscr{I}}(A_1) = T$ and $v_{\mathscr{I}}(A_2) = T$
$v_{\mathscr{I}}(A_1 \wedge A_2) = F$	otherwise
$v_{\mathscr{I}}(A_1 \rightarrow A_2) = F$	if $v_{\mathscr{I}}(A_1) = T$ and $v_{\mathscr{I}}(A_2) = F$
$v_{\mathscr{I}}(A_1 \rightarrow A_2) = T$	otherwise
$v_{\mathscr{I}}(A_1 \uparrow A_2) = F$	if $v_{\mathscr{I}}(A_1) = T$ and $v_{\mathscr{I}}(A_2) = T$
$v_{\mathscr{I}}(A_1 \uparrow A_2) = T$	otherwise
$v_{\mathscr{I}}(A_1 \downarrow A_2) = T$	if $v_{\mathscr{I}}(A_1) = F$ and $v_{\mathscr{I}}(A_2) = F$
$v_{\mathscr{I}}(A_1 \downarrow A_2) = F$	otherwise
$v_{\mathscr{I}}(A_1 \leftrightarrow A_2) = T$	if $v_{\mathscr{I}}(A_1) = v_{\mathscr{I}}(A_2)$
$v_{\mathscr{I}}(A_1 \leftrightarrow A_2) = F$	if $v_{\mathscr{I}}(A_1) \neq v_{\mathscr{I}}(A_2)$
$v_{\mathscr{I}}(A_1 \oplus A_2) = T$	if $v_{\mathscr{I}}(A_1) \neq v_{\mathscr{I}}(A_2)$
$v_{\mathscr{I}}(A_1 \oplus A_2) = F$	if $v_{\mathscr{I}}(A_1) = v_{\mathscr{I}}(A_2)$

Fig. 2.3 Truth values of formulas

2.2 Interpretations

We now define the semantics—the meaning—of formulas. Consider again arithmetic expressions. Given an expression E such as $a * b + 2$, we can *assign* values to a and b and then *evaluate* the expression. For example, if $a = 2$ and $b = 3$ then E evaluates to 8. In propositional logic, truth values are assigned to the atoms of a formula in order to evaluate the truth value of the formula.

2.2.1 The Definition of an Interpretation

Definition 2.15 Let $A \in \mathscr{F}$ be a formula and let $\mathscr{P}_A$ be the set of atoms appearing in A. An *interpretation* for A is a total function $\mathscr{I}_A : \mathscr{P}_A \mapsto \{T, F\}$ that assigns one of the *truth values* T or F to *every* atom in $\mathscr{P}_A$. ∎

Definition 2.16 Let $\mathscr{I}_A$ be an interpretation for $A \in \mathscr{F}$. $v_{\mathscr{I}_A}(A)$, the *truth value of A under $\mathscr{I}_A$* is defined inductively on the structure of A as shown in Fig. 2.3. ∎

In Fig. 2.3, we have abbreviated $v_{\mathscr{I}_A}(A)$ by $v_{\mathscr{I}}(A)$. The abbreviation $\mathscr{I}$ for $\mathscr{I}_A$ will be used whenever the formula is clear from the context.

Example 2.17 Let $A = (p \rightarrow q) \leftrightarrow (\neg q \rightarrow \neg p)$ and let $\mathscr{I}_A$ be the interpretation:

$$\mathscr{I}_A(p) = F, \qquad \mathscr{I}_A(q) = T.$$

The truth value of A can be evaluated inductively using Fig. 2.3:

$$
\begin{aligned}
v_{\mathscr{I}}(p) &= \mathscr{I}_A(p) = F \\
v_{\mathscr{I}}(q) &= \mathscr{I}_A(q) = T \\
v_{\mathscr{I}}(p \to q) &= T \\
v_{\mathscr{I}}(\neg q) &= F \\
v_{\mathscr{I}}(\neg p) &= T \\
v_{\mathscr{I}}(\neg q \to \neg p) &= T \\
v_{\mathscr{I}}((p \to q) \leftrightarrow (\neg q \to \neg p)) &= T.
\end{aligned}
$$

∎

Partial Interpretations *

We will later need the following definition, but you can skip it for now:

Definition 2.18 Let $A \in \mathscr{F}$. A *partial interpretation* for A is a partial function $\mathscr{I}_A : \mathscr{P}_A \mapsto \{T, F\}$ that assigns one of the *truth values* T or F to *some* of the atoms in $\mathscr{P}_A$. ∎

It is possible that the truth value of a formula can be determined in a partial interpretation.

Example 2.19 Consider the formula $A = p \wedge q$ and the partial interpretation that assigns F to p. Clearly, the truth value of A is F. If the partial interpretation assigned T to p, we cannot compute the truth value of A. ∎

2.2.2 Truth Tables

A truth table is a convenient format for displaying the semantics of a formula by showing its truth value for every possible interpretation of the formula.

Definition 2.20 Let $A \in \mathscr{F}$ and supposed that there are n atoms in $\mathscr{P}_A$. A *truth table* is a table with $n + 1$ columns and 2^n rows. There is a column for each atom in $\mathscr{P}_A$, plus a column for the formula A. The first n columns specify the interpretation $\mathscr{I}$ that maps atoms in $\mathscr{P}_A$ to $\{T, F\}$. The last column shows $v_{\mathscr{I}}(A)$, the truth value of A for the interpretation $\mathscr{I}$. ∎

Since each of the n atoms can be assigned T or F independently, there are 2^n interpretations and thus 2^n rows in a truth table.

Example 2.21 Here is the truth table for the formula $p \to q$:

p	q	$p \to q$
T	T	T
T	F	F
F	T	T
F	F	T

∎

When the formula A is complex, it is easier to build a truth table by adding columns that show the truth value for subformulas of A.

Example 2.22 Here is a truth table for the formula $(p \to q) \leftrightarrow (\neg q \to \neg p)$ from Example 2.17:

p	q	$p \to q$	$\neg p$	$\neg q$	$\neg q \to \neg p$	$(p \to q) \leftrightarrow (\neg q \to \neg p)$
T	T	T	F	F	T	T
T	F	F	F	T	F	T
F	T	T	T	F	T	T
F	F	T	T	T	T	T

∎

A convenient way of computing the truth value of a formula for a specific interpretation $\mathscr{I}$ is to write the value T or F of $\mathscr{I}(p_i)$ under each atom p_i and then to write down the truth values incrementally under each operator as you perform the computation. Each step of the computation consists of choosing an innermost subformula and evaluating it.

Example 2.23 The computation of the truth value of $(p \to q) \leftrightarrow (\neg q \to \neg p)$ for the interpretation $\mathscr{I}(p) = T$ and $\mathscr{I}(q) = F$ is:

(p	$\to$	q)	$\leftrightarrow$	($\neg$	q	$\to$	$\neg$	p)
T		F			F			T
T		F		T	F			T
T		F		T	F		F	T
T		F		T	F	F	F	T
T	F	F		T	F	F	F	T
T	F	F	T	T	F	F	F	T

If the computations for all subformulas are written on the same line, the truth table from Example 2.22 can be written as follows:

p	q	(p	$\rightarrow$	q)	$\leftrightarrow$	($\neg$	q	$\rightarrow$	$\neg$	p)
T	T	T	T	T	T	F	T	T	F	T
T	F	T	F	F	T	T	F	F	F	T
F	T	F	T	T	T	F	T	T	T	F
F	F	F	T	F	T	T	F	T	T	F

■

2.2.3 Understanding the Boolean Operators

The natural reading of the Boolean operators $\neg$ and $\wedge$ correspond with their formal semantics as defined in Fig. 2.3. The operators $\uparrow$ and $\downarrow$ are simply negations of $\wedge$ and $\vee$. Here we comment on the operators $\vee$, $\oplus$ and $\rightarrow$, whose formal semantics can be the source of confusion.

Inclusive or *vs.* Exclusive or

Disjunction $\vee$ is *inclusive or* and is a distinct operator from $\oplus$ which is *exclusive or*. Consider the compound statement:

At eight o'clock 'I will go to the movies' *or* 'I will go to the theater'.

The intended meaning is 'movies' $\oplus$ 'theater', because I can't be in both places at the same time. This contrasts with the disjunctive operator $\vee$ which evaluates to true when either or both of the statements are true:

Do you want 'popcorn' or 'candy'?

This can be denoted by 'popcorn' $\vee$ 'candy', because it is possible to want both of them at the same time.

For $\vee$, it is sufficient for one statement to be true for the compound statement to be true. Thus, the following strange statement is true because the truth of the first statement by itself is sufficient to ensure the truth of the compound statement:

'Earth is farther from the sun than Venus' $\vee$ '1 + 1 = 3'.

The difference between $\vee$ and $\oplus$ is seen when both subformulas are true:

'Earth is farther from the sun than Venus' $\vee$ '1 + 1 = 2'.
'Earth is farther from the sun than Venus' $\oplus$ '1 + 1 = 2'.

The first statement is true but the second is false.

Inclusive or *vs.* Exclusive or in Programming Languages

When *or* is used in the context of programming languages, the intention is usually inclusive or:

```
if (index < min || index > max) /* There is an error */
```

The truth of one of the two subexpressions causes the following statements to be executed. The operator `||` is not really a Boolean operator because it uses *short-circuit evaluation*: if the first subexpression is true, the second subexpression is not evaluated, because its truth value cannot change the decision to execute the following statements. There is an operator `|` that performs true Boolean evaluation; it is usually used when the operands are bit vectors:

```
mask1 = 0xA0;
mask2 = 0x0A;
mask  = mask1 | mask2;
```

Exclusive or `^` is used to implement encoding and decoding in error-correction and cryptography. The reason is that when used twice, the original value can be recovered. Suppose that we encode bit of data with a secret key:

```
codedMessage = data ^ key;
```

The recipient of the message can decode it by computing:

```
clearMessage = codedMessage ^ key;
```

as shown by the following computation:

```
clearMessage == codedMessage ^ key
             == (data ^ key) ^ key
             == data ^ (key ^ key)
             == data ^ false
             == data
```

Implication

The operator of $p \to q$ is called *material implication*; p is the *antecedent* and q is the *consequent*. Material implication does not claim causation; that is, it does not assert there the antecedent *causes* the consequent (or is even related to the consequent in any way). A material implication merely states that if the antecedent is true the consequent must be true (see Fig. 2.3), so it can be falsified only if the antecedent is true and the consequent is false. Consider the following two compound statements:

'Earth is farther from the sun than Venus' $\to$ '$1 + 1 = 3$'.

is false since the antecedent is true and the consequent is false, but:

'Earth is farther from the sun than Mars' $\rightarrow$ '$1 + 1 = 3$'.

is true! The falsity of the antecedent by itself is sufficient to ensure the truth of the implication.

2.2.4 An Interpretation for a Set of Formulas

Definition 2.24 Let $S = \{A_1, \ldots\}$ be a set of formulas and let $\mathscr{P}_S = \bigcup_i \mathscr{P}_{A_i}$, that is, $\mathscr{P}_S$ is the set of all the atoms that appear in the formulas of S. An *interpretation* for S is a function $\mathscr{I}_S : \mathscr{P}_S \mapsto \{T, F\}$. For any $A_i \in S$, $v_{\mathscr{I}_S}(A_i)$, the *truth value of A_i under $\mathscr{I}_S$*, is defined as in Definition 2.16. ∎

The definition of $\mathscr{P}_S$ as the union of the sets of atoms in the formulas of S ensures that each atom is assigned exactly one truth value.

Example 2.25 Let $S = \{p \rightarrow q,\ p,\ q \wedge r,\ p \vee s \leftrightarrow s \wedge q\}$ and let $\mathscr{I}_S$ be the interpretation:

$$\mathscr{I}_S(p) = T, \qquad \mathscr{I}_S(q) = F, \qquad \mathscr{I}_S(r) = T, \qquad \mathscr{I}_S(s) = T.$$

The truth values of the elements of S can be evaluated as:

$$
\begin{aligned}
v_{\mathscr{I}}(p \rightarrow q) &= F \\
v_{\mathscr{I}}(p) &= \mathscr{I}_S(p) = T \\
v_{\mathscr{I}}(q \wedge r) &= F \\
v_{\mathscr{I}}(p \vee s) &= T \\
v_{\mathscr{I}}(s \wedge q) &= F \\
v_{\mathscr{I}}(p \vee s \leftrightarrow s \wedge q) &= F.
\end{aligned}
$$

∎

2.3 Logical Equivalence

Definition 2.26 Let $A_1, A_2 \in \mathscr{F}$. If $v_{\mathscr{I}}(A_1) = v_{\mathscr{I}}(A_2)$ for all interpretations $\mathscr{I}$, then A_1 is *logically equivalent* to A_2, denoted $A_1 \equiv A_2$. ∎

Example 2.27 Is the formula $p \vee q$ logically equivalent to $q \vee p$? There are four distinct interpretations that assign to the atoms p and q:

$\mathscr{I}(p)$	$\mathscr{I}(q)$	$v_{\mathscr{I}}(p \vee q)$	$v_{\mathscr{I}}(q \vee p)$
T	T	T	T
T	F	T	T
F	T	T	T
F	F	F	F

Since $p \vee q$ and $q \vee p$ agree on all the interpretations, $p \vee q \equiv q \vee p$.   ∎

This example can be generalized to arbitrary *formulas*:

Theorem 2.28 *Let* $A_1, A_2 \in \mathscr{F}$. *Then* $A_1 \vee A_2 \equiv A_2 \vee A_1$.

Proof Let $\mathscr{I}$ be an arbitrary interpretation for $A_1 \vee A_2$. Obviously, $\mathscr{I}$ is also an interpretation for $A_2 \vee A_1$ since $\mathscr{P}_{A_1} \cup \mathscr{P}_{A_2} = \mathscr{P}_{A_2} \cup \mathscr{P}_{A_1}$.

Since $\mathscr{P}_{A_1} \subseteq \mathscr{P}_{A_1} \cup \mathscr{P}_{A_2}$, $\mathscr{I}$ assigns truth values to all atoms in A_1 and can be considered to be an interpretation for A_1. Similarly, $\mathscr{I}$ can be considered to be an interpretation for A_2.

Now $v_{\mathscr{I}}(A_1 \vee A_2) = T$ if and only if either $v_{\mathscr{I}}(A_1) = T$ or $v_{\mathscr{I}}(A_2) = T$, and $v_{\mathscr{I}}(A_2 \vee A_1) = T$ if and only if either $v_{\mathscr{I}}(A_2) = T$ or $v_{\mathscr{I}}(A_1) = T$. If $v_{\mathscr{I}}(A_1) = T$, then:

$$v_{\mathscr{I}}(A_1 \vee A_2) = T = v_{\mathscr{I}}(A_2 \vee A_1),$$

and similarly if $v_{\mathscr{I}}(A_2) = T$. Since $\mathscr{I}$ was arbitrary, $A_1 \vee A_2 \equiv A_2 \vee A_1$.   ∎

This type of argument will be used frequently. In order to prove that something is true of *all* interpretations, we let $\mathscr{I}$ be an *arbitrary* interpretation and then write a proof without using any property that distinguishes one interpretation from another.

2.3.1 The Relationship Between ↔ and ≡

Equivalence, ↔, is a Boolean operator in propositional logic and can appear in formulas of the logic. Logical equivalence, ≡, is not a Boolean operator; instead, is a notation for a *property* of pairs of formulas in propositional logic. There is potential for confusion because we are using a similar vocabulary both for the *object language*, in this case the language of propositional logic, and for the *metalanguage* that we use reason about the object language.

Equivalence and logical equivalence are, nevertheless, closely related as shown by the following theorem:

Theorem 2.29 $A_1 \equiv A_2$ *if and only if* $A_1 \leftrightarrow A_2$ *is true in every interpretation.*

Proof Suppose that $A_1 \equiv A_2$ and let $\mathscr{I}$ be an arbitrary interpretation; then $v_{\mathscr{I}}(A_1) = v_{\mathscr{I}}(A_2)$ by definition of logical equivalence. From Fig. 2.3, $v_{\mathscr{I}}(A_1 \leftrightarrow A_2) = T$. Since $\mathscr{I}$ was arbitrary, $v_{\mathscr{I}}(A_1 \leftrightarrow A_2) = T$ in all interpretations. The proof of the converse is similar.   ∎

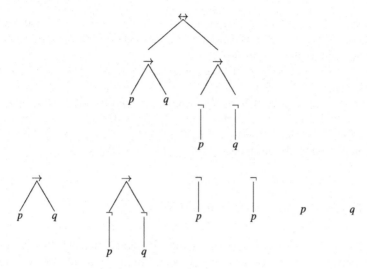

Fig. 2.4 Subformulas

2.3.2 Substitution

Logical equivalence justifies substitution of one formula for another.

Definition 2.30 A is a *subformula* of B if A is a subtree of B. If A is not the same as B, it is a *proper subformula* of B. ∎

Example 2.31 Figure 2.4 shows a formula (the left formula from Fig. 2.1) and its proper subformulas. Represented as strings, $(p \rightarrow q) \leftrightarrow (\neg p \rightarrow \neg q)$ contains the proper subformulas: $p \rightarrow q, \neg p \rightarrow \neg q, \neg p, \neg q, p, q$. ∎

Definition 2.32 Let A be a subformula of B and let A' be any formula. $B\{A \leftarrow A'\}$, the *substitution of A' for A in B*, is the formula obtained by replacing all occurrences of the subtree for A in B by A'. ∎

Example 2.33 Let $B = (p \rightarrow q) \leftrightarrow (\neg p \rightarrow \neg q)$, $A = p \rightarrow q$ and $A' = \neg p \vee q$.

$$B\{A \leftarrow A'\} = (\neg p \vee q) \leftrightarrow (\neg q \rightarrow \neg p).$$

∎

Given a formula A, substitution of a logically equivalent formula for a subformula of A does not change its truth value under any interpretation.

Theorem 2.34 *Let A be a subformula of B and let A' be a formula such that $A \equiv A'$. Then $B \equiv B\{A \leftarrow A'\}$.*

Proof Let $\mathscr{I}$ be an arbitrary interpretation. Then $v_{\mathscr{I}}(A) = v_{\mathscr{I}}(A')$ and we must show that $v_{\mathscr{I}}(B) = v_{\mathscr{I}}(B')$. The proof is by induction on the depth d of the highest occurrence of the subtree A in B.

If $d = 0$, there is only one occurrence of A, namely B itself. Obviously, $v_{\mathscr{I}}(B) = v_{\mathscr{I}}(A) = v_{\mathscr{I}}(A') = v_{\mathscr{I}}(B')$.

If $d \neq 0$, then B is $\neg B_1$ or $B_1 \, op \, B_2$ for some formulas B_1, B_2 and operator op. In B_1, the depth of A is less than d. By the inductive hypothesis, $v_{\mathscr{I}}(B_1) = v_{\mathscr{I}}(B_1') = v_{\mathscr{I}}(B_1\{A \leftarrow A'\})$, and similarly $v_{\mathscr{I}}(B_2) = v_{\mathscr{I}}(B_2') = v_{\mathscr{I}}(B_2\{A \leftarrow A'\})$. By the definition of v on the Boolean operators, $v_{\mathscr{I}}(B) = v_{\mathscr{I}}(B')$.   ∎

2.3.3 Logically Equivalent Formulas

Substitution of logically equivalence formulas is frequently done, for example, to simplify a formula, and it is essential to become familiar with the common equivalences that are listed in this subsection. Their proofs are elementary from the definitions and are left as exercises.

Absorption of Constants

Let us extend the syntax of Boolean formulas to include the two constant atomic propositions *true* and *false*. (Another notation is ⊤ for *true* and ⊥ for *false*.) Their semantics are defined by $\mathscr{I}(true) = T$ and $\mathscr{I}(false) = F$ for any interpretation. Do not confuse these symbols in the object language of propositional logic with the truth values T and F used to define interpretations. Alternatively, it is possible to regard *true* and *false* as abbreviations for the formulas $p \vee \neg p$ and $p \wedge \neg p$, respectively.

The appearance of a constant in a formula can collapse the formula so that the binary operator is no longer needed; it can even make a formula become a constant whose truth value no longer depends on the non-constant subformula.

$$
\begin{array}{llcl}
A \vee true & \equiv & true & \qquad A \wedge true \equiv A \\
A \vee false & \equiv & A & \qquad A \wedge false \equiv false \\
A \rightarrow true & \equiv & true & \qquad true \rightarrow A \equiv A \\
A \rightarrow false & \equiv & \neg A & \qquad false \rightarrow A \equiv true \\
A \leftrightarrow true & \equiv & A & \qquad A \oplus true \equiv \neg A \\
A \leftrightarrow false & \equiv & \neg A & \qquad A \oplus false \equiv A
\end{array}
$$

Identical Operands

Collapsing can also occur when both operands of an operator are the same or one is the negation of another.

$$
\begin{array}{llllll}
A & \equiv & \neg\,\neg\,A \\
A & \equiv & A \wedge A & \quad A & \equiv & A \vee A \\
A \vee \neg A & \equiv & true & \quad A \wedge \neg A & \equiv & false \\
A \rightarrow A & \equiv & true \\
A \leftrightarrow A & \equiv & true & \quad A \oplus A & \equiv & false \\
\neg A & \equiv & A \uparrow A \, \neg A & \quad \equiv & & A \downarrow A
\end{array}
$$

Commutativity, Associativity and Distributivity

The binary Boolean operators are commutative, except for implication.

$$
\begin{array}{llllll}
A \vee B & \equiv & B \vee A & \quad A \wedge B & \equiv & B \wedge A \\
A \leftrightarrow B & \equiv & B \leftrightarrow A & \quad A \oplus B & \equiv & B \oplus A \\
A \uparrow B & \equiv & B \uparrow A & \quad A \downarrow B & \equiv & B \downarrow A
\end{array}
$$

If negations are added, the direction of an implication can be reversed:

$$
A \rightarrow B \equiv \neg B \rightarrow \neg A
$$

The formula $\neg B \rightarrow \neg A$ is the *contrapositive* of $A \rightarrow B$.

Disjunction, conjunction, equivalence and non-equivalence are associative.

$$
\begin{array}{llllll}
A \vee (B \vee C) & \equiv & (A \vee B) \vee C & \quad A \wedge (B \wedge C) & \equiv & (A \wedge B) \wedge C \\
A \leftrightarrow (B \leftrightarrow C) & \equiv & (A \leftrightarrow B) \leftrightarrow C & \quad A \oplus (B \oplus C) & \equiv & (A \oplus B) \oplus C
\end{array}
$$

Implication, *nor* and *nand* are not associative.

Disjunction and conjunction distribute over each other.

$$
\begin{array}{lll}
A \vee (B \wedge C) & \equiv & (A \vee B) \wedge (A \vee C) \\
A \wedge (B \vee C) & \equiv & (A \wedge B) \vee (A \wedge C)
\end{array}
$$

Defining One Operator in Terms of Another

When proving theorems *about* propositional logic using structural induction, we have to prove the inductive step for each of the binary operators. It will simplify proofs if we can eliminate some of the operators by replacing subformulas with formulas that use another operator. For example, equivalence can be eliminated because it can be defined in terms of conjunction and implication. Another reason for eliminating operators is that many algorithms on propositional formulas require that the formulas be in a *normal form*, using a specified subset of the Boolean operators. Here is a list of logical equivalences that can be used to eliminate operators.

$$A \leftrightarrow B \ \equiv \ (A \rightarrow B) \wedge (B \rightarrow A) \qquad A \oplus B \ \equiv \ \neg(A \rightarrow B) \vee \neg(B \rightarrow A)$$
$$A \rightarrow B \ \equiv \ \neg A \vee B \qquad\qquad\qquad A \rightarrow B \ \equiv \ \neg(A \wedge \neg B)$$
$$A \vee B \ \equiv \ \neg(\neg A \wedge \neg B) \qquad\qquad A \wedge B \ \equiv \ \neg(\neg A \vee \neg B)$$
$$A \vee B \ \equiv \ \neg A \rightarrow B \qquad\qquad\quad A \wedge B \ \equiv \ \neg(A \rightarrow \neg B)$$

The definition of conjunction in terms of disjunction and negation, and the definition of disjunction in terms of conjunction and negation are called *De Morgan's laws*.

2.4 Sets of Boolean Operators *

From our earliest days in school, we are taught that there are four basic operators in arithmetic: addition, subtraction, multiplication and division. Later on, we learn about additional operators like modulo and absolute value. On the other hand, multiplication and division are theoretically redundant because they can be defined in terms of addition and subtraction.

In this section, we will look at two issues: What Boolean operators are there? What sets of operators are adequate, meaning that all other operators can be defined using just the operators in the set?

2.4.1 Unary and Binary Boolean Operators

Since there are only two Boolean values T and F, the number of possible n-place operators is 2^{2^n}, because for each of the n arguments we can choose either of the two values T and F and for each of these 2^n n-tuples of arguments we can choose the value of the operator to be either T or F. We will restrict ourselves to one- and two-place operators.

The following table shows the $2^{2^1} = 4$ possible one-place operators, where the first column gives the value of the operand x and the other columns give the value of the nth operator $o_n(x)$:

x	o_1	o_2	o_3	o_4
T	T	T	F	F
F	T	F	T	F

x_1	x_2	$\circ_1$	$\circ_2$	$\circ_3$	$\circ_4$	$\circ_5$	$\circ_6$	$\circ_7$	$\circ_8$
T	T	T	T	T	T	T	T	T	T
T	F	T	T	T	T	F	F	F	F
F	T	T	T	F	F	T	T	F	F
F	F	T	F	T	F	T	F	T	F

x_1	x_2	$\circ_9$	$\circ_{10}$	$\circ_{11}$	$\circ_{12}$	$\circ_{13}$	$\circ_{14}$	$\circ_{15}$	$\circ_{16}$
T	T	F	F	F	F	F	F	F	F
T	F	T	T	T	T	F	F	F	F
F	T	T	T	F	F	T	T	F	F
F	F	T	F	T	F	T	F	T	F

Fig. 2.5 Two-place Boolean operators

Of the four one-place operators, three are trivial: $\circ_1$ and $\circ_4$ are the constant operators, and $\circ_2$ is the identity operator which simply maps the operand to itself. The only non-trivial one-place operator is $\circ_3$ which is negation.

There are $2^{2^2} = 16$ two-place operators (Fig. 2.5). Several of the operators are trivial: $\circ_1$ and $\circ_{16}$ are constant; $\circ_4$ and $\circ_6$ are projection operators, that is, their value is determined by the value of only one of operands; $\circ_{11}$ and $\circ_{13}$ are the negations of the projection operators.

The correspondence between the operators in the table and those we defined in Definition 2.1 are shown in the following table, where the operators in the right-hand column are the negations of those in the left-hand column.

op	name	symbol	op	name	symbol
$\circ_2$	disjunction	$\vee$	$\circ_{15}$	nor	$\downarrow$
$\circ_8$	conjunction	$\wedge$	$\circ_9$	nand	$\uparrow$
$\circ_5$	implication	$\rightarrow$			
$\circ_7$	equivalence	$\leftrightarrow$	$\circ_{10}$	exclusive or	$\oplus$

The operator $\circ_{12}$ is the negation of implication and is not used. Reverse implication, $\circ_3$, is used in logic programming (Chap. 11); its negation, $\circ_{14}$, is not used.

2.4.2 Adequate Sets of Operators

Definition 2.35 A binary operator $\circ$ *is defined from* a set of operators $\{\circ_1, \dots, \circ_n\}$ iff there is a logical equivalence $A_1 \circ A_2 \equiv A$, where A is a formula constructed from occurrences of A_1 and A_2 using the operators $\{\circ_1, \dots, \circ_n\}$. The unary operator $\neg$

is defined by a formula $\neg A_1 \equiv A$, where A is constructed from occurrences of A_1 and the operators in the set.   ∎

Theorem 2.36 *The Boolean operators* $\vee, \wedge, \rightarrow, \leftrightarrow, \oplus, \uparrow, \downarrow$ *can be defined from negation and one of* $\vee, \wedge, \rightarrow$.

Proof The theorem follows by using the logical equivalences in Sect. 2.3.3. The *nand* and *nor* operators are the negations of conjunction and disjunction, respectively. Equivalence can be defined from implication and conjunction and non-equivalence can be defined using these operators and negation. Therefore, we need only $\rightarrow, \vee, \wedge$, but each of these operators can be defined by one of the others and negation as shown by the equivalences on page 26.   ∎

It may come as a surprise that it is possible to define all Boolean operators from either *nand* or *nor* alone. The equivalence $\neg A \equiv A \uparrow A$ is used to define negation from *nand* and the following sequence of equivalences shows how conjunction can be defined:

$$
\begin{array}{lll}
(A \uparrow B) \uparrow (A \uparrow B) & \equiv & \text{by the definition of } \uparrow \\
\neg((A \uparrow B) \wedge (A \uparrow B)) & \equiv & \text{by idempotence} \\
\neg(A \uparrow B) & \equiv & \text{by the definition of } \uparrow \\
\neg\neg(A \wedge B) & \equiv & \text{by double negation} \\
A \wedge B. & &
\end{array}
$$

From the formulas for negation and conjunction, all other operators can be defined. Similarly definitions are possible using *nor*.

In fact it can be proved that only *nand* and *nor* have this property.

Theorem 2.37 *Let* $\circ$ *be a binary operator that can define negation and all other binary operators by itself. Then* $\circ$ *is either nand or nor.*

Proof We give an outline of the proof and leave the details as an exercise.

Suppose that $\circ$ is an operator that can define all the other operators. Negation must be defined by an equivalence of the form:

$$\neg A \equiv A \circ \cdots \circ A.$$

Any binary operator *op* must be defined by an equivalence:

$$A_1 \ op \ A_2 \equiv B_1 \circ \cdots \circ B_n,$$

where each B_i is either A_1 or A_2. (If $\circ$ is not associative, add parentheses as necessary.) We will show that these requirements impose restrictions on $\circ$ so that it must be *nand* or *nor*.

Let $\mathscr{I}$ be any interpretation such that $v_{\mathscr{I}}(A) = T$; then

$$F = v_{\mathscr{I}}(\neg A) = v_{\mathscr{I}}(A \circ \cdots \circ A).$$

Prove by induction on the number of occurrences of ∘ that $v_{\mathscr{I}}(A_1 \circ A_2) = F$ when $v_{\mathscr{I}}(A_1) = T$ and $v_{\mathscr{I}}(A_2) = T$. Similarly, if $\mathscr{I}$ is an interpretation such that $v_{\mathscr{I}}(A) = F$, prove that $v_{\mathscr{I}}(A_1 \circ A_2) = T$.

Thus the only freedom we have in defining ∘ is in the case where the two operands are assigned different truth values:

A_1	A_2	$A_1 \circ A_2$
T	T	F
T	F	T or F
F	T	T or F
F	F	T

If ∘ is defined to give the same truth value T for these two lines then ∘ is *nand*, and if ∘ is defined to give the same truth value F then ∘ is *nor*.

The remaining possibility is that ∘ is defined to give different truth values for these two lines. Prove by induction that only projection and negated projection are definable in the sense that:

$$B_1 \circ \cdots \circ B_n \equiv \neg \cdots \neg B_i$$

for some i and zero or more negations. ∎

2.5 Satisfiability, Validity and Consequence

We now define the fundamental concepts of the semantics of formulas:

Definition 2.38 Let $A \in \mathscr{F}$.

- A is *satisfiable* iff $v_{\mathscr{I}}(A) = T$ for *some* interpretation $\mathscr{I}$.
  A satisfying interpretation is a *model* for A.
- A is *valid*, denoted $\models A$, iff $v_{\mathscr{I}}(A) = T$ for *all* interpretations $\mathscr{I}$.
  A valid propositional formula is also called a *tautology*.
- A is *unsatisfiable* iff it is not satisfiable, that is, if $v_{\mathscr{I}}(A) = F$ for *all* interpretations $\mathscr{I}$.
- A is *falsifiable*, denoted $\not\models A$, iff it is not valid, that is, if $v_{\mathscr{I}}(A) = F$ for *some* interpretation v. ∎

These concepts are illustrated in Fig. 2.6.

The four semantical concepts are closely related.

Theorem 2.39 Let $A \in \mathscr{F}$. *A is valid if and only if* $\neg A$ *is unsatisfiable. A is satisfiable if and only if* $\neg A$ *is falsifiable.*

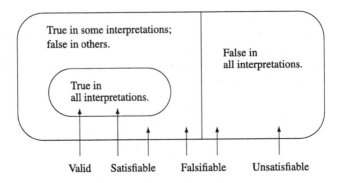

Fig. 2.6 Satisfiability and validity of formulas

Proof Let $\mathscr{I}$ be an arbitrary interpretation. $v_{\mathscr{I}}(A) = T$ if and only if $v_{\mathscr{I}}(\neg A) = F$ by the definition of the truth value of a negation. Since $\mathscr{I}$ was arbitrary, A is true in all interpretations if and only if $\neg A$ is false in all interpretations, that is, iff $\neg A$ is unsatisfiable.

If A is satisfiable then for *some* interpretation $\mathscr{I}$, $v_{\mathscr{I}}(A) = T$. By definition of the truth value of a negation, $v_{\mathscr{I}}(\neg A) = F$ so that $\neg A$ is falsifiable. Conversely, if $v_{\mathscr{I}}(\neg A) = F$ then $v_{\mathscr{I}}(A) = T$.   ∎

2.5.1 Decision Procedures in Propositional Logic

Definition 2.40 Let $\mathscr{U} \subseteq \mathscr{F}$ be a set of formulas. An algorithm is a *decision procedure* for $\mathscr{U}$ if given an arbitrary formula $A \in \mathscr{F}$, it terminates and returns the answer *yes* if $A \in \mathscr{U}$ and the answer *no* if $A \notin \mathscr{U}$.   ∎

If $\mathscr{U}$ is the set of satisfiable formulas, a decision procedure for $\mathscr{U}$ is called a decision procedure for satisfiability, and similarly for validity.

By Theorem 2.39, a decision procedure for satisfiability can be used as a decision procedure for validity. To decide if A is valid, apply the decision procedure for satisfiability to $\neg A$. If it reports that $\neg A$ is satisfiable, then A is not valid; if it reports that $\neg A$ is not satisfiable, then A is valid. Such an decision procedure is called a *refutation procedure*, because we prove the validity of a formula by refuting its negation. Refutation procedures can be efficient algorithms for deciding validity, because instead of checking that the formula is always true, we need only search for a falsifying counterexample.

The existence of a decision procedure for satisfiability in propositional logic is trivial, because we can build a truth table for any formula. The truth table in Example 2.21 shows that $p \rightarrow q$ is satisfiable, but not valid; Example 2.22 shows that $(p \rightarrow q) \leftrightarrow (\neg q \rightarrow \neg p)$ is valid. The following example shows an unsatisfiable formula.

Example 2.41 The formula $(p \vee q) \wedge \neg p \wedge \neg q$ is unsatisfiable because all lines of its truth table evaluate to F.

p	q	$p \vee q$	$\neg p$	$\neg q$	$(p \vee q) \wedge \neg p \wedge \neg q$
T	T	T	F	F	F
T	F	T	F	T	F
F	T	T	T	F	F
F	F	F	T	T	F

■

The method of truth tables is a very inefficient decision procedure because we need to evaluate a formula for each of 2^n possible interpretations, where n is the number of distinct atoms in the formula. In later chapters we will discuss more efficient decision procedures for satisfiability, though it is extremely unlikely that there is a decision procedure that is efficient for all formulas (see Sect. 6.7).

2.5.2 Satisfiability of a Set of Formulas

The concept of satisfiability can be extended to a set of formulas.

Definition 2.42 A set of formulas $U = \{A_1, \ldots\}$ is *(simultaneously) satisfiable* iff there exists an interpretation $\mathscr{I}$ such that $v_{\mathscr{I}}(A_i) = T$ for all i. The satisfying interpretation is a *model* of U. U is *unsatisfiable* iff for every interpretation $\mathscr{I}$, there exists an i such that $v_{\mathscr{I}}(A_i) = F$. ■

Example 2.43 The set $U_1 = \{p, \neg p \vee q, q \wedge r\}$ is simultaneously satisfiable by the interpretation which assigns T to each atom, while the set $U_2 = \{p, \neg p \vee q, \neg p\}$ is unsatisfiable. Each formula in U_2 is satisfiable by itself, but the set is not simultaneously satisfiable. ■

The proofs of the following elementary theorems are left as exercises.

Theorem 2.44 *If U is satisfiable, then so is $U - \{A_i\}$ for all i.*

Theorem 2.45 *If U is satisfiable and B is valid, then $U \cup \{B\}$ is satisfiable.*

Theorem 2.46 *If U is unsatisfiable, then for any formula B, $U \cup \{B\}$ is unsatisfiable.*

Theorem 2.47 *If U is unsatisfiable and for some i, A_i is valid, then $U - \{A_i\}$ is unsatisfiable.*

2.5.3 Logical Consequence

Definition 2.48 Let U be a set of formulas and A a formula. A is a *logical consequence* of U, denoted $U \models A$, iff every model of U is a model of A.   ∎

The formula A need not be true in every possible interpretation, only in those interpretations which satisfy U, that is, those interpretations which satisfy every formula in U. If U is empty, logical consequence is the same as validity.

Example 2.49 Let $A = (p \vee r) \wedge (\neg q \vee \neg r)$. Then A is a logical consequence of $\{p, \neg q\}$, denoted $\{p, \neg q\} \models A$, since A is true in all interpretations $\mathscr{I}$ such that $\mathscr{I}(p) = T$ and $\mathscr{I}(q) = F$. However, A is not valid, since it is not true in the interpretation $\mathscr{I}'$ where $\mathscr{I}'(p) = F$, $\mathscr{I}'(q) = T$, $\mathscr{I}'(r) = T$.   ∎

The caveat concerning $\leftrightarrow$ and $\equiv$ also applies to $\rightarrow$ and $\models$. Implication, $\rightarrow$, is an operator in the object language, while $\models$ is a symbol for a concept in the metalanguage. However, as with equivalence, the two concepts are related:

Theorem 2.50 $U \models A$ *if and only if* $\models \bigwedge_i A_i \rightarrow A$.

Definition 2.51 $\bigwedge_{i=1}^{i=n} A_i$ is an abbreviation for $A_1 \wedge \cdots \wedge A_n$. The notation $\bigwedge_i$ is used if the bounds are obvious from the context or if the set of formulas is infinite. A similar notation $\bigvee$ is used for disjunction.   ∎

Example 2.52 From Example 2.49, $\{p, \neg q\} \models (p \vee r) \wedge (\neg q \vee \neg r)$, so by Theorem 2.50, $\models (p \wedge \neg q) \rightarrow (p \vee r) \wedge (\neg q \vee \neg r)$.   ∎

The proof of Theorem 2.50, as well as the proofs of the following two theorems are left as exercises.

Theorem 2.53 *If* $U \models A$ *then* $U \cup \{B\} \models A$ *for any formula* B.

Theorem 2.54 *If* $U \models A$ *and* B *is valid then* $U - \{B\} \models A$.

2.5.4 Theories *

Logical consequence is the central concept in the foundations of mathematics. Valid logical formulas such as $p \vee q \leftrightarrow q \vee p$ are of little mathematical interest. It is much more interesting to assume that a set of formulas is true and then to investigate the consequences of these assumptions. For example, Euclid assumed five formulas about geometry and deduced an extensive set of logical consequences. The formal definition of a mathematical theory is as follows.

Definition 2.55 Let $\mathscr{T}$ be a set of formulas. $\mathscr{T}$ is *closed under logical consequence* iff for all formulas A, if $\mathscr{T} \models A$ then $A \in \mathscr{T}$. A set of formulas that is closed under logical consequence is a *theory*. The elements of $\mathscr{T}$ are *theorems*.   ∎

Theories are constructed by selecting a set of formulas called *axioms* and deducing their logical consequences.

Definition 2.56 Let $\mathscr{T}$ be a theory. $\mathscr{T}$ is said to be *axiomatizable* iff there exists a set of formulas U such that $\mathscr{T} = \{A \mid U \models A\}$. The set of formulas U are the *axioms* of $\mathscr{T}$. If U is finite, $\mathscr{T}$ is said to be *finitely axiomatizable*. ∎

Arithmetic is axiomatizable: There is a set of axioms developed by Peano whose logical consequences are theorems of arithmetic. Arithmetic is not finitely axiomatizable, because the induction axiom is not by a single axiom but an axiom scheme with an instance for each property in arithmetic.

2.6 Semantic Tableaux

The method of *semantic tableaux* is an efficient decision procedure for satisfiability (and by duality validity) in propositional logic. We will use semantic tableaux extensively in the next chapter to prove important theorems about deductive systems. The principle behind semantic tableaux is very simple: search for a model (satisfying interpretation) by decomposing the formula into sets of atoms and negations of atoms. It is easy to check if there is an interpretation for each set: a set of atoms and negations of atoms is satisfiable iff the set does not contain an atom p and its negation $\neg p$. The formula is satisfiable iff one of these sets is satisfiable.

We begin with some definitions and then analyze the satisfiability of two formulas to motivate the construction of semantic tableaux.

2.6.1 Decomposing Formulas into Sets of Literals

Definition 2.57 A *literal* is an atom or the negation of an atom. An atom is a *positive literal* and the negation of an atom is a *negative literal*. For any atom p, $\{p, \neg p\}$ is a *complementary pair of literals*.

For any formula A, $\{A, \neg A\}$ is a *complementary pair of formulas*. A is the *complement* of $\neg A$ and $\neg A$ is the *complement* of A. ∎

Example 2.58 In the set of literals $\{\neg p, q, r, \neg r\}$, q and r are positive literals, while $\neg p$ and $\neg r$ are negative literals. The set contains the complementary pair of literals $\{r, \neg r\}$. ∎

Example 2.59 Let us analyze the satisfiability of the formula:

$$A = p \wedge (\neg q \vee \neg p)$$

in an arbitrary interpretation $\mathscr{I}$, using the inductive rules for the evaluation of the truth value of a formula.

- The principal operator of A is conjunction, so $v_\mathscr{I}(A) = T$ if and only if both $v_\mathscr{I}(p) = T$ and $v_\mathscr{I}(\neg q \vee \neg p) = T$.
- The principal operator of $\neg q \vee \neg p$ is disjunction, so $v_\mathscr{I}(\neg q \vee \neg p) = T$ if and only if either $v_\mathscr{I}(\neg q) = T$ or $v_\mathscr{I}(\neg p) = T$.
- Integrating the information we have obtained from this analysis, we conclude that $v_\mathscr{I}(A) = T$ if and only if either:

  1. $v_\mathscr{I}(p) = T$ and $v_\mathscr{I}(\neg q) = T$, or
  2. $v_\mathscr{I}(p) = T$ and $v_\mathscr{I}(\neg p) = T$.

A is satisfiable if and only if there is an interpretation such that (1) holds or an interpretation such that (2) holds.   ∎

We have reduced the question of the satisfiability of A to a question about the satisfiability of sets of literals.

Theorem 2.60 *A set of literals is satisfiable if and only if it does* not *contain a complementary pair of literals.*

Proof Let L be a set of literals that does not contain a complementary pair. Define the interpretation $\mathscr{I}$ by:

$$\mathscr{I}(p) = T \quad \text{if } p \in L,$$
$$\mathscr{I}(p) = F \quad \text{if } \neg p \in L.$$

The interpretation is well-defined—there is only one value assigned to each atom in L—since there is no complementary pair of literals in L. Each literal in L evaluates to T so L is satisfiable.

Conversely, if $\{p, \neg p\} \subseteq L$, then for any interpretation $\mathscr{I}$ for the atoms in L, either $v_\mathscr{I}(p) = F$ or $v_\mathscr{I}(\neg p) = F$, so L is not satisfiable.   ∎

Example 2.61 Continuing the analysis of the formula $A = p \wedge (\neg q \vee \neg p)$ from Example 2.59, A is satisfiable if and only at least one of the sets $\{p, \neg p\}$ and $\{p, \neg q\}$ does not contain a complementary pair of literals. Clearly, only the second set does not contain a complementary pair of literals. Using the method described in Theorem 2.60, we obtain the interpretation:

$$\mathscr{I}(p) = T, \qquad \mathscr{I}(q) = F.$$

We leave it to the reader to check that for this interpretation, $v_\mathscr{I}(A) = T$.   ∎

The following example shows what happens if a formula is unsatisfiable.

Example 2.62 Consider the formula:

$$B = (p \vee q) \wedge (\neg p \wedge \neg q).$$

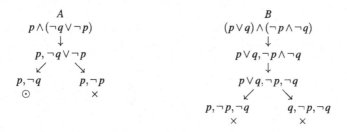

Fig. 2.7 Semantic tableaux

The analysis of the formula proceeds as follows:

- $v_{\mathscr{I}}(B) = T$ if and only if $v_{\mathscr{I}}(p \vee q) = T$ and $v_{\mathscr{I}}(\neg p \wedge \neg q) = T$.
- Decomposing the conjunction, $v_{\mathscr{I}}(B){=}T$ if and only if $v_{\mathscr{I}}(p \vee q) = T$ and $v_{\mathscr{I}}(\neg p) = v_{\mathscr{I}}(\neg q) = T$.
- Decomposing the disjunction, $v_{\mathscr{I}}(B) = T$ if and only if either:

   1. $v_{\mathscr{I}}(p) = v_{\mathscr{I}}(\neg p) = v_{\mathscr{I}}(\neg q) = T$, or
   2. $v_{\mathscr{I}}(q) = v_{\mathscr{I}}(\neg p) = v_{\mathscr{I}}(\neg q) = T$.

Both sets of literals $\{p, \neg p, \neg q\}$ and $\{q, \neg p, \neg q\}$ contain complementary pairs, so by Theorem 2.60, both set of literals are unsatisfiable. We conclude that it is impossible to find a model for B; in other words, B is unsatisfiable.   ∎

2.6.2 Construction of Semantic Tableaux

The decomposition of a formula into sets of literals is rather difficult to follow when expressed textually, as we did in Examples 2.59 and 2.62. In the method of semantic tableaux, sets of formulas label nodes of a tree, where each path in the tree represents the formulas that must be satisfied in one possible interpretation.

The initial formula labels the root of the tree; each node has one or two child nodes depending on how a formula labeling the node is decomposed. The leaves are labeled by the sets of literals. A leaf labeled by a set of literals containing a complementary pair of literals is marked ×, while a leaf labeled by a set not containing a complementary pair is marked ⊙.

Figure 2.7 shows semantic tableaux for the formulas from the examples.

The tableau construction is not unique; here is another tableau for B:

$$(p \vee q) \wedge (\neg p \wedge \neg q)$$
$$\downarrow$$
$$p \vee q, \neg p \wedge \neg q$$

$$p, \neg p \wedge \neg q \qquad\qquad q, \neg p \wedge \neg q$$
$$\downarrow \qquad\qquad\qquad \downarrow$$
$$p, \neg p, \neg q \qquad\qquad q, \neg p, \neg q$$
$$\times \qquad\qquad\qquad \times$$

α	α_1	α_2	β	β_1	β_2
$\neg\neg A_1$	A_1				
$A_1 \wedge A_2$	A_1	A_2	$\neg(B_1 \wedge B_2)$	$\neg B_1$	$\neg B_2$
$\neg(A_1 \vee A_2)$	$\neg A_1$	$\neg A_2$	$B_1 \vee B_2$	B_1	B_2
$\neg(A_1 \rightarrow A_2)$	A_1	$\neg A_2$	$B_1 \rightarrow B_2$	$\neg B_1$	B_2
$\neg(A_1 \uparrow A_2)$	A_1	A_2	$B_1 \uparrow B_2$	$\neg B_1$	$\neg B_2$
$A_1 \downarrow A_2$	$\neg A_1$	$\neg A_2$	$\neg(B_1 \downarrow B_2)$	B_1	B_2
$A_1 \leftrightarrow A_2$	$A_1 \rightarrow A_2$	$A_2 \rightarrow A_1$	$\neg(B_1 \leftrightarrow B_2)$	$\neg(B_1 \rightarrow B_2)$	$\neg(B_2 \rightarrow B_1)$
$\neg(A_1 \oplus A_2)$	$A_1 \rightarrow A_2$	$A_2 \rightarrow A_1$	$B_1 \oplus B_2$	$\neg(B_1 \rightarrow B_2)$	$\neg(B_2 \rightarrow B_1)$

Fig. 2.8 Classification of α- and β-formulas

It is constructed by branching to search for a satisfying interpretation for $p \vee q$ before searching for one for $\neg p \wedge \neg q$. The first tableau contains fewer nodes, showing that it is preferable to decompose conjunctions before disjunctions.

A concise presentation of the rules for creating a semantic tableau can be given if formulas are classified according to their principal operator (Fig. 2.8). If the formula is a negation, the classification takes into account both the negation and the principal operator. α-formulas are conjunctive and are satisfiable only if both subformulas α_1 and α_2 are satisfied, while β-formulas are disjunctive and are satisfied even if only one of the subformulas β_1 or β_2 is satisfiable.

Example 2.63 The formula $p \wedge q$ is classified as an α-formula because it is true if and only if both p and q are true. The formula $\neg(p \wedge q)$ is classified as a β-formula. It is logically equivalent to $\neg p \vee \neg q$ and is true if and only if either $\neg p$ is true or $\neg q$ is true. ∎

We now give the algorithm for the construction of a semantic tableau for a formula in propositional logic.

Algorithm 2.64 (Construction of a semantic tableau)
Input: A formula ϕ of propositional logic.
Output: A semantic tableau $\mathscr{T}$ for ϕ all of whose leaves are marked.

Initially, $\mathscr{T}$ is a tree consisting of a single root node labeled with the singleton set $\{\phi\}$. This node is not marked.

Repeat the following step as long as possible: *Choose* an unmarked leaf l labeled with a set of formulas $U(l)$ and apply one of the following rules.

- $U(l)$ is a set of literals. Mark the leaf *closed* $\times$ if it contains a complementary pair of literals. If not, mark the leaf *open* $\odot$.
- $U(l)$ is not a set of literals. *Choose* a formula in $U(l)$ which is not a literal. Classify the formula as an α-formula A or as a β-formula B and perform one of the following steps according to the classification:

– A is an α-formula. Create a new node l' as a child of l and label l' with:

$$U(l') = (U(l) - \{A\}) \cup \{A_1, A_2\}.$$

(In the case that A is $\neg\neg A_1$, there is no A_2.)
– B is a β-formula. Create two new nodes l' and l'' as children of l. Label l' with:

$$U(l') = (U(l) - \{B\}) \cup \{B_1\},$$

and label l'' with:

$$U(l'') = (U(l) - \{B\}) \cup \{B_2\}.$$

∎

Definition 2.65 A tableau whose construction has terminated is a *completed tableau*. A completed tableau is *closed* if all its leaves are marked closed. Otherwise (if some leaf is marked open), it is *open*. ∎

2.6.3 Termination of the Tableau Construction

Since each step of the algorithm decomposes one formula into one or two simpler formulas, it is clear that the construction of the tableau for any formula terminates, but it is worth proving this claim.

Theorem 2.66 *The construction of a tableau for any formula ϕ terminates. When the construction terminates, all the leaves are marked $\times$ or $\odot$.*

Proof Let us assume that $\leftrightarrow$ and $\oplus$ do not occur in the formula ϕ; the extension of the proof for these cases is left as an exercise.

Consider an unmarked leaf l that is chosen to be expanded during the construction of the tableau. Let $b(l)$ be the total number of binary operators in all formulas in $U(l)$ and let $n(l)$ be the total number of negations in $U(l)$. Define:

$$W(l) = 3 \cdot b(l) + n(l).$$

For example, if $U(l) = \{p \vee q, \neg p \wedge \neg q\}$, then $W(l) = 3 \cdot 2 + 2 = 8$.

Each step of the algorithm adds either a new node l' or a pair of new nodes l', l'' as children of l. We claim that $W(l') < W(l)$ and, if there is a second node, $W(l'') < W(l)$.

Suppose that $A = \neg(A_1 \vee A_2)$ and that the rule for this α-formula is applied at l to obtain a new leaf l' labeled:

$$U(l') = (U(l) - \{\neg(A_1 \vee A_2)\}) \cup \{\neg A_1, \neg A_2\}.$$

Then:

$$W(l') = W(l) - (3 \cdot 1 + 1) + 2 = W(l) - 2 < W(l),$$

because one binary operator and one negation are removed, while two negations are added.

Suppose now that $B = B_1 \vee B_2$ and that the rule for this β-formula is applied at l to obtain two new leaves l', l'' labeled:

$$U(l') = (U(l) - \{B_1 \vee B_2\}) \cup \{B_1\},$$
$$U(l'') = (U(l) - \{B_1 \vee B_2\}) \cup \{B_2\}.$$

Then:

$$W(l') \leq W(l) - (3 \cdot 1) < W(l), \qquad W(l'') \leq W(l) - (3 \cdot 1) < W(l).$$

We leave it to the reader to prove that $W(l)$ decreases for the other α- and β-formulas.

The value of $W(l)$ decreases as each branch in the tableau is extended. Since, obviously, $W(l) \geq 0$, no branch can be extended indefinitely and the construction of the tableau must eventually terminate.

A branch can always be extended if its leaf is labeled with a set of formulas that is not a set of literals. Therefore, when the construction of the tableau terminates, all leaves are labeled with sets of literals and each is marked open or closed by the first rule of the algorithm.   ∎

2.6.4 Improving the Efficiency of the Algorithm *

The algorithm for constructing a tableau is not deterministic: at most steps, there is a choice of which leaf to extend and if the leaf contains more than one formula which is not a literal, there is a choice of which formula to decompose. This opens the possibility of applying heuristics in order to cause the tableau to be completed quickly. We saw in Sect. 2.6.2 that it is better to decompose α-formulas before β-formulas to avoid duplication.

Tableaux can be shortened by closing a branch if it contains a formula and its negation and not just a pair of complementary literals. Clearly, there is no reason to continue expanding a node containing:

$$(p \wedge (q \vee r)), \quad \neg(p \wedge (q \vee r)).$$

We leave it as an exercise to prove that this modification preserves the correctness of the algorithm.

There is a lot of redundancy in copying formulas from one node to another:

$$U(l') = (U(l) - \{A\}) \cup \{A_1, A_2\}.$$

In a variant of semantic tableaux called *analytic tableaux* (Smullyan, 1968), when a new node is created, it is labeled only with the new formulas:

$$U(l') = \{A_1, A_2\}.$$

The algorithm is changed so that the formula to be decomposed is selected from the set of formulas labeling the nodes on the branch from the root to a leaf (provided, of course, that the formula has not already been selected). A leaf is marked closed if two complementary literals (or formulas) appear in the labels of one or two nodes on a branch, and a leaf is marked open if is not closed but there are no more formulas to decompose.

Here is an analytic tableau for the formula B from Example 2.62, where the formula $p \vee q$ is not copied from the second node to the third when $p \wedge q$ is decomposed:

$$(p \vee q) \wedge (\neg p \wedge \neg q)$$
$$\downarrow$$
$$p \vee q, \neg p \wedge \neg q$$
$$\downarrow$$
$$\neg p, \neg q$$

$$\swarrow \qquad \searrow$$

$$p \qquad\qquad q$$
$$\times \qquad\qquad \times$$

We prefer to use semantic tableaux because it is easy to see which formulas are candidates for decomposition and how to mark leaves.

2.7 Soundness and Completeness

The construction of a semantic tableau is a purely formal. The decomposition of a formula depends solely on its syntactical properties: its principal operator and—if it is a negation—the principal operator of the formula that is negated. We gave several examples to motivate semantic tableau, but we have not yet proven that the algorithm is correct. We have not connected the syntactical outcome of the algorithm (Is the tableau closed or not?) with the semantical concept of truth value. In this section, we prove that the algorithm is correct in the sense that it reports that a formula is satisfiable or unsatisfiable if and only if there exists or does not exist a model for the formula.

The proof techniques of this section should be studied carefully because they will be used again and again in other logical systems.

Theorem 2.67 Soundness and completeness *Let $\mathscr{T}$ be a completed tableau for a formula A. A is unsatisfiable if and only if $\mathscr{T}$ is closed.*

Here are some corollaries that follow from the theorem.

Corollary 2.68 *A is satisfiable if and only if $\mathscr{T}$ is open.*

Proof A is satisfiable iff (by definition) A is not unsatisfiable iff (by Theorem 2.67) $\mathscr{T}$ is not closed iff (by definition) $\mathscr{T}$ is open. ∎

Corollary 2.69 *A is valid if and only if the tableau for $\neg A$ closes.*

Proof A is valid iff $\neg A$ is unsatisfiable iff the tableau for $\neg A$ closes. ∎

Corollary 2.70 *The method of semantic tableaux is a decision procedure for validity in propositional logic.*

Proof Let A be a formula of propositional logic. By Theorem 2.66, the construction of the semantic tableau for $\neg A$ terminates in a completed tableau. By the previous corollary, A is valid if and only if the completed tableau is closed. ∎

The forward direction of Corollary 2.69 is called *completeness*: if A is valid, we can discover this fact by constructing a tableau for $\neg A$ and the tableau will close. The converse direction is called *soundness*: any formula A that the tableau construction claims valid (because the tableau for $\neg A$ closes) actually is valid. Invariably in logic, soundness is easier to show than completeness. The reason is that while we only include in a formal system rules that are obviously sound, it is hard to be sure that we haven't forgotten some rule that may be needed for completeness. At the extreme, the following vacuous algorithm is sound but far from complete!

Algorithm 2.71 (Incomplete decision procedure for validity)
Input: A formula A of propositional logic.
Output: A is not valid. ∎

Example 2.72 If the rule for $\neg (A_1 \vee A_2)$ is omitted, the construction of the tableau is still sound, but it is not complete, because it is impossible to construct a closed tableau for the obviously valid formula $A = \neg p \vee p$. Label the root of the tableau with the negation $\neg A = \neg (\neg p \vee p)$; there is now no rule that can be used to decompose the formula. ∎

2.7.1 Proof of Soundness

The theorem to be proved is: if the tableau $\mathscr{T}$ for a formula A closes, then A is unsatisfiable. We will prove a more general theorem: if $\mathscr{T}_n$, the subtree rooted at node n of $\mathscr{T}$, closes then the set of formulas $U(n)$ labeling n is unsatisfiable. Soundness is the special case for the root.

To make the proof easier to follow, we will use $A_1 \wedge A_2$ and $B_1 \vee B_2$ as representatives of the classes of α- and β-formulas, respectively.

Proof of Soundness The proof is by induction on the height h_n of the node n in $\mathscr{T}_n$. Clearly, a closed leaf is labeled by an unsatisfiable set of formulas. Recall (Definition 2.42) that a set of formulas is unsatisfiable iff for any interpretation the truth value of at least one formula is false. In the inductive step, if the children of a node n are labeled by an unsatisfiable set of formulas, then: (a) either the unsatisfiable formula also appears in the label of n, or (b) the unsatisfiable formulas in the labels of the children were used to construct an unsatisfiable formula in the label of n. Let us write out the formal proof.

For the base case, $h_n = 0$, assume that $\mathscr{T}_n$ closes. Since $h_n = 0$ means that n is a leaf, $U(n)$ must contain a complementary set of literals so it is unsatisfiable.

For the inductive step, let n be a node such that $h_n > 0$ in $\mathscr{T}_n$. We need to show that $\mathscr{T}_n$ is closed implies that $U(n)$ is unsatisfiable. By the inductive hypothesis, we can assume that for any node m of height $h_m < h_n$, if $\mathscr{T}_m$ closes, then $U(m)$ is unsatisfiable.

Since $h_n > 0$, the rule for some α- or β-formula was used to create the children of n:

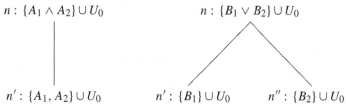

Case 1: $U(n) = \{A_1 \wedge A_2\} \cup U_0$ and $U(n') = \{A_1, A_2\} \cup U_0$ for some (possibly empty) set of formulas U_0.

Clearly, $\mathscr{T}_{n'}$ is also a closed tableau and since $h_{n'} = h_n - 1$, by the inductive hypothesis $U(n')$ is unsatisfiable. Let $\mathscr{I}$ be an arbitrary interpretation. There are two possibilities:

- $v_{\mathscr{I}}(A_0) = F$ for some formula $A_0 \in U_0$. But $U_0 \subset U(n)$ so $U(n)$ is also unsatisfiable.
- Otherwise, $v_{\mathscr{I}}(A_0) = T$ for all $A_0 \in U_0$, so $v_{\mathscr{I}}(A_1) = F$ or $v_{\mathscr{I}}(A_2) = F$. Suppose that $v_{\mathscr{I}}(A_1) = F$. By the definition of the semantics of $\wedge$, this implies that $v_{\mathscr{I}}(A_1 \wedge A_2) = F$. Since $A_1 \wedge A_2 \in U(n)$, $U(n)$ is unsatisfiable. A similar argument holds if $v_{\mathscr{I}}(A_2) = F$.

Case 2: $U(n) = \{B_1 \vee B_2\} \cup U_0$, $U(n') = \{B_1\} \cup U_0$, and $U(n'') = \{B_2\} \cup U_0$ for some (possibly empty) set of formulas U_0.

Clearly, $\mathscr{T}_{n'}$ and $\mathscr{T}_{n''}$ are also closed tableaux and since $h_{n'} \leq h_n - 1$ and $h_{n''} \leq h_n - 1$, by the inductive hypothesis $U(n')$ and $U(n'')$ are both unsatisfiable. Let $\mathscr{I}$ be an arbitrary interpretation. There are two possibilities:

- $v_{\mathscr{I}}(B_0) = F$ for some formula $B_0 \in U_0$. But $U_0 \subset U(n)$ so $U(n)$ is also unsatisfiable.
- Otherwise, $v_{\mathscr{I}}(B_0) = T$ for all $B_0 \in U_0$, so $v_{\mathscr{I}}(B_1) = F$ (since $U(n')$ is unsatisfiable) *and* $v_{\mathscr{I}}(B_2) = F$ (since $U(n'')$ is unsatisfiable). By the definition of the semantics of $\vee$, this implies that $v_{\mathscr{I}}(B_1 \vee B_2) = F$. Since $B_1 \vee B_2 \in U(n)$, $U(n)$ is unsatisfiable. ∎

2.7.2 Proof of Completeness

The theorem to be proved is: if A is unsatisfiable then *every* tableau for A closes. Completeness is much more difficult to prove than soundness. For soundness, we had a single (though arbitrary) closed tableau for a formula A and we proved that A is unsatisfiable by induction on the structure of a tableau. Here we need to prove that no matter how the tableau for A is constructed, it must close.

Rather than prove that every tableau must close, we prove the contrapositive (Corollary 2.68): if some tableau for A is open (has an open branch), then A is satisfiable. Clearly, there is a model for the set of literals labeling the leaf of an open branch. We extend this to an interpretation for A and then prove by induction on the length of the branch that the interpretation is a model of the sets of formulas labeling the nodes on the branch, including the singleton set $\{A\}$ that labels the root.

Let us look at some examples.

Example 2.73 Let $A = p \wedge (\neg q \vee \neg p)$. We have already constructed the tableau for A which is reproduced here:

$$p \wedge (\neg q \vee \neg p)$$
$$\downarrow$$
$$p, \neg q \vee \neg p$$
$$\swarrow \qquad \searrow$$
$$p, \neg q \qquad\qquad p, \neg p$$
$$\odot \qquad\qquad\quad \times$$

The interpretation $\mathscr{I}(p) = T$, $\mathscr{I}(q) = F$ defined by assigning T to the literals labeling the leaf of the open branch is clearly a model for A.   ∎

Example 2.74 Now let $A = p \vee (q \wedge \neg q)$; here is a tableau for A:

$$p \vee (q \wedge \neg q)$$
$$\swarrow \qquad \searrow$$
$$p \qquad\quad q \wedge \neg q$$
$$\odot \qquad\qquad \downarrow$$
$$\qquad\qquad q, \neg q$$
$$\qquad\qquad \times$$

The open branch of the tableau terminates in a leaf labeled with the singleton set of literals $\{p\}$. We can conclude that any model for A must define $\mathscr{I}(p) = T$. However, an interpretation for A must also define an assignment to q and the leaf gives us no guidance as to which value to choose for $\mathscr{I}(q)$. But it is obvious that it doesn't matter what value is assigned to q; in either case, the interpretation will be a model of A.   ∎

To prove completeness we need to show that the assignment of T to the literals labeling the leaf of an open branch can be extended to a model of the formula labeling the root. There are four steps in the proof:

1. Define a property of sets of formulas;
2. Show that the union of the formulas labeling nodes in an open branch has this property;
3. Prove that any set having this property is satisfiable;
4. Note that the formula labeling the root is in the set.

Definition 2.75 Let U be a set of formulas. U is a *Hintikka set* iff:

1. For all atoms p appearing in a formula of U, either $p \notin U$ or $\neg p \notin U$.
2. If $A \in U$ is an α-formula, then $A_1 \in U$ *and* $A_2 \in U$.
3. If $B \in U$ is a β-formula, then $B_1 \in U$ *or* $B_2 \in U$.   ∎

Example 2.76 U, the union of the set of formulas labeling the nodes in the open branch of Example 2.74, is $\{p, \ p \vee (q \wedge \neg q)\}$. We claim that U is a Hintikka set. Condition (1) obviously holds since there is only one literal p in U and $\neg p \notin U$. Condition (2) is vacuous. For Condition (3), $B = p \vee (q \wedge \neg q) \in U$ is a β-formula and $B_1 = p \in U$.   ∎

Condition (1) requires that a Hintikka set not contain a complementary pair of literals, which to be expected on an open branch of a tableau. Conditions (2) and (3) ensure that U is *downward saturated*, that is, U contains sufficient subformulas so that the decomposition of the formula to be satisfied will not take us out of U. In turn, this ensures that an interpretation defined by the set of literals in U will make all formulas in U true.

The second step of the proof of completeness is to show that the set of formulas labeling the nodes in an open branch is a Hintikka set.

Theorem 2.77 *Let l be an open leaf in a completed tableau $\mathcal{T}$. Let $U = \bigcup_i U(i)$, where i runs over the set of nodes on the branch from the root to l. Then U is a Hintikka set.*

Proof In the construction of the semantic tableau, there are no rules for decomposing a literal p or $\neg p$. Thus if a literal p or $\neg p$ appears for the first time in $U(n)$ for some n, the literal will be copied into $U(k)$ for all nodes k on the branch from n to l, in particular, $p \in U(l)$ or $\neg p \in U(l)$. This means that all literals in U appear in $U(l)$. Since the branch is open, no complementary pair of literals appears in $U(l)$, so Condition (1) holds for U.

Suppose that $A \in U$ is an α-formula. Since the tableau is completed, A was the formula selected for decomposing at some node n in the branch from the root to l. Then $\{A_1, A_2\} \subseteq U(n') \subseteq U$, so Condition (2) holds.

Suppose that $B \in U$ is a β-formula. Since the tableau is completed, B was the formula selected for decomposing at some node n in the branch from the root to l. Then either $B_1 \in U(n') \subseteq U$ or $B_2 \in U(n') \subseteq U$, so Condition (3) holds.   ∎

The third step of the proof is to show that a Hintikka set is satisfiable.

Theorem 2.78 (Hintikka's Lemma) *Let U be a Hintikka set. Then U is satisfiable.*

Proof We define an interpretation and then show that the interpretation is a model of U. Let $\mathscr{P}_U$ be set of all atoms appearing in all formulas of U. Define an interpretation $\mathscr{I} : \mathscr{P}_U \mapsto \{T, F\}$ as follows:

$$
\begin{aligned}
\mathscr{I}(p) &= T &&\text{if } p \in U, \\
\mathscr{I}(p) &= F &&\text{if } \neg\, p \in U, \\
\mathscr{I}(p) &= T &&\text{if } p \notin U \text{ and } \neg\, p \notin U.
\end{aligned}
$$

Since U is a Hintikka set, by Condition (1) $\mathscr{I}$ is well-defined, that is, every atom in $\mathscr{P}_U$ is given exactly one value. Example 2.74 demonstrates the third case: the atom q appears in a *formula* of U so $q \in \mathscr{P}_U$, but neither the literal q nor its complement $\neg\, q$ appear in U. The atom is arbitrarily mapped to the truth value T.

We show by structural induction that for any $A \in U$, $v_{\mathscr{I}}(A) = T$.

- If A is an atom p, then $v_{\mathscr{I}}(A) = v_{\mathscr{I}}(p) = \mathscr{I}(p) = T$ since $p \in U$.
- If A is a negated atom $\neg\, p$, then since $\neg\, p \in U$, $\mathscr{I}(p) = F$, so $v_{\mathscr{I}}(A) = v_{\mathscr{I}}(\neg\, p) = T$.
- If A is an α-formula, by Condition (2) $A_1 \in U$ and $A_2 \in U$. By the inductive hypothesis, $v_{\mathscr{I}}(A_1) = v_{\mathscr{I}}(A_2) = T$, so $v_{\mathscr{I}}(A) = T$ by definition of the conjunctive operators.
- If A is β-formula B, by Condition (3) $B_1 \in U$ or $B_2 \in U$. By the inductive hypothesis, either $v_{\mathscr{I}}(B_1) = T$ or $v_{\mathscr{I}}(B_2) = T$, so $v_{\mathscr{I}}(A) = v_{\mathscr{I}}(B) = T$ by definition of the disjunctive operators.   ∎

Proof of Completeness Let $\mathscr{T}$ be a completed open tableau for A. Then U, the union of the labels of the nodes on an open branch, is a Hintikka set by Theorem 2.77. Theorem 2.78 shows an interpretation $\mathscr{I}$ can be found such that U is simultaneously satisfiable in $\mathscr{I}$. A, the formula labeling the root, is an element of U so $\mathscr{I}$ is a model of A.   ∎

2.8 Summary

The presentation of propositional logic was carried out in a manner that we will use for all systems of logic. First, the syntax of formulas is given. The formulas are defined as trees, which avoids ambiguity and simplifies the description of structural induction.

The second step is to define the semantics of formulas. An interpretation is a mapping of atomic propositions to the values $\{T, F\}$. An interpretation is used to give a truth value to any formula by induction on the structure of the formula, starting from atoms and proceeding to more complex formulas using the definitions of the Boolean operators.

A formula is satisfiable iff it is true in *some* interpretation and it is valid iff is true in *all* interpretations. Two formulas whose values are the same in all interpretations are logically equivalent and can be substituted for each other. This can be used to show that for any formula, there exists a logically equivalent formula that uses only negation and either conjunction or disjunction.

While truth tables can be used as a decision procedure for the satisfiability or validity of formulas of propositional logic, semantic tableaux are usually much more efficient. In a semantic tableau, a tree is constructed during a search for a model of a formula; the construction is based upon the structure of the formula. A semantic tableau is closed if the formula is unsatisfiable and open if it is satisfiable.

We proved that the algorithm for semantic tableaux is sound and complete as a decision procedure for satisfiability. This theorem connects the syntactical aspect of a formula that guides the construction of the tableau with its meaning. The central concept in the proof is that of a Hintikka set, which gives conditions that ensure that a model can be found for a set of formulas.

2.9 Further Reading

The presentation of semantic tableaux follows that of Smullyan (1968) although he uses analytic tableaux. Advanced textbooks that also use tableaux are Nerode and Shore (1997) and Fitting (1996).

2.10 Exercises

2.1 Draw formation trees and construct truth tables for

$$(p \to (q \to r)) \to ((p \to q) \to (p \to r)),$$
$$(p \to q) \to p,$$
$$((p \to q) \to p) \to p.$$

2.2 Prove that there is a unique formation tree for every derivation tree.

2.3 Prove the following logical equivalences:

$$A \wedge (B \vee C) \equiv (A \wedge B) \vee (A \wedge C),$$
$$A \vee B \equiv \neg(\neg A \wedge \neg B),$$
$$A \wedge B \equiv \neg(\neg A \vee \neg B),$$
$$A \to B \equiv \neg A \vee B,$$
$$A \to B \equiv \neg(A \wedge \neg B).$$

2.4 Prove $((A \oplus B) \oplus B) \equiv A$ and $((A \leftrightarrow B) \leftrightarrow B) \equiv A$.

2.5 Simplify $A \wedge (A \vee B)$ and $A \vee (A \wedge B)$.

2.6 Prove the following logical equivalences using truth tables, semantic tableaux or Venn diagrams:

$$
\begin{aligned}
A \to B &\equiv A \leftrightarrow (A \wedge B), \\
A \to B &\equiv B \leftrightarrow (A \vee B), \\
A \wedge B &\equiv (A \leftrightarrow B) \leftrightarrow (A \vee B), \\
A \leftrightarrow B &\equiv (A \vee B) \to (A \wedge B).
\end{aligned}
$$

2.7 Prove $\models (A \to B) \vee (B \to C)$.

2.8 Prove or disprove:

$$
\models ((A \to B) \to B) \to B,
$$

$$
\models (A \leftrightarrow B) \leftrightarrow (A \leftrightarrow (B \leftrightarrow A)).
$$

2.9 Prove:

$$
\models ((A \wedge B) \to C) \to ((A \to C) \vee (B \to C)).
$$

This formula may seem strange since it could be misinterpreted as saying that if C follows from $A \wedge B$, then it follows from one or the other of A or B. To clarify this, show that:

$$
\{A \wedge B \to C\} \models (A \to C) \vee (B \to C),
$$

but:

$$
\{A \wedge B \to C\} \not\models A \to C,
$$

$$
\{A \wedge B \to C\} \not\models B \to C.
$$

2.10 Complete the proof that $\uparrow$ and $\downarrow$ can each define all unary and binary Boolean operators (Theorem 2.37).

2.11 Prove that $\wedge$ and $\vee$ cannot define all Boolean operators.

2.12 Prove that $\{\neg, \leftrightarrow\}$ *cannot* define all Boolean operators.

2.13 Prove that $\uparrow$ and $\downarrow$ are not associative.

2.14 Prove that if U is satisfiable then $U \cup \{B\}$ is not necessarily satisfiable.

2.15 Prove Theorems 2.44–2.47 on the satisfiability of sets of formulas.

2.16 Prove Theorems 2.50–2.54 on logical consequence.

2.17 Prove that for a set of axioms U, $\mathcal{T}(U)$ is closed under logical consequence (see Definition 2.55).

2.18 Complete the proof that the construction of a semantic tableau terminates (Theorem 2.66).

2.19 Prove that the method of semantic tableaux remains sound and complete if a tableau can be closed non-atomically.

2.20 Manna (1974) Let *ifte* be a *tertiary* (3-place) operator defined by:

A	B	C	$ifte(A, B, C)$
T	T	T	T
T	T	F	T
T	F	T	F
T	F	F	F
F	T	T	T
F	T	F	F
F	F	T	T
F	F	F	F

The operator can be defined using infix notation as:

$$if\ A\ then\ B\ else\ C.$$

1. Prove that *if then else* by itself forms an adequate sets of operators if the use of the constant formulas *true* and *false* is allowed.
2. Prove: $\models if\ A\ then\ B\ else\ C \equiv (A \rightarrow B) \wedge (\neg A \rightarrow C)$.
3. Add a rule for the operator *if then else* to the algorithm for semantic tableaux.

References

M. Fitting. *First-Order Logic and Automated Theorem Proving (Second Edition)*. Springer, 1996.
J.E. Hopcroft, R. Motwani, and J.D. Ullman. *Introduction to Automata Theory, Languages and Computation (Third Edition)*. Addison-Wesley, 2006.
Z. Manna. *Mathematical Theory of Computation*. McGraw-Hill, New York, NY, 1974. Reprinted by Dover, 2003.
A. Nerode and R.A. Shore. *Logic for Applications (Second Edition)*. Springer, 1997.
R.M. Smullyan. *First-Order Logic*. Springer-Verlag, 1968. Reprinted by Dover, 1995.

Chapter 3
Propositional Logic: Deductive Systems

The concept of deducing theorems from a set of axioms and rules of inference is very old and is familiar to every high-school student who has studied Euclidean geometry. Modern mathematics is expressed in a style of reasoning that is not far removed from the reasoning used by Greek mathematicians. This style can be characterized as 'formalized informal reasoning', meaning that while the proofs are expressed in natural language rather than in a formal system, there are conventions among mathematicians as to the forms of reasoning that are allowed. The deductive systems studied in this chapter were developed in an attempt to formalize mathematical reasoning.

We present two deductive systems for propositional logic. The second one $\mathscr{H}$ will be familiar because it is a formalization of step-by-step proofs in mathematics: It contains a set of three axioms and one rule of inference; proofs are constructed as a sequence of formulas, each of which is either an axiom (or a formula that has been previously proved) or a derivation of a formula from previous formulas in the sequence using the rule of inference. The system $\mathscr{G}$ will be less familiar because it has one axiom and many rules of inference, but we present it first because it is almost trivial to prove the soundness and completeness of $\mathscr{G}$ from its relationship with semantic tableaux. The proof of the soundness and completeness of $\mathscr{H}$ is then relatively easy to show by using $\mathscr{G}$. The chapter concludes with three short sections: the definition of an important property called *consistency*, a generalization to infinite sets of formulas, and a survey of other deductive systems for propositional logic.

3.1 Why Deductive Proofs?

Let $U = \{A_1, \ldots, A_n\}$. Theorem 2.50 showed that $U \models A$ if and only if $\models A_1 \wedge \cdots \wedge A_n \to A$. Therefore, if U is a set of axioms, we can use the completeness of the method of semantic tableaux to determine if A follows from U (see Sect. 2.5.4 for precise definitions). Why would we want to go through the trouble of searching for a mathematical proof when we can easily compute if a formula is valid?

M. Ben-Ari, *Mathematical Logic for Computer Science*,
DOI 10.1007/978-1-4471-4129-7_3, © Springer-Verlag London 2012

There are several problems with a purely semantical approach:

- The set of axioms may be infinite. For example, the axiom of induction in arithmetic is really an infinite set of axioms, one for each property to be proved. For semantic tableaux in propositional logic, the only formulas that appear in the tableaux are subformulas of the formula being checked or their negations, and there are only a finite number of such formulas.
- Very few logics have decision procedures like propositional logic.
- A decision procedure may not give insight into the relationship between the axioms and the theorem. For example, in proofs of theorems about prime numbers, we would want to know exactly where primality is used (Velleman, 2006, Sect. 3.7). This understanding can also help us propose other formulas that might be theorems.
- A decision procedure produces a 'yes/no' answer, so it is difficult to recognize intermediate results (lemmas). Clearly, the millions of mathematical theorems in existence could not have been inferred directly from axioms.

Definition 3.1 A *deductive system* is a set of formulas called *axioms* and a set of *rules of inference*. A *proof* in a deductive system is a sequence of formulas $S = \{A_1, \ldots, A_n\}$ such that each formula A_i is either an axiom or it can be inferred from previous formulas of the sequence $A_{j_1}, \ldots, A_{j_k}$, where $j_1 < \cdots < j_k < i$, using a rule of inference. For A_n, the last formula in the sequence, we say that A_n is a *theorem*, the sequence S is a *proof* of A_n, and A_n is *provable*, denoted $\vdash A_n$. If $\vdash A$, then A may be used like an axiom in a subsequent proof.   ∎

The deductive approach can overcome the problems described above:

- There may be an infinite number of axioms, but only a finite number will appear in any proof.
- Although a proof is not a decision procedure, it can be mechanically *checked*; that is, given a sequence of formulas, an syntax-based algorithm can easily check whether the sequence is a proof as defined above.
- The proof of a formula clearly shows which axioms, theorems and rules are used and for what purposes.
- Once a theorem has been proved, it can be used in proofs like an axiom.

Deductive proofs are not generated by decision procedures because the formulas that appear in a proof are not limited to subformulas of the theorem and because there is no algorithm telling us how to generate the next formula in the sequence forming a proof. Nevertheless, algorithms and heuristics can be used to build software systems called *automatic theorem provers* which search for proofs. In Chap. 4, we will study a deductive system that has been successfully used in automatic theorem provers. Another promising approach is to use a *proof assistant* which performs administrative tasks such as proof checking, bookkeeping and cataloging previously proved theorems, but a person guides the search by suggesting lemmas that are likely to lead to a proof.

α	α_1	α_2	β	β_1	β_2
$\neg\neg A$	A				
$\neg(A_1 \wedge A_2)$	$\neg A_1$	$\neg A_2$	$B_1 \wedge B_2$	B_1	B_2
$A_1 \vee A_2$	A_1	A_2	$\neg(B_1 \vee B_2)$	$\neg B_1$	$\neg B_2$
$A_1 \rightarrow A_2$	$\neg A_1$	A_2	$\neg(B_1 \rightarrow B_2)$	B_1	$\neg B_2$
$A_1 \uparrow A_2$	$\neg A_1$	$\neg A_2$	$\neg(B_1 \uparrow B_2)$	B_1	B_2
$\neg(A_1 \downarrow A_2)$	A_1	A_2	$B_1 \downarrow B_2$	$\neg B_1$	$\neg B_2$
$\neg(A_1 \leftrightarrow A_2)$	$\neg(A_1 \rightarrow A_2)$	$\neg(A_2 \rightarrow A_1)$	$B_1 \leftrightarrow B_2$	$B_1 \rightarrow B_2$	$B_2 \rightarrow B_1$
$A_1 \oplus A_2$	$\neg(A_1 \rightarrow A_2)$	$\neg(A_2 \rightarrow A_1)$	$\neg(B_1 \oplus B_2)$	$B_1 \rightarrow B_2$	$B_2 \rightarrow B_1$

Fig. 3.1 Classification of α- and β-formulas

3.2 Gentzen System $\mathscr{G}$

The first deductive system that we study is based on a system proposed by Gerhard Gentzen in the 1930s. The system itself will seem unfamiliar because it has one type of axiom and many rules of inference, unlike familiar mathematical theories which have multiple axioms and only a few rules of inference. Furthermore, deductions in the system can be naturally represented as trees rather in the linear format characteristic of mathematical proofs. However, it is this property that makes it easy to relate Gentzen systems to semantic tableaux.

Definition 3.2 (Gentzen system $\mathscr{G}$) An *axiom* of $\mathscr{G}$ is a set of literals U containing a complementary pair. *Rule of inference* are used to infer a set of formulas U from one or two other sets of formulas U_1 and U_2; there are two types of rules, defined with reference to Fig. 3.1:

- Let $\{\alpha_1, \alpha_2\} \subseteq U_1$ and let $U_1' = U_1 - \{\alpha_1, \alpha_2\}$. Then $U = U_1' \cup \{\alpha\}$ can be inferred.
- Let $\{\beta_1\} \subseteq U_1$, $\{\beta_2\} \subseteq U_2$ and let $U_1' = U_1 - \{\beta_1\}$, $U_2' = U_2 - \{\beta_2\}$. Then $U = U_1' \cup U_2' \cup \{\beta\}$ can be inferred.

The set or sets of formulas U_1, U_2 are the *premises* and set of formulas U that is inferred is the *conclusion*. A set of formulas U that is an axiom or a conclusion is said to be *proved*, denoted $\vdash U$. The following notation is used for rules of inference:

$$\frac{\vdash U_1' \cup \{\alpha_1, \alpha_2\}}{\vdash U_1' \cup \{\alpha\}} \qquad \frac{\vdash U_1' \cup \{\beta_1\} \qquad \vdash U_2' \cup \{\beta_2\}}{\vdash U_1' \cup U_2' \cup \{\beta\}}.$$

Braces can be omitted with the understanding that a sequence of formulas is to be interpreted as a set (with no duplicates). ∎

Example 3.3 The following set of formulas is an axiom because it contains the complementary pair $\{r, \neg r\}$:

$$\vdash p \wedge q, q, r, \neg r, q \vee \neg r.$$

The disjunction rule for $A_1 = q$, $A_2 = \neg r$ can be used to deduce:

$$\frac{\vdash p \wedge q, q, r, \neg r, q \vee \neg r}{\vdash p \wedge q, r, q \vee \neg r, q \vee \neg r}.$$

Removing the duplicate formula $q \vee \neg r$ gives:

$$\frac{\vdash p \wedge q, q, r, \neg r, q \vee \neg r}{\vdash p \wedge q, r, q \vee \neg r}.$$

Note that the premises $\{q, \neg r\}$ are no longer elements of the conclusion. ∎

A proof can be written as a sequence of sets of formulas, which are numbered for convenient reference. On the right of each line is its *justification*: either the set of formulas is an axiom, or it is the conclusion of a rule of inference applied to a set or sets of formulas earlier in the sequence. A rule of inference is identified by the rule used for the α- or β-formula on the principal operator of the conclusion and by the number or numbers of the lines containing the premises.

Example 3.4 Prove $\vdash (p \vee q) \to (q \vee p)$ in $\mathcal{G}$.

Proof

1.	$\vdash \neg p, q, p$	Axiom
2.	$\vdash \neg q, q, p$	Axiom
3.	$\vdash \neg (p \vee q), q, p$	$\beta \vee, 1, 2$
4.	$\vdash \neg (p \vee q), (q \vee p)$	$\alpha \vee, 3$
5.	$\vdash (p \vee q) \to (q \vee p)$	$\alpha \to, 4$

∎

Example 3.5 Prove $\vdash p \vee (q \wedge r) \to (p \vee q) \wedge (p \vee r)$ in $\mathcal{G}$.

Proof

1.	$\vdash \neg p, p, q$	Axiom
2.	$\vdash \neg p, (p \vee q)$	$\alpha \vee, 1$
3.	$\vdash \neg p, p, r$	Axiom
4.	$\vdash \neg p, (p \vee r)$	$\alpha \vee, 3$
5.	$\vdash \neg p, (p \vee q) \wedge (p \vee r)$	$\beta \wedge, 2, 4$

6.	$\vdash \neg q, \neg r, p, q$	Axiom
7.	$\vdash \neg q, \neg r, (p \vee q)$	$\alpha \vee, 6$
8.	$\vdash \neg q, \neg r, p, r$	Axiom
9.	$\vdash \neg q, \neg r, (p \vee r)$	$\alpha \vee, 8$
10.	$\vdash \neg q, \neg r, (p \vee q) \wedge (p \vee r)$	$\beta \wedge, 7, 9$
11.	$\vdash \neg (q \wedge r), (p \vee q) \wedge (p \vee r)$	$\alpha \wedge, 10$

12.	$\vdash \neg (p \vee (q \wedge r)), (p \vee q) \wedge (p \vee r)$	$\beta \vee, 5, 11$
13.	$\vdash p \vee (q \wedge r) \to (p \vee q) \wedge (p \vee r)$	$\alpha \to, 12$

∎

3.2.1 The Relationship Between $\mathscr{G}$ and Semantic Tableaux

It might seem that we have been rather clever to arrange all the inferences in these proofs so that everything comes out exactly right in the end. In fact, no cleverness was required. Let us rearrange the Gentzen proof into a tree format rather than a linear sequence of sets of formulas. Let the axioms be the leaves of the tree, and let the inference rules define the interior nodes. The root at the bottom will be labeled with the formula that is proved.

The proof from Example 3.4 is displayed in tree form on the left below:

$$\neg p, q, p \qquad\qquad \neg q, q, p \qquad\qquad\qquad \neg[(p \vee q) \to (q \vee p)]$$

$$\searrow \qquad \swarrow \qquad\qquad\qquad\qquad \downarrow$$

$$\neg(p \vee q), q, p \qquad\qquad\qquad p \vee q, \neg(q \vee p)$$

$$\downarrow \qquad\qquad\qquad\qquad\qquad \downarrow$$

$$\neg(p \vee q), (q \vee p) \qquad\qquad\qquad p \vee q, \neg q, \neg p$$

$$\downarrow \qquad\qquad\qquad\qquad\qquad \swarrow \qquad\qquad \searrow$$

$$(p \vee q) \to (q \vee p) \qquad\qquad p, \neg q, \neg p \qquad\qquad q, \neg q, \neg p$$

$$\times \qquad\qquad\qquad \times$$

If this looks familiar, it should. The semantic tableau on the right results from turning the derivation in $\mathscr{G}$ upside down and replacing each formula in the labels on the nodes by its complement (Definition 2.57).

A set of formulas labeling a node in a semantic tableau is an *implicit conjunction*, that is, all the formulas in the set must evaluate to true for the set to be true. By taking complements, a set of formulas labeling a node in a derivation in $\mathscr{G}$ is an *implicit disjunction*.

An axiom in $\mathscr{G}$ is valid: Since it contains a complementary pair of literals, as a disjunction it is:

$$\cdots \vee p \vee \cdots \vee \neg p \vee \cdots,$$

which is valid.

Consider a rule applied to obtain an α-formula, for example, $A_1 \vee A_2$; when the rule is written using disjunctions it becomes:

$$\frac{\vdash \bigvee U_1' \vee A_1 \vee A_2}{\vdash \bigvee U_1' \vee (A_1 \vee A_2)},$$

and this is a valid inference in propositional logic that follows immediately from associativity.

Similarly, when a rule is applied to obtain a β-formula, we have:

$$\frac{\vdash \bigvee U_1' \vee B_1 \qquad\qquad \bigvee U_2' \vee B_2}{\vdash \bigvee U_1' \vee \bigvee U_2' \vee (B_1 \wedge B_2)},$$

which follows by the distribution of disjunction over conjunction. This inference simply says that if we can prove both B_1 and B_2 then we can prove $B_1 \wedge B_2$.

The relationship between semantic tableaux and Gentzen systems is formalized in the following theorem.

Theorem 3.6 *Let A be a formula in propositional logic. Then $\vdash A$ in $\mathscr{G}$ if and only if there is a closed semantic tableau for $\neg A$.*

This follows immediately from a more general theorem on sets of formulas.

Theorem 3.7 *Let U be a set of formulas and let $\bar{U}$ be the set of complements of formulas in U. Then $\vdash U$ in $\mathscr{G}$ if and only if there is a closed semantic tableau for $\bar{U}$.*

Proof Let $\mathscr{T}$ be a closed semantic tableau for $\bar{U}$. We prove $\vdash U$ by induction on h, the height of $\mathscr{T}$. The other direction is left as an exercise.

If $h = 0$, then $\mathscr{T}$ consists of a single node labeled by $\bar{U}$. By assumption, $\mathscr{T}$ is closed, so it contains a complementary pair of literals $\{p, \neg p\}$, that is, $\bar{U} = \bar{U}' \cup \{p, \neg p\}$. Obviously, $U = U' \cup \{\neg p, p\}$ is an axiom in $\mathscr{G}$, hence $\vdash U$.

If $h > 0$, then some tableau rule was used on an α- or β-formula at the root of $\mathscr{T}$ on a formula $\bar{\phi} \in \bar{U}$, that is, $\bar{U} = \bar{U}' \cup \{\bar{\phi}\}$. The proof proceeds by cases, where you must be careful to distinguish between applications of the tableau rules and applications of the Gentzen rules of the same name.

Case 1: $\bar{\phi}$ is an α-formula (such as) $\neg(A_1 \vee A_2)$. The tableau rule created a child node labeled by the set of formulas $\bar{U}' \cup \{\neg A_1, \neg A_2\}$. By assumption, the subtree rooted at this node is a closed tableau, so by the inductive hypothesis, $\vdash U' \cup \{A_1, A_2\}$. Using the appropriate rule of inference from $\mathscr{G}$, we obtain $\vdash U' \cup \{A_1 \vee A_2\}$, that is, $\vdash U' \cup \{\phi\}$, which is $\vdash U$.

Case 2: $\bar{\phi}$ is a β-formula (such as) $\neg(B_1 \wedge B_2)$. The tableau rule created two child nodes labeled by the sets of formulas $\bar{U}' \cup \{\neg B_1\}$ and $\bar{U}' \cup \{\neg B_2\}$. By assumption, the subtrees rooted at this node are closed, so by the inductive hypothesis $\vdash U' \cup \{B_1\}$ and $\vdash U' \cup \{B_2\}$. Using the appropriate rule of inference from $\mathscr{G}$, we obtain $\vdash U' \cup \{B_1 \wedge B_2\}$, that is, $\vdash U' \cup \{\phi\}$, which is $\vdash U$. ∎

Theorem 3.8 (Soundness and completeness of $\mathscr{G}$)
$\models A$ *if and only if* $\vdash A$ *in* $\mathscr{G}$.

Proof A is valid iff $\neg A$ is unsatisfiable iff there is a closed semantic tableau for $\neg A$ iff there is a proof of A in $\mathscr{G}$. ∎

The proof is very simple because we did all the hard work in the proof of the soundness and completeness of tableaux.

The Gentzen system $\mathscr{G}$ described in this section is not very useful; other versions (surveyed in Sect. 3.9) are more convenient for proving theorems and are closer to Gentzen's original formulation. We introduced $\mathscr{G}$ as a theoretical stepping stone to Hilbert systems which we now describe.

3.3 Hilbert System $\mathscr{H}$

In Gentzen systems there is one axiom and many rules of inference, while in a Hilbert system there are several axioms but only one rule of inference. In this section, we define the deductive system $\mathscr{H}$ and use it to prove many theorems. Actually, only one theorem (Theorem 3.10) will be proved directly from the axioms and the rule of inference; practical use of the system depends on the use of derived rules, especially the deduction rule.

Notation: Capital letters $A, B, C, \ldots$ represent arbitrary formulas in propositional logic. For example, the notation $\vdash A \rightarrow A$ means: for *any* formula A of propositional logic, the formula $A \rightarrow A$ can be proved.

Definition 3.9 (Deductive system $\mathscr{H}$) The *axioms* of $\mathscr{H}$ are:

   Axiom 1 $\vdash (A \rightarrow (B \rightarrow A))$,

   Axiom 2 $\vdash (A \rightarrow (B \rightarrow C)) \rightarrow ((A \rightarrow B) \rightarrow (A \rightarrow C))$,

   Axiom 3 $\vdash (\neg B \rightarrow \neg A) \rightarrow (A \rightarrow B)$.

The *rule of inference* is **modus ponens** (*MP* for short):

$$\frac{\vdash A \qquad\qquad \vdash A \rightarrow B}{\vdash B}.$$

In words: the formula B can be inferred from A and $A \rightarrow B$.
 The terminology used for $\mathscr{G}$—*premises, conclusion, theorem, proved*— carries over to $\mathscr{H}$, as does the symbol $\vdash$ meaning that a formula is proved. ■

Theorem 3.10 $\vdash A \rightarrow A$.

Proof

1.	$\vdash (A \rightarrow ((A \rightarrow A) \rightarrow A)) \rightarrow ((A \rightarrow (A \rightarrow A)) \rightarrow (A \rightarrow A))$	Axiom 2
2.	$\vdash A \rightarrow ((A \rightarrow A) \rightarrow A)$	Axiom 1
3.	$\vdash (A \rightarrow (A \rightarrow A)) \rightarrow (A \rightarrow A)$	MP 1, 2
4.	$\vdash A \rightarrow (A \rightarrow A)$	Axiom 1
5.	$\vdash A \rightarrow A$	MP 3, 4

                                      ■

When an axiom is given as the justification, identify which formulas are substituted for the formulas A, B, C in the definition of the axioms above.

3.3.1 Axiom Schemes and Theorem Schemes *

As we noted above, a capital letter can be replaced by any formula of propositional logic, so, strictly speaking, $\vdash A \rightarrow (B \rightarrow A)$ is not an axiom, and similarly, $\vdash A \rightarrow A$

is not a theorem. A more precise terminology would be to say that $\vdash A \rightarrow (B \rightarrow A)$ is an *axiom scheme* that is a shorthand for an infinite number of axioms obtained by replacing the 'variables' A and B with actual formulas, for example:

$$\overbrace{((p \vee \neg q) \leftrightarrow r)}^{A} \rightarrow (\overbrace{\neg (q \wedge \neg r)}^{B} \rightarrow \overbrace{((p \vee \neg q) \leftrightarrow r)}^{A}).$$

Similarly, $\vdash A \rightarrow A$ is a *theorem scheme* that is a shorthand for an infinite number of theorems that can be proved in $\mathscr{H}$, including, for example:

$$\vdash ((p \vee \neg q) \leftrightarrow r) \rightarrow ((p \vee \neg q) \leftrightarrow r).$$

We will not retain this precision in our presentation because it will always clear if a given formula is an instance of a particular axiom scheme or theorem scheme. For example, a formula ϕ is an instance of Axiom 1 if it is of the form:

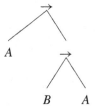

where there are subtrees for the formulas represented by A and B. There is a simple and efficient algorithm that checks if ϕ is of this form and if the two subtrees A are identical.

3.3.2 The Deduction Rule

The proof of Theorem 3.10 is rather complicated for such a trivial formula. In order to formalize the powerful methods of inference used in mathematics, we introduce new rules of inference called *derived rules*. The most important derived rule is the *deduction rule*. Suppose that you want to prove $A \rightarrow B$. *Assume* that A has already been proved and use it in the proof of B. This is not a proof of B unless A is an axiom or theorem that has been previously proved, in which case it can be used directly in the proof. However, we claim that the proof can be mechanically transformed into a proof of $A \rightarrow B$.

Example 3.11 The deduction rule is used frequently in mathematics. Suppose that you want to prove that *the sum of any two odd integer numbers is even*, expressed formally as:

$$odd(x) \wedge odd(y) \rightarrow even(x + y),$$

for every x and y. To prove this formula, let us *assume* the formula $odd(x) \wedge odd(y)$ as if it were an additional axiom. We have available all the theorems we have already deduced about odd numbers, in particular, the theorem that any odd number can be expressed as $2k + 1$. Computing:

$$x + y = 2k_1 + 1 + 2k_2 + 1 = 2(k_1 + k_2 + 1),$$

we obtain that $x + y$ is a multiple of 2, that is, $even(x + y)$. The theorem now follows from the deduction rule which *discharges* the assumption.   ∎

To express the deduction rule, we extend the definition of *proof*.

Definition 3.12 Let U be a set of formulas and A a formula. The notation $U \vdash A$ means that the formulas in U are *assumptions* in the proof of A. A *proof* is a sequence of lines $U_i \vdash \phi_i$, such that for each i, $U_i \subseteq U$, and ϕ_i is an axiom, a previously proved theorem, a member of U_i or can be derived by *MP* from previous lines $U_{i'} \vdash \phi_{i'}, U_{i''} \vdash \phi_{i''}$, where $i', i'' < i$.   ∎

Rule 3.13 (Deduction rule)

$$\frac{U \cup \{A\} \vdash B}{U \vdash A \rightarrow B}.$$

We must show that this derived rule is *sound*, that is, that the use of the derived rule does not increase the set of provable theorems in $\mathscr{H}$. This is done by showing how to transform any proof using the rule into one that does not use the rule. Therefore, in principle, any proof that uses the derived rule could be transformed to one that uses only the three axioms and *MP*.

Theorem 3.14 (Deduction theorem) *The deduction rule is a sound derived rule.*

Proof We show by induction on the length n of the proof of $U \cup \{A\} \vdash B$ how to obtain a proof of $U \vdash A \rightarrow B$ that does not use the deduction rule.

For $n = 1$, B is proved in one step, so B must be either an element of $U \cup \{A\}$ or an axiom of $\mathscr{H}$ or a previously proved theorem:

- If B is A, then $\vdash A \rightarrow A$ by Theorem 3.10, so certainly $U \vdash A \rightarrow A$.
- Otherwise (B is an axiom or a previously proved theorem), here is a proof of $U \vdash A \rightarrow B$ that does not use the deduction rule or the assumption A:

  1. $U \vdash B$ Axiom or theorem
  2. $U \vdash B \rightarrow (A \rightarrow B)$ Axiom 1
  3. $U \vdash A \rightarrow B$ MP 1, 2

If $n > 1$, the last step in the proof of $U \cup \{A\} \vdash B$ is either a one-step inference of B or an inference of B using *MP*. In the first case, the result holds by the proof for $n = 1$. Otherwise, *MP* was used, so there is a formula C and lines $i, j < n$ in the proof such that line i in the proof is $U \cup \{A\} \vdash C$ and line j is $U \cup \{A\} \vdash C \rightarrow B$. By the inductive hypothesis, $U \vdash A \rightarrow C$ and $U \vdash A \rightarrow (C \rightarrow B)$. A proof of $U \vdash A \rightarrow B$ is given by:

1. $U \vdash A \rightarrow C$ Inductive hypothesis
2. $U \vdash A \rightarrow (C \rightarrow B)$ Inductive hypothesis
3. $U \vdash (A \rightarrow (C \rightarrow B)) \rightarrow ((A \rightarrow C) \rightarrow (A \rightarrow B))$ Axiom 2
4. $U \vdash (A \rightarrow C) \rightarrow (A \rightarrow B)$ MP 2, 3
5. $U \vdash A \rightarrow B$ MP 1, 4
∎

3.4 Derived Rules in $\mathscr{H}$

The general form of a derived rule will be one of:

$$\frac{U \vdash \phi_1}{U \vdash \phi}, \qquad \frac{U \vdash \phi_1 \qquad U \vdash \phi_2}{U \vdash \phi}.$$

The first form is justified by proving the formula $U \vdash \phi_1 \rightarrow \phi$ and the second by $U \vdash \phi_1 \rightarrow (\phi_2 \rightarrow \phi)$; the formula $U \vdash \phi$ that is the conclusion of the rule follows immediately by one or two applications of *MP*. For example, from Axiom 3 we immediately have the following rule:

Rule 3.15 (Contrapositive rule)

$$\frac{U \vdash \neg B \rightarrow \neg A}{U \vdash A \rightarrow B}.$$

The contrapositive is used extensively in mathematics. We showed the completeness of the method of semantic tableaux by proving: *If a tableau is open, the formula is satisfiable*, which is the contrapositive of the theorem that we wanted to prove: *If a formula is unsatisfiable (not satisfiable), the tableau is closed (not open)*.

Theorem 3.16 $\vdash (A \rightarrow B) \rightarrow [(B \rightarrow C) \rightarrow (A \rightarrow C)]$.

Proof
1. $\{A \rightarrow B, B \rightarrow C, A\} \vdash A$ Assumption
2. $\{A \rightarrow B, B \rightarrow C, A\} \vdash A \rightarrow B$ Assumption
3. $\{A \rightarrow B, B \rightarrow C, A\} \vdash B$ MP 1, 2
4. $\{A \rightarrow B, B \rightarrow C, A\} \vdash B \rightarrow C$ Assumption
5. $\{A \rightarrow B, B \rightarrow C, A\} \vdash C$ MP 3, 4
6. $\{A \rightarrow B, B \rightarrow C\} \vdash A \rightarrow C$ Deduction 5
7. $\{A \rightarrow B\} \vdash [(B \rightarrow C) \rightarrow (A \rightarrow C)]$ Deduction 6
8. $\vdash (A \rightarrow B) \rightarrow [(B \rightarrow C) \rightarrow (A \rightarrow C)]$ Deduction 7
∎

Rule 3.17 (Transitivity rule)

$$\frac{U \vdash A \to B \qquad\qquad U \vdash B \to C}{U \vdash A \to C}.$$

The transitivity rule justifies the step-by-step development of a mathematical theorem $\vdash A \to C$ through a series of *lemmas*. The antecedent A of the theorem is used to prove a lemma $\vdash A \to B_1$ whose consequent is used to prove the next lemma $\vdash B_1 \to B_2$ and so on until the consequent of the theorem appears as $\vdash B_n \to C$. Repeated use of the transitivity rule enables us to deduce $\vdash A \to C$.

Theorem 3.18 $\vdash [A \to (B \to C)] \to [B \to (A \to C)]$.

Proof

1.	$\{A \to (B \to C), B, A\} \vdash A$	Assumption
2.	$\{A \to (B \to C), B, A\} \vdash A \to (B \to C)$	Assumption
3.	$\{A \to (B \to C), B, A\} \vdash B \to C$	MP 1, 2
4.	$\{A \to (B \to C), B, A\} \vdash B$	Assumption
5.	$\{A \to (B \to C), B, A\} \vdash C$	MP 3, 4
6.	$\{A \to (B \to C), B\} \vdash A \to C$	Deduction 5
7.	$\{A \to (B \to C)\} \vdash B \to (A \to C)$	Deduction 6
8.	$\vdash [A \to (B \to C)] \to [B \to (A \to C)]$	Deduction 7

∎

Rule 3.19 (Exchange of antecedent rule)

$$\frac{U \vdash A \to (B \to C)}{U \vdash B \to (A \to C)}.$$

Exchanging the antecedent simply means that it doesn't matter in which order we use the lemmas necessary in a proof.

Theorem 3.20 $\vdash \neg A \to (A \to B)$.

Proof

1.	$\{\neg A\} \vdash \neg A \to (\neg B \to \neg A)$	Axiom 1
2.	$\{\neg A\} \vdash \neg A$	Assumption
3.	$\{\neg A\} \vdash \neg B \to \neg A$	MP 1, 2
4.	$\{\neg A\} \vdash (\neg B \to \neg A) \to (A \to B)$	Axiom 3
5.	$\{\neg A\} \vdash A \to B$	MP 3, 4
6.	$\vdash \neg A \to (A \to B)$	Deduction 5

∎

Theorem 3.21 $\vdash A \to (\neg A \to B)$.

Proof

1.	$\vdash \neg A \to (A \to B)$	Theorem 3.20
2.	$\vdash A \to (\neg A \to B)$	Exchange 1

∎

These two theorems are of major theoretical importance. They say that if you can prove some formula A and its negation $\neg A$, then you can prove *any* formula B! If you can prove any formula then there are no unprovable formulas so the concept of proof becomes meaningless.

Theorem 3.22 $\vdash \neg\neg A \to A$.

Proof

1.	$\{\neg\neg A\} \vdash \neg\neg A \to (\neg\neg\neg\neg A \to \neg\neg A)$	Axiom 1
2.	$\{\neg\neg A\} \vdash \neg\neg A$	Assumption
3.	$\{\neg\neg A\} \vdash \neg\neg\neg\neg A \to \neg\neg A$	MP 1, 2
4.	$\{\neg\neg A\} \vdash \neg A \to \neg\neg\neg A$	Contrapositive 3
5.	$\{\neg\neg A\} \vdash \neg\neg A \to A$	Contrapositive 4
6.	$\{\neg\neg A\} \vdash A$	MP 2, 5
7.	$\vdash \neg\neg A \to A$	Deduction 6

∎

Theorem 3.23 $\vdash A \to \neg\neg A$.

Proof

1.	$\vdash \neg\neg\neg A \to \neg A$	Theorem 3.22
2.	$\vdash A \to \neg\neg A$	Contrapositive 1

∎

Rule 3.24 (Double negation rule)

$$\frac{U \vdash \neg\neg A}{U \vdash A}, \qquad \frac{U \vdash A}{U \vdash \neg\neg A}.$$

Double negation is a very intuitive rule. We expect that 'it is raining' and 'it is not true that it is not raining' will have the same truth value, and that the second formula can be simplified to the first. Nevertheless, some logicians reject the rule because it is not constructive. Suppose that we can prove for some number n, 'it is not true that n is prime' which is the same as 'it is not true that n is not composite'. This double negation could be reduced by the rule to 'n is composite', but we have not actually demonstrated any factors of n.

Theorem 3.25 $\vdash (A \to B) \to (\neg B \to \neg A)$.

Proof

1.	$\{A \to B\} \vdash A \to B$	Assumption
2.	$\{A \to B\} \vdash \neg\neg A \to A$	Theorem 3.22
3.	$\{A \to B\} \vdash \neg\neg A \to B$	Transitivity 2, 1
4.	$\{A \to B\} \vdash B \to \neg\neg B$	Theorem 3.23
5.	$\{A \to B\} \vdash \neg\neg A \to \neg\neg B$	Transitivity 3, 4
6.	$\{A \to B\} \vdash \neg B \to \neg A$	Contrapositive 5
7.	$\vdash (A \to B) \to (\neg B \to \neg A)$	Deduction 6

∎

Rule 3.26 (Contrapositive rule)

$$\frac{U \vdash A \rightarrow B}{U \vdash \neg B \rightarrow \neg A}.$$

This is the other direction of the contrapositive rule shown earlier.

Recall from Sect. 2.3.3 the definition of the logical constants *true* as an abbreviation for $p \vee \neg p$ and *false* as an abbreviation for $p \wedge \neg p$. These can be expressed using implication and negation alone as $p \rightarrow p$ and $\neg (p \rightarrow p)$.

Theorem 3.27

$$\vdash true,$$
$$\vdash \neg false.$$

Proof $\vdash$ *true* is an instance of Theorem 3.10. $\vdash \neg false$, which is $\vdash \neg\neg(p \rightarrow p)$, follows by double negation. ∎

Theorem 3.28 $\vdash (\neg A \rightarrow false) \rightarrow A$.

Proof

1.	$\{\neg A \rightarrow false\} \vdash \neg A \rightarrow false$	Assumption
2.	$\{\neg A \rightarrow false\} \vdash \neg false \rightarrow \neg\neg A$	Contrapositive
3.	$\{\neg A \rightarrow false\} \vdash \neg false$	Theorem 3.27
4.	$\{\neg A \rightarrow false\} \vdash \neg\neg A$	MP 2, 3
5.	$\{\neg A \rightarrow false\} \vdash A$	Double negation 4
6.	$\vdash (\neg A \rightarrow false) \rightarrow A$	Deduction 5

∎

Rule 3.29 (Reductio ad absurdum)

$$\frac{U \vdash \neg A \rightarrow false}{U \vdash A}.$$

Reductio ad absurdum is a very useful rule in mathematics: Assume the negation of what you wish to prove and show that it leads to a contradiction. This rule is also controversial because proving that $\neg A$ leads to a contradiction provides no reason that directly justifies A.

Here is an example of the use of this rule:

Theorem 3.30 $\vdash (A \rightarrow \neg A) \rightarrow \neg A$.

Proof

1.	$\{A \rightarrow \neg A, \neg\neg A\} \vdash \neg\neg A$	Assumption
2.	$\{A \rightarrow \neg A, \neg\neg A\} \vdash A$	Double negation 1
3.	$\{A \rightarrow \neg A, \neg\neg A\} \vdash A \rightarrow \neg A$	Assumption
4.	$\{A \rightarrow \neg A, \neg\neg A\} \vdash \neg A$	MP 2, 3
5.	$\{A \rightarrow \neg A, \neg\neg A\} \vdash A \rightarrow (\neg A \rightarrow false)$	Theorem 3.21
6.	$\{A \rightarrow \neg A, \neg\neg A\} \vdash \neg A \rightarrow false$	MP 2, 5
7.	$\{A \rightarrow \neg A, \neg\neg A\} \vdash false$	MP 4, 6
8.	$\{A \rightarrow \neg A\} \vdash \neg\neg A \rightarrow false$	Deduction 7
9.	$\{A \rightarrow \neg A\} \vdash \neg A$	Reductio ad absurdum 8
10.	$\vdash (A \rightarrow \neg A) \rightarrow \neg A$	Deduction 9

∎

We leave the proof of the following theorem as an exercise.

Theorem 3.31 $\vdash (\neg A \rightarrow A) \rightarrow A$.

These two theorems may seem strange, but they can be understood on the semantic level. For the implication of Theorem 3.31 to be false, the antecedent $\neg A \rightarrow A$ must be true and the consequent A false. But if A is false, then so is $\neg A \rightarrow A \equiv A \vee A$, so the formula is true.

3.5 Theorems for Other Operators

So far we have worked with only negation and implication as operators. These two operators are adequate for defining all others (Sect. 2.4), so we can use these definitions to prove theorems using other operators. Recall that $A \wedge B$ is defined as $\neg (A \rightarrow \neg B)$, and $A \vee B$ is defined as $\neg A \rightarrow B$.

Theorem 3.32 $\vdash A \to (B \to (A \land B))$.

Proof

1.	$\{A, B\} \vdash (A \to \neg B) \to (A \to \neg B)$	Theorem 3.10
2.	$\{A, B\} \vdash A \to ((A \to \neg B) \to \neg B)$	Exchange 1
3.	$\{A, B\} \vdash A$	Assumption
4.	$\{A, B\} \vdash (A \to \neg B) \to \neg B$	MP 2, 3
5.	$\{A, B\} \vdash \neg\neg B \to \neg(A \to \neg B)$	Contrapositive 4
6.	$\{A, B\} \vdash B$	Assumption
7.	$\{A, B\} \vdash \neg\neg B$	Double negation 6
8.	$\{A, B\} \vdash \neg(A \to \neg B)$	MP 5, 7
9.	$\{A\} \vdash B \to \neg(A \to \neg B)$	Deduction 8
10.	$\vdash A \to (B \to \neg(A \to \neg B))$	Deduction 9
11.	$\vdash A \to (B \to (A \land B))$	Definition of $\land$

∎

Theorem 3.33 (Commutativity) $\vdash A \lor B \leftrightarrow B \lor A$.

Proof

1.	$\{\neg A \to B, \neg B\} \vdash \neg A \to B$	Assumption
2.	$\{\neg A \to B, \neg B\} \vdash \neg B \to \neg\neg A$	Contrapositive 1
3.	$\{\neg A \to B, \neg B\} \vdash \neg B$	Assumption
4.	$\{\neg A \to B, \neg B\} \vdash \neg\neg A$	MP 2, 3
5.	$\{\neg A \to B, \neg B\} \vdash A$	Double negation 4
6.	$\{\neg A \to B\} \vdash \neg B \to A$	Deduction 5
7.	$\vdash (\neg A \to B) \to (\neg B \to A)$	Deduction 6
8.	$\vdash A \lor B \to B \lor A$	Def. of $\lor$

The other direction is similar.

∎

The proofs of the following theorems are left as exercises.

Theorem 3.34 (Weakening)

$\vdash A \to A \lor B$,
$\vdash B \to A \lor B$,
$\vdash (A \to B) \to ((C \lor A) \to (C \lor B))$.

Theorem 3.35 (Associativity)

$\vdash A \lor (B \lor C) \leftrightarrow (A \lor B) \lor C$.

Theorem 3.36 (Distributivity)

$\vdash A \lor (B \land C) \leftrightarrow (A \lor B) \land (A \lor C)$,
$\vdash A \land (B \lor C) \leftrightarrow (A \land B) \lor (A \land C)$.

3.6 Soundness and Completeness of $\mathcal{H}$

We now prove the soundness and completeness of the Hilbert system $\mathcal{H}$. As usual, soundness is easy to prove. Proving completeness will not be too difficult because we already know that the Gentzen system $\mathcal{G}$ is complete so it is sufficient to show how to transform any proof in $\mathcal{G}$ into a proof in $\mathcal{H}$.

Theorem 3.37 *The Hilbert system $\mathcal{H}$ is sound: If $\vdash A$ then $\models A$.*

Proof The proof is by structural induction. First we show that the axioms are valid, and then we show that *MP* preserves validity. Here are closed semantic tableaux for the *negations* of Axioms 1 and 3:

$$\neg [A \rightarrow (B \rightarrow A)] \qquad\qquad \neg [(\neg B \rightarrow \neg A) \rightarrow (A \rightarrow B)]$$
$$\downarrow \qquad\qquad\qquad\qquad\qquad \downarrow$$
$$A, \neg (B \rightarrow A) \qquad\qquad \neg B \rightarrow \neg A, \neg (A \rightarrow B)$$
$$\downarrow \qquad\qquad\qquad\qquad\qquad \downarrow$$
$$A, B, \neg A \qquad\qquad\qquad \neg B \rightarrow \neg A, A, \neg B$$
$$\times \qquad\qquad\qquad\qquad \swarrow \qquad\qquad \searrow$$
$$\neg\neg B, A, \neg B \qquad\qquad \neg A, A, \neg B$$
$$\downarrow \qquad\qquad\qquad\qquad\qquad \times$$
$$B, A, \neg B$$
$$\times$$

The construction of a tableau for the negation of Axiom 2 is left as an exercise.

Suppose that *MP* were not sound. There would be a set of formulas $\{A, A \rightarrow B, B\}$ such that A and $A \rightarrow B$ are valid, but B is not valid. Since B is not valid, there is an interpretation $\mathscr{I}$ such that $v_{\mathscr{I}}(B) = F$. Since A and $A \rightarrow B$ are valid, for *any* interpretation, in particular for $\mathscr{I}$, $v_{\mathscr{I}}(A) = v_{\mathscr{I}}(A \rightarrow B) = T$. By definition of $v_{\mathscr{I}}$ for implication, $v_{\mathscr{I}}(B) = T$, contradicting $v_{\mathscr{I}}(B) = F$. ∎

There is no circularity in the final sentence of the proof: We are not using the syntactical proof rule *MP*, but, rather, the semantic definition of truth value in the presence of the implication operator.

Theorem 3.38 *The Hilbert system $\mathcal{H}$ is complete: If $\models A$ then $\vdash A$.*

By the completeness of the Gentzen system $\mathcal{G}$ (Theorem 3.8), if $\models A$, then $\vdash A$ in $\mathcal{G}$. The proof of the theorem showed how to construct the proof of A by first constructing a semantic tableau for $\neg A$; the tableau is guaranteed to close since A is valid. The completeness of $\mathcal{H}$ is proved by showing how to transform a proof in $\mathcal{G}$ into a proof in $\mathcal{H}$. Note that all three steps can be carried out algorithmically: Given an arbitrary valid formula in propositional logic, a computer can generate its proof.

We need a more general result because a proof in $\mathscr{G}$ is a sequence of *sets* of formulas, while a proof in $\mathscr{H}$ is a sequence of formulas.

Theorem 3.39 *If* $\vdash U$ *in* $\mathscr{G}$, *then* $\vdash \bigvee U$ *in* $\mathscr{H}$.

The difficulty arises from the clash of the data structures used: U is a set while $\bigvee U$ is a single formula. To see why this is a problem, consider the base case of the induction. The set $\{\neg p, p\}$ is an axiom in $\mathscr{G}$ and we immediately have $\vdash \neg p \vee p$ in $\mathscr{H}$ since this is simply $\vdash p \to p$. But if the axiom in $\mathscr{G}$ is $\{q, \neg p, r, p, s\}$, we can't immediately conclude that $\vdash q \vee \neg p \vee r \vee p \vee s$ in $\mathscr{H}$.

Lemma 3.40 *If* $U' \subseteq U$ *and* $\vdash \bigvee U'$ *in* $\mathscr{H}$ *then* $\vdash \bigvee U$ *in* $\mathscr{H}$.

Proof The proof is by induction using weakening, commutativity and associativity of disjunction (Theorems 3.34–3.35). We give the outline here and leave it as an exercise to fill in the details.

Suppose we have a proof of $\bigvee U'$. By repeated application of Theorem 3.34, we can transform this into a proof of $\bigvee U''$, where U'' is a permutation of the elements of U. By repeated applications of commutativity and associativity, we can move the elements of U'' to their proper places.   ∎

Example 3.41 Let $U' = \{A, C\} \subset \{A, B, C\} = U$ and suppose we have a proof of $\vdash \bigvee U' = A \vee C$. This can be transformed into a proof of $\vdash \bigvee U = A \vee (B \vee C)$ as follows, where Theorems 3.34–3.35 are used as derived rules:

1.	$\vdash A \vee C$	Assumption
2.	$\vdash (A \vee C) \vee B$	Weakening, 1
3.	$\vdash A \vee (C \vee B)$	Associativity, 2
4.	$\vdash (C \vee B) \to (B \vee C)$	Commutativity
5.	$\vdash A \vee (C \vee B) \to A \vee (B \vee C)$	Weakening, 4
6.	$\vdash A \vee (B \vee C)$	MP 3, 5

∎

Proof of Theorem 3.39 The proof is by induction on the structure of the proof in $\mathscr{G}$. If U is an axiom, it contains a pair of complementary literals and $\vdash \neg p \vee p$ can be proved in $\mathscr{H}$. By Lemma 3.40, this can be transformed into a proof of $\bigvee U$.

Otherwise, the last step in the proof of U in $\mathscr{G}$ is the application of a rule to an α- or β-formula. As usual, we will use disjunction and conjunction as representatives of α- and β-formulas.

Case 1: A rule in $\mathscr{G}$ was applied to obtain an α-formula $\vdash U_1 \cup \{A_1 \vee A_2\}$ from $\vdash U_1 \cup \{A_1, A_2\}$. By the inductive hypothesis, $\vdash ((\bigvee U_1) \vee A_1) \vee A_2$ in $\mathscr{H}$ from which we infer $\vdash \bigvee U_1 \vee (A_1 \vee A_2)$ by associativity.

Case 2: A rule in $\mathscr{G}$ was applied to obtain a β-formula $\vdash U_1 \cup U_2 \cup \{A_1 \wedge A_2\}$ from $\vdash U_1 \cup \{A_1\}$ and $\vdash U_2 \cup \{A_2\}$. By the inductive hypothesis, $\vdash (\bigvee U_1) \vee A_1$ and $\vdash (\bigvee U_2) \vee A_2$ in $\mathscr{H}$. We leave it to the reader to justify each step of the following deduction of $\vdash \bigvee U_1 \vee \bigvee U_2 \vee (A_1 \wedge A_2)$:

1. $\vdash \bigvee U_1 \vee A_1$
2. $\vdash \neg \bigvee U_1 \rightarrow A_1$
3. $\vdash A_1 \rightarrow (A_2 \rightarrow (A_1 \wedge A_2))$
4. $\vdash \neg \bigvee U_1 \rightarrow (A_2 \rightarrow (A_1 \wedge A_2))$
5. $\vdash A_2 \rightarrow (\neg \bigvee U_1 \rightarrow (A_1 \wedge A_2))$
6. $\vdash \bigvee U_2 \vee A_2$
7. $\vdash \neg \bigvee U_2 \rightarrow A_2$
8. $\vdash \neg \bigvee U_2 \rightarrow (\neg \bigvee U_1 \rightarrow (A_1 \wedge A_2))$
9. $\vdash \bigvee U_1 \vee \bigvee U_2 \vee (A_1 \wedge A_2)$

∎

Proof of Theorem 3.38 If $\models A$ then $\vdash A$ in $\mathcal{G}$ by Theorem 3.8. By the remark at the end of Definition 3.2, $\vdash A$ is an abbreviation for $\vdash \{A\}$. By Theorem 3.39, $\vdash \bigvee\{A\}$ in $\mathcal{H}$. Since A is a single formula, $\vdash A$ in $\mathcal{H}$. ∎

3.7 Consistency

What would mathematics be like if both $1 + 1 = 2$ and $\neg (1 + 1 = 2) \equiv 1 + 1 \neq 2$ could be proven? An inconsistent deductive system is useless, because *all* formulas are provable and the concept of proof becomes meaningless.

Definition 3.42 A set of formulas U is *inconsistent* iff for some formula A, both $U \vdash A$ and $U \vdash \neg A$. U is *consistent* iff it is not inconsistent. A deductive system is *inconsistent* iff it contains an inconsistent set of formulas. ∎

Theorem 3.43 *U is inconsistent iff for all A, $U \vdash A$.*

Proof Let A be an arbitrary formula. If U is inconsistent, for some formula B, $U \vdash B$ and $U \vdash \neg B$. By Theorem 3.21, $\vdash B \rightarrow (\neg B \rightarrow A)$. Using *MP* twice, $U \vdash A$. The converse is trivial. ∎

Corollary 3.44 *U is consistent if and only if for some A, $U \nvdash A$.*

If a deductive system is sound, then $\vdash A$ implies $\models A$, and, conversely, $\not\models A$ implies $\nvdash A$. Therefore, if there is even a single falsifiable formula A in a sound system, the system must be consistent! Since $\not\models \textit{false}$ (where *false* is an abbreviation for $\neg (p \rightarrow p)$), by the soundness of $\mathcal{H}$, $\nvdash \textit{false}$. By Corollary 3.44, $\mathcal{H}$ is consistent.

The following theorem is another way of characterizing inconsistency.

Theorem 3.45 $U \vdash A$ *if and only if* $U \cup \{\neg A\}$ *is inconsistent.*

Proof If $U \vdash A$, obviously $U \cup \{\neg A\} \vdash A$, since the extra assumption will not be used in the proof. $U \cup \{\neg A\} \vdash \neg A$ because $\neg A$ is an assumption. By Definition 3.42, $U \cup \{\neg A\}$ is inconsistent.

Conversely, if $U \cup \{\neg A\}$ is inconsistent, then $U \cup \{\neg A\} \vdash A$ by Theorem 3.43. By the deduction theorem, $U \vdash \neg A \rightarrow A$, and $U \vdash A$ follows by *MP* from $\vdash (\neg A \rightarrow A) \rightarrow A$ (Theorem 3.31). ∎

3.8 Strong Completeness and Compactness *

The construction of a semantic tableau can be generalized to an infinite set of formulas $S = \{A_1, A_2, \ldots\}$. The label of the root is $\{A_1\}$. Whenever a rule is applied to a leaf of depth n, A_{n+1} will be added to the label(s) of its child(ren) in addition to the α_i or β_i.

Theorem 3.46 *A set of formulas* $S = \{A_1, A_2, \ldots\}$ *is unsatisfiable if and only if a semantic tableau for S closes.*

Proof Here is an outline of the proof that is given in detail in Smullyan (1968, Chap. III).

If the tableau closes, there is only a finite subset $S_0 \subset S$ of formulas on each closed branch, and S_0 is unsatisfiable. By a generalization of Theorem 2.46 to an infinite set of formulas, it follows that $S = S_0 \cup (S - S_0)$ is unsatisfiable.

Conversely, if the tableau is open, it can be shown that there must be an infinite branch containing all formulas in S, and the union of formulas in the labels of nodes on the branch forms a Hintikka set, from which a satisfying interpretation can be found. ∎

The completeness of propositional logic now generalizes to:

Theorem 3.47 (Strong completeness) *Let U be a finite or countably infinite set of formulas and let A be a formula. If* $U \models A$ *then* $U \vdash A$.

The same construction proves the following important theorem.

Theorem 3.48 (Compactness) *Let S be a countably infinite set of formulas, and suppose that every finite subset of S is satisfiable. Then S is satisfiable.*

Proof Suppose that S were unsatisfiable. Then a semantic tableau for S must close. There are only a finite number of formulas labeling nodes on each closed branch. Each such set of formulas is a finite unsatisfiable subset of S, contracting the assumption that all finite subsets are satisfiable. ∎

3.9 Variant Forms of the Deductive Systems *

$\mathcal{G}$ and $\mathcal{H}$, the deductive systems that we presented in detail, are two of many possible deductive systems for propositional logic. Different systems are obtained by changing the operators, the axioms or the representations of proofs. In propositional logic, all these systems are equivalent in the sense that they are sound and complete. In this section, we survey some of these variants.

3.9.1 Hilbert Systems

Hilbert systems almost invariably have *MP* as the only rule. They differ in the choice of primitive operators and axioms. For example, $\mathcal{H}'$ is an Hilbert system where Axiom 3 is replaced by:

$$\text{Axiom } 3' \quad \vdash (\neg B \to \neg A) \to ((\neg B \to A) \to B).$$

Theorem 3.49 $\mathcal{H}$ and $\mathcal{H}'$ are equivalent in the sense that a proof in one system can be transformed into a proof in the other.

Proof We prove Axiom 3' in $\mathcal{H}$. It follows that any proof in $\mathcal{H}'$ can be transformed into a proof in $\mathcal{H}$, by starting with this proof of the new axiom and using it as a previously proved theorem.

1.	$\{\neg B \to \neg A, \neg B \to A, \neg B\} \vdash \neg B$	Assumption
2.	$\{\neg B \to \neg A, \neg B \to A, \neg B\} \vdash \neg B \to A$	Assumption
3.	$\{\neg B \to \neg A, \neg B \to A, \neg B\} \vdash A$	MP 1, 2
4.	$\{\neg B \to \neg A, \neg B \to A, \neg B\} \vdash \neg B \to \neg A$	Assumption
5.	$\{\neg B \to \neg A, \neg B \to A, \neg B\} \vdash A \to B$	Contrapositive 4
6.	$\{\neg B \to \neg A, \neg B \to A, \neg B\} \vdash B$	MP 3, 5
7.	$\{\neg B \to \neg A, \neg B \to A\} \vdash \neg B \to B$	Deduction 7
8.	$\{\neg B \to \neg A, \neg B \to A\} \vdash (\neg B \to B) \to B$	Theorem 3.31
9.	$\{\neg B \to \neg A, \neg B \to A\} \vdash B$	MP 8, 9
10.	$\{\neg B \to \neg A\} \vdash (\neg B \to A) \to B$	Deduction 9
11.	$\vdash (\neg B \to \neg A) \to ((\neg B \to A) \to B)$	Deduction 10

The use of the deduction theorem is legal because its proof in $\mathcal{H}$ does not use Axiom 3, so the identical proof can be done in $\mathcal{H}'$.

We leave it as an exercise to prove Axiom 3 in $\mathcal{H}'$.   ∎

Either conjunction or disjunction may replace implication as the binary operator in the formulation of a Hilbert system. Implication can then be defined by $\neg (A \wedge \neg B)$ or $\neg A \vee B$, respectively, and *MP* is still the only inference rule. For disjunction, a set of axioms is:

$$\textbf{Axiom 1} \quad \vdash A \vee A \to A,$$
$$\textbf{Axiom 2} \quad \vdash A \to A \vee B,$$
$$\textbf{Axiom 3} \quad \vdash A \vee B \to B \vee A,$$
$$\textbf{Axiom 4} \quad \vdash (B \to C) \to (A \vee B \to A \vee C).$$

The steps needed to show the equivalence of this system with $\mathscr{H}$ are given in Mendelson (2009, Exercise 1.54).

Finally, Meredith's axiom:

$$\vdash (\{[(A \to B) \to (\neg C \to \neg D)] \to C\} \to E) \to [(E \to A) \to (D \to A)],$$

together with MP as the rule of inference is a complete deductive system for propositional logic. Adventurous readers are invited to prove the axioms of $\mathscr{H}$ from Meredith's axiom following the 37-step plan given in Monk (1976, Exercise 8.50).

3.9.2 Gentzen Systems

$\mathscr{G}$ was constructed in order to simplify the theoretical treatment by using a notation that is identical to that of semantic tableaux. We now present a deductive system similar to the one that Gentzen originally proposed; this system is taken from Smullyan (1968, Chap. XI).

Definition 3.50 If U and V are (possibly empty) sets of formulas, then $U \Rightarrow V$ is a *sequent*. ∎

Intuitively, a sequent represents 'provable from' in the sense that the formulas in U are assumptions for the set of formulas V that are to be proved. The symbol $\Rightarrow$ is similar to the symbol $\vdash$ in Hilbert systems, except that $\Rightarrow$ is part of the object language of the deductive system being formalized, while $\vdash$ is a metalanguage notation used to reason about deductive systems.

Definition 3.51 Axioms in the Gentzen sequent system $\mathscr{S}$ are sequents of the form:

$$U \cup \{A\} \Rightarrow V \cup \{A\}.$$

The rules of inference are shown in Fig. 3.2. ∎

The semantics of the sequent system $\mathscr{S}$ are defined as follows:

Definition 3.52 Let $S = U \Rightarrow V$ be a sequent where $U = \{U_1, \ldots, U_n\}$ and $V = \{V_1, \ldots, V_m\}$, and let $\mathscr{I}$ be an interpretation for $U \cup V$. Then $v_{\mathscr{I}}(S) = T$ if and only if $v_{\mathscr{I}}(U_1) = \cdots = v_{\mathscr{I}}(U_n) = T$ implies that for some i, $v_{\mathscr{I}}(V_i) = T$. ∎

This definition relates sequents to formulas: Given an interpretation $\mathscr{I}$ for $U \cup V$, $v_{\mathscr{I}}(U \Rightarrow V) = T$ if and only if $v_{\mathscr{I}}(\bigwedge U \to \bigvee V) = T$.

op	Introduction into consequent	Introduction into antecedent
$\wedge$	$$\dfrac{U \Rightarrow V \cup \{A\} \qquad U \Rightarrow V \cup \{B\}}{U \Rightarrow V \cup \{A \wedge B\}}$$	$$\dfrac{U \cup \{A, B\} \Rightarrow V}{U \cup \{A \wedge B\} \Rightarrow V}$$
$\vee$	$$\dfrac{U \Rightarrow V \cup \{A, B\}}{U \Rightarrow V \cup \{A \vee B\}}$$	$$\dfrac{U \cup \{A\} \Rightarrow V \qquad U \cup \{B\} \Rightarrow V}{U \cup \{A \vee B\} \Rightarrow V}$$
$\rightarrow$	$$\dfrac{U \cup \{A\} \Rightarrow V \cup \{B\}}{U \Rightarrow V \cup \{A \rightarrow B\}}$$	$$\dfrac{U \Rightarrow V \cup \{A\} \qquad U \cup \{B\} \Rightarrow V}{U \cup \{A \rightarrow B\} \Rightarrow V}$$
$\neg$	$$\dfrac{U \cup \{A\} \Rightarrow V}{U \Rightarrow V \cup \{\neg A\}}$$	$$\dfrac{U \Rightarrow V \cup \{A\}}{U \cup \{\neg A\} \Rightarrow V}$$

Fig. 3.2 Rules of inference for sequents

3.9.3 Natural Deduction

The advantage of working with sequents is that the deduction theorem is a rule of inference: introduction into the consequent of $\rightarrow$. The convenience of Gentzen systems is apparent when proofs are presented in a format called *natural deduction* that emphasizes the role of assumptions.

Look at the proof of Theorem 3.30, for example. The assumptions are dragged along throughout the entire deduction, even though each is used only twice, once as an assumption and once in the deduction rule. The way we reason in mathematics is to set out the assumptions once when they are first needed and then to *discharge* them by using the deduction rule. A natural deduction proof of Theorem 3.30 is shown in Fig. 3.3.

The boxes indicate the scope of assumptions. Just as in programming where local variables in procedures can only be used within the procedure and disappear when the procedure is left, an assumption can only be used within the scope of its box, and once it is discharged by using it in a deduction, it is no longer available.

3.9.4 Subformula Property

Definition 3.53 A deductive system has the *subformula property* iff any formula appearing in a proof of A is either a subformula of A or the negation of a subformula of A. ∎

The systems $\mathcal{G}$ and $\mathcal{S}$ have the subformula property while $\mathcal{H}$ does not. For example, in the proof of the theorem of double negation $\vdash \neg \neg A \rightarrow A$, the formula $\vdash \neg \neg \neg \neg A \rightarrow \neg \neg A$ appeared even though it is obviously not a subformula of the theorem.

Gentzen proposed his deductive system in order to obtain a system with the subformula property. Then he defined the system $\mathcal{S}'$ by adding an additional rule of inference, the *cut rule*:

$$\frac{U, A \Rightarrow V \qquad U \Rightarrow V, A}{U \Rightarrow V}$$

1.	$A \to \neg A$	Assumption
2.	$\neg \neg A$	Assumption
3.	A	Double negation 2
4.	$\neg A$	MP 1, 3
5.	$A \to (\neg A \to false)$	Theorem 3.21
6.	$\neg A \to false$	MP 3, 5
7.	$false$	MP 4, 6
8.	$\neg \neg A \to false$	Deduction 2, 7
9.	$\neg A$	Reductio ad absurdum 8
10.	$(A \to \neg A) \to \neg A$	Deduction 1, 9

Fig. 3.3 A natural deduction proof

to the system $\mathscr{S}$ and showed that proofs in $\mathscr{S}'$ can be mechanically transformed into proofs in $\mathscr{S}$. See Smullyan (1968, Chap. XII) for a proof of the following theorem.

Theorem 3.54 (Gentzen's Hauptsatz) *Any proof in $\mathscr{S}'$ can be transformed into a proof in $\mathscr{S}$ not using the cut rule.*

3.10 Summary

Deductive systems were developed to formalize mathematical reasoning. The structure of Hilbert systems such as $\mathscr{H}$ imitates the style of mathematical theories: a small number of axioms, *modus ponens* as the sole rule of inference and proofs as linear sequences of formulas. The problem with Hilbert systems is that they offer no guidance on how to find a proof of a formula. Gentzen systems such as $\mathscr{G}$ (and variants that use sequents or natural deduction) facilitate finding proofs because all formulas that appear are subformulas of the formula to be proved or their negations.

Both the deductive systems $\mathscr{G}$ and $\mathscr{H}$ are sound and complete. Completeness of $\mathscr{G}$ follows directly from the completeness of the method of semantic tableaux as a decision procedure for satisfiability and validity in propositional logic. However, the method of semantic tableaux is not very efficient. Our task in the next chapters is to study more efficient algorithms for satisfiability and validity.

3.11 Further Reading

Our presentation is based upon Smullyan (1968) who showed how Gentzen systems are closely related to tableaux. The deductive system $\mathscr{H}$ is from Mendelson (2009); he develops the theory of $\mathscr{H}$ (and later its generalization to first-order logic) without recourse to tableaux. Huth and Ryan (2004) base their presentation of logic on natural deduction. Velleman (2006) will help you learn how to prove theorems in mathematics.

3.12 Exercises

3.1 Prove in $\mathcal{G}$:

$$\vdash (A \to B) \to (\neg B \to \neg A),$$
$$\vdash (A \to B) \to ((\neg A \to B) \to B),$$
$$\vdash ((A \to B) \to A) \to A.$$

3.2 Prove that if $\vdash U$ in $\mathcal{G}$ then there is a closed semantic tableau for $\bar{U}$ (the forward direction of Theorem 3.7).

3.3 Prove the derived rule *modus tollens*:

$$\frac{\vdash \neg B \qquad\qquad \vdash A \to B}{\vdash \neg A}.$$

3.4 Give proofs in $\mathcal{G}$ for each of the three axioms of $\mathcal{H}$.

3.5 Prove $\vdash (\neg A \to A) \to A$ (Theorem 3.31) in $\mathcal{H}$.

3.6 Prove $\vdash (A \to B) \vee (B \to C)$ in $\mathcal{H}$.

3.7 Prove $\vdash ((A \to B) \to A) \to A$ in $\mathcal{H}$.

3.8 Prove $\{\neg A\} \vdash (\neg B \to A) \to B$ in $\mathcal{H}$.

3.9 Prove Theorem 3.34 in $\mathcal{H}$:

$$\vdash A \to A \vee B,$$
$$\vdash B \to A \vee B,$$
$$\vdash (A \to B) \to ((C \vee A) \to (C \vee B)).$$

3.10 Prove Theorem 3.35 in $\mathcal{H}$:

$$\vdash A \vee (B \vee C) \leftrightarrow (A \vee B) \vee C.$$

3.11 Prove Theorem 3.36 in $\mathcal{H}$:

$$\vdash A \vee (B \wedge C) \leftrightarrow (A \vee B) \wedge (A \vee C),$$
$$\vdash A \wedge (B \vee C) \leftrightarrow (A \wedge B) \vee (A \wedge C).$$

3.12 Prove that Axiom 2 of $\mathcal{H}$ is valid by constructing a semantic tableau for its negation.

3.13 Complete the proof that if $U' \subseteq U$ and $\vdash \bigvee U'$ then $\vdash \bigvee U$ (Lemma 3.40).

3.14 Prove the last two formulas of Exercise 3.1 in $\mathcal{H}$.

3.15 * Prove Axiom 3 of $\mathcal{H}$ in $\mathcal{H}'$.

3.16 * Prove that the Gentzen sequent system $\mathcal{S}$ is sound and complete.

3.17 * Prove that a set of formulas U is inconsistent if and only if there is a finite set of formulas $\{A_1, \ldots, A_n\} \subseteq U$ such that $\vdash \neg A_1 \vee \cdots \vee \neg A_n$.

3.18 A set of formulas U is *maximally consistent* iff every proper superset of U is not consistent. Let S be a countable, consistent set of formulas. Prove:

1. Every finite subset of S is satisfiable.
2. For every formula A, at least one of $S \cup \{A\}$, $S \cup \{\neg A\}$ is consistent.
3. S can be extended to a maximally consistent set.

References

M. Huth and M.D. Ryan. *Logic in Computer Science: Modelling and Reasoning about Systems (Second Edition)*. Cambridge University Press, 2004.

E. Mendelson. *Introduction to Mathematical Logic (Fifth Edition)*. Chapman & Hall/CRC, 2009.

J.D. Monk. *Mathematical Logic*. Springer, 1976.

R.M. Smullyan. *First-Order Logic*. Springer-Verlag, 1968. Reprinted by Dover, 1995.

D.J. Velleman. *How to Prove It: A Structured Approach (Second Edition)*. Cambridge University Press, 2006.

Chapter 4
Propositional Logic: Resolution

The method of resolution, invented by J.A. Robinson in 1965, is an efficient method for searching for a proof. In this section, we introduce resolution for the propositional logic, though its advantages will not become apparent until it is extended to first-order logic. It is important to become familiar with resolution, because it is widely used in automatic theorem provers and it is also the basis of logic programming (Chap. 11).

4.1 Conjunctive Normal Form

Definition 4.1 A formula is in *conjunctive normal form (CNF)* iff it is a conjunction of disjunctions of literals. ∎

Example 4.2 The formula:

$$(\neg p \vee q \vee r) \wedge (\neg q \vee r) \wedge (\neg r)$$

is in CNF while the formula:

$$(\neg p \vee q \vee r) \wedge ((p \wedge \neg q) \vee r) \wedge (\neg r)$$

is not in CNF, because $(p \wedge \neg q) \vee r$ is not a disjunction.
The formula:

$$(\neg p \vee q \vee r) \wedge \neg (\neg q \vee r) \wedge (\neg r)$$

is not in CNF because the second disjunction is negated. ∎

Theorem 4.3 *Every formula in propositional logic can be transformed into an equivalent formula in CNF.*

Proof To convert an arbitrary formula to a formula in CNF perform the following steps, each of which preserves logical equivalence:

M. Ben-Ari, *Mathematical Logic for Computer Science*,
DOI 10.1007/978-1-4471-4129-7_4, © Springer-Verlag London 2012

1. Eliminate all operators except for negation, conjunction and disjunction by substituting logically equivalent formulas:

$$A \leftrightarrow B \ \equiv \ (A \rightarrow B) \wedge (B \rightarrow A),$$
$$A \oplus B \ \equiv \ \neg (A \rightarrow B) \vee \neg (B \rightarrow A),$$
$$A \rightarrow B \ \equiv \ \neg A \vee B,$$
$$A \uparrow B \ \equiv \ \neg (A \wedge B),$$
$$A \downarrow B \ \equiv \ \neg (A \vee B).$$

2. Push negations inward using De Morgan's laws:

$$\neg (A \wedge B) \ \equiv \ (\neg A \vee \neg B),$$
$$\neg (A \vee B) \ \equiv \ (\neg A \wedge \neg B),$$

   until they appear only before atomic propositions or atomic propositions preceded by negations.
3. Eliminate sequences of negations by deleting double negation operators:

$$\neg \neg A \equiv A.$$

4. The formula now consists of disjunctions and conjunctions of literals. Use the
   distributive laws:

$$A \vee (B \wedge C) \ \equiv \ (A \vee B) \wedge (A \vee C),$$
$$(A \wedge B) \vee C \ \equiv \ (A \vee C) \wedge (B \vee C)$$

   to eliminate conjunctions within disjunctions.   ∎

Example 4.4 The following sequence of formulas shows the four steps applied to
the formula $(\neg p \rightarrow \neg q) \rightarrow (p \rightarrow q)$:

$$\begin{aligned}
(\neg p \rightarrow \neg q) \rightarrow (p \rightarrow q) \ &\equiv \ \neg (\neg \neg p \vee \neg q) \vee (\neg p \vee q) \\
&\equiv \ (\neg \neg \neg p \wedge \neg \neg q) \vee (\neg p \vee q) \\
&\equiv \ (\neg p \wedge q) \vee (\neg p \vee q) \\
&\equiv \ (\neg p \vee \neg p \vee q) \wedge (q \vee \neg p \vee q).
\end{aligned}$$

∎

4.2 Clausal Form

The clausal form of formula is a notational variant of CNF. Recall (Definition 2.57) that a *literal* is an atom or the negation of an atom.

Definition 4.5

- A *clause* is a set of literals.
- A clause is considered to be an implicit disjunction of its literals.
- A *unit clause* is a clause consisting of exactly one literal.
- The empty set of literals is the *empty clause*, denoted by □.
- A formula in *clausal form* is a set of clauses.
- A formula is considered to be an implicit conjunction of its clauses.
- The formula that is the *empty set of clauses* is denoted by ∅.   ∎

The only significant difference between clausal form and the standard syntax is that clausal form is defined in terms of sets, while our standard syntax was defined in terms of trees. A node in a tree may have multiple children that are identical subtrees, but a set has only one occurrence of each of its elements. However, this difference is of no logical significance.

Corollary 4.6 *Every formula ϕ in propositional logic can be transformed into an logically equivalent formula in clausal form.*

Proof By Theorem 4.3, ϕ can be transformed into a logically equivalent formula ϕ' in CNF. Transform each disjunction in ϕ' into a clause (a set of literals) and ϕ' itself into the set of these clauses. Clearly, the transformation into sets will cause multiple occurrences of literals and clauses to collapse into single occurrences. Logical equivalence is preserved by idempotence: $A \wedge A \equiv A$ and $A \vee A \equiv A$.   ∎

Example 4.7 The CNF formula:

$$(p \vee r) \wedge (\neg q \vee \neg p \vee q) \wedge (p \vee \neg p \vee q \vee p \vee \neg p) \wedge (r \vee p)$$

is logically equivalent to its clausal form:

$$\{\{p, r\}, \{\neg q, \neg p, q\}, \{p, \neg p, q\}\}.$$

The clauses corresponding to the first and last disjunctions collapse into a single set, while in the third disjunction multiple occurrences of p and $\neg p$ have been collapsed to obtain the third clause.   ∎

Trivial Clauses

A formula in clausal form can be simplified by removing trivial clauses.

Definition 4.8 A clause if *trivial* if it contains a pair of clashing literals.   ∎

Since a trivial clause is valid ($p \vee \neg p \equiv true$), it can be removed from a set of clauses without changing the truth value of the formula.

Lemma 4.9 *Let S be a set of clauses and let $C \in S$ be a trivial clause. Then $S - \{C\}$ is logically equivalent to S.*

Proof Since a clause is an implicit disjunction, C is logically equivalent to a formula obtained by weakening, commutativity and associativity of a valid disjunction $p \vee \neg p$ (Theorems 3.34–3.35). Let $\mathscr{I}$ be any interpretation for $S - \{C\}$. Since $S - \{C\}$ is an implicit conjunction, the value $v_{\mathscr{I}}(S - \{C\})$ is not changed by adding the clause C, since $v_{\mathscr{I}}(C) = T$ and $A \wedge T \equiv A$. Therefore, $v_{\mathscr{I}}(S - \{C\}) = v_{\mathscr{I}}(S)$. Since $\mathscr{I}$ was arbitrary, it follows that $S - \{C\} \equiv S$. ∎

Henceforth, we will assume that all trivial clauses have been deleted from formulas in clausal form.

The Empty Clause and the Empty Set of Clauses

The following results may be a bit hard to understand at first, but they are very important. The proof uses reasoning about vacuous sets.

Lemma 4.10

$\square$, *the empty clause, is unsatisfiable.* $\emptyset$, *the empty set of clauses, is valid.*

Proof A clause is satisfiable iff there is *some* interpretation under which *at least one literal* in the clause is true. Let $\mathscr{I}$ be an arbitrary interpretation. Since there are no literals in $\square$, there are *no* literals whose value is true under $\mathscr{I}$. But $\mathscr{I}$ was an arbitrary interpretation, so $\square$ is unsatisfiable.

A set of clauses is valid iff *every* clause in the set is true in *every* interpretation. But there are no clauses in $\emptyset$ that need be true, so $\emptyset$ is valid. ∎

Notation

When working with clausal form, the following additional notational conventions will be used:

- An abbreviated notation will be used for a formula in clausal form. The set delimiters { and } are removed from each clause and a negated literal is denoted by a bar over the atomic proposition. In this notation, the formula in Example 4.7 becomes:

$$\{pr, \bar{q}\bar{p}q, p\bar{p}q\}.$$

- S is a formula in clausal form, C is a clause and l is a literal. The symbols will be subscripted and primed as necessary.
- If l is a literal l^c is its complement: if $l = p$ then $l^c = \bar{p}$ and if $l = \bar{p}$ then $l^c = p$.

- The concept of an interpretation is generalized to literals. Let l be a literal defined on the atomic proposition p, that is, l is p or l is $\bar{p}$. Then an interpretation $\mathscr{I}$ for a set of atomic propositions including p is extended to l as follows:

  - $\mathscr{I}(l) = T$, if $l = p$ and $\mathscr{I}(p) = T$,
  - $\mathscr{I}(l) = F$, if $l = p$ and $\mathscr{I}(p) = F$,
  - $\mathscr{I}(l) = T$, if $l = \bar{p}$ and $\mathscr{I}(p) = F$,
  - $\mathscr{I}(l) = F$, if $l = \bar{p}$ and $\mathscr{I}(p) = T$.

The Restriction of CNF to 3CNF *

Definition 4.11 A formula is in *3CNF* iff it is in CNF and each disjunction has exactly three literals. ∎

The problem of finding a model for a formula in CNF belongs to an important class of problems called $\mathscr{N}\mathscr{P}$-complete problems (Sect. 6.7). This important theoretical result holds even if the formulas are restricted to 3CNF. To prove this, an efficient algorithm is needed to transform a CNF formula into one in 3CNF.

Algorithm 4.12 (CNF to 3CNF)
Input: A formula in CNF.
Output: A formula in 3CNF.
   For each disjunction $C_i = l_i^1 \vee l_i^2 \vee \cdots \vee l_i^{n_i}$, perform the appropriate transformation depending of the value of n_i:

- If $n_i = 1$, create two new atoms p_i^1, p_i^2 and replace C_i by:

$$(l_i^1 \vee p_i^1 \vee p_i^2) \wedge (l_i^1 \vee \neg p_i^1 \vee p_i^2) \wedge (l_i^1 \vee p_i^1 \vee \neg p_i^2) \wedge (l_i^1 \vee \neg p_i^1 \vee \neg p_i^2).$$

- If $n_i = 2$, create one new atom p_i^1 and replace C_i by:

$$(l_i^1 \vee l_i^2 \vee p_i^1) \wedge (l_i^1 \vee l_i^2 \vee \neg p_i^1).$$

- If $n_i = 3$, do nothing.
- If $n_i > 3$, create $n - 3$ new atoms $p_i^1, p_i^2, \ldots, p_i^{n-3}$ and replace C_i by:

$$(l_i^1 \vee l_i^2 \vee p_i^1) \wedge (\neg p_i^1 \vee l_i^3 \vee p_i^2) \wedge \cdots \wedge (\neg p_i^{n-3} \vee l_i^{n-1} \vee l_i^n).$$

∎

We leave the proof of the following theorem as an exercise.

Theorem 4.13 *Let A be a formula in CNF and let A' be the formula in 3CNF constructed from A by Algorithm 4.12. Then A is satisfiable if and only if A' is satisfiable. The length of A' (the number of symbols in A') is a polynomial in the length of A.*

4.3 Resolution Rule

Resolution is a refutation procedure used to check if a formula in clausal form is unsatisfiable. The resolution procedure consists of a sequence of applications of the resolution rule to a set of clauses. The rule maintains satisfiability: if a set of clauses is satisfiable, so is the set of clauses produced by an application of the rule. Therefore, if the (unsatisfiable) empty clause is ever obtained, the original set of clauses must have been unsatisfiable.

Rule 4.14 (Resolution rule) *Let C_1, C_2 be clauses such that $l \in C_1$, $l^c \in C_2$. The clauses C_1, C_2 are said to be* clashing clauses *and to* clash *on the complementary pair of literals l, l^c. C, the* resolvent *of C_1 and C_2, is the clause:*

$$Res(C_1, C_2) = (C_1 - \{l\}) \cup (C_2 - \{l^c\}).$$

C_1 and C_2 are the parent clauses *of C.*   ∎

Example 4.15 The pair of clauses $C_1 = ab\bar{c}$ and $C_2 = bc\bar{e}$ clash on the pair of complementary literals c, $\bar{c}$. The resolvent is:

$$C = (ab\bar{c} - \{\bar{c}\}) \cup (bc\bar{e} - \{c\}) = ab \cup b\bar{e} = ab\bar{e}.$$

Recall that a clause is a set so duplicate literals are removed when taking the union: $\{a, b\} \cup \{b, \bar{e}\} = \{a, b, \bar{e}\}$.   ∎

Resolution is only performed if the pair of clauses clash on *exactly* one pair of complementary literals.

Lemma 4.16 *If two clauses clash on more than one literal, their resolvent is a trivial clause (Definition 4.8).*

Proof Consider a pair of clauses:

$$\{l_1, l_2\} \cup C_1, \qquad \{l_1^c, l_2^c\} \cup C_2,$$

and suppose that we perform the resolution rule because the clauses clash on the pair of literals $\{l_1, l_1^c\}$. The resolvent is the trivial clause:

$$\{l_2, l_2^c\} \cup C_1 \cup C_2.$$

∎

It is not strictly incorrect to perform resolution on such clauses, but since trivial clauses contribute nothing to the satisfiability or unsatisfiability of a set of clauses (Theorem 4.9), we agree to delete them from any set of clauses and not to perform resolution on clauses with two clashing pairs of literals.

Theorem 4.17 *The resolvent C is satisfiable if and only if the parent clauses C_1 and C_2 are both satisfiable.*

Proof Let C_1 and C_2 be satisfiable under an interpretation $\mathscr{I}$. Since l, l^c are complementary, either $\mathscr{I}(l) = T$ or $\mathscr{I}(l^c) = T$. Suppose that $\mathscr{I}(l) = T$; then $\mathscr{I}(l^c) = F$ and C_2, the clause containing l^c, can be satisfied only if $\mathscr{I}(l') = T$ for some *other* literal $l' \in C_2, l' \neq l^c$. By construction in the resolution rule, $l' \in C$, so $\mathscr{I}$ is also a model for C. A symmetric argument holds if $\mathscr{I}(l^c) = T$.

Conversely, let $\mathscr{I}$ be an interpretation which satisfies C; then $\mathscr{I}(l') = T$ for at least one literal $l' \in C$. By the resolution rule, $l' \in C_1$ or $l' \in C_2$ (or both). If $l' \in C_1$, then $v_{\mathscr{I}}(C_1) = T$. Since neither $l \in C$ nor $l^c \in C$, $\mathscr{I}$ is not defined on either l or l^c, and we can extend $\mathscr{I}$ to an interpretation $\mathscr{I}'$ by defining $\mathscr{I}(l^c) = T$. Since $l^c \in C_2$, $v_{\mathscr{I}'}(C_2) = T$ and $v_{\mathscr{I}'}(C_1) = v_{\mathscr{I}}(C_1) = T$ (because $\mathscr{I}$ is an extension of v) so $\mathscr{I}'$ is a model for both C_1 and C_2. A symmetric argument holds if $l' \in C_2$. $\blacksquare$

Algorithm 4.18 (Resolution procedure)
Input: A set of clauses S.
Output: S is satisfiable or unsatisfiable.

Let S be a set of clauses and define $S_0 = S$.

Repeat the following steps to obtain S_{i+1} from S_i until the procedure terminates as defined below:

- *Choose* a pair of clashing clauses $\{C_1, C_2\} \subseteq S_i$ that has not been chosen before.
- Compute $C = Res(C_1, C_2)$ according to the resolution rule.
- If C is not a trivial clause, let $S_{i+1} = S_i \cup \{C\}$; otherwise, $S_{i+1} = S_i$.

Terminate the procedure if:

- $C = \square$.
- All pairs of clashing clauses have be resolved. $\blacksquare$

Example 4.19 Consider the set of clauses:

$$S = \{(1)\, p,\ (2)\, \bar{p}q,\ (3)\, \bar{r},\ (4)\, \bar{p}\bar{q}r\},$$

where the clauses have been numbered. Here is a resolution derivation of $\square$ from S, where the justification for each line is the pair of the numbers of the parent clauses that have been resolved to give the resolvent clause:

5.	$\bar{p}\bar{q}$	3, 4
6.	$\bar{p}$	5, 2
7.	$\square$	6, 1

$\blacksquare$

It is easier to read a resolution derivation if it is presented as a tree. Figure 4.1 shows the tree that represents the derivation of Example 4.19. The clauses of S label leaves, and the resolvents label interior nodes whose children are the parent clauses used in the resolution.

Fig. 4.1 A resolution
refutation represented
as a tree

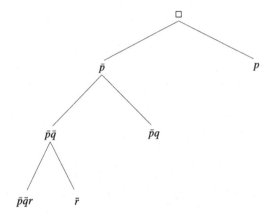

Definition 4.20 A derivation of □ from a set of clauses S is a *refutation by resolution* of S or a *resolution refutation* of S. ∎

Since □ is unsatisfiable, by Theorem 4.17 if there exists a refutation of S by resolution then S is unsatisfiable.

In Example 4.19, we derived the unsatisfiable clause □, so we conclude that the set of clauses S is unsatisfiable. We leave it to the reader to check that S is the clausal form of $\neg A$ where A is an instance of Axiom 2 of $\mathscr{H}$ $(p \to (q \to r)) \to ((p \to q) \to (p \to r))$. Since $\neg A$ is unsatisfiable, A is valid.

4.4 Soundness and Completeness of Resolution *

The soundness of resolution follows easily from Theorem 4.17, but completeness is rather difficult to prove, so you may want to skip the this section on your first reading.

Theorem 4.21 *If the set of clauses labeling the leaves of a resolution tree is satisfiable then the clause at the root is satisfiable.*

The proof is by induction using Theorem 4.17 and is left as an exercise.

The converse to Theorem 4.21 is not true because we have no way of ensuring that the extensions made to $\mathscr{I}$ on all branches are consistent. In the tree in Fig. 4.2, the set of clauses on the leaves $S = \{r, pq\bar{r}, \bar{r}, p\bar{q}r\}$ is not satisfiable even though the clause p at the root is satisfiable. Since S is unsatisfiable, it has a refutation: whenever the pair of clashing clauses r and $\bar{r}$ is chosen, the resolvent will be □.

Resolution is a refutation procedure, so soundness and completeness are better expressed in terms of unsatisfiability, rather than validity.

Fig. 4.2 Incomplete
resolution tree

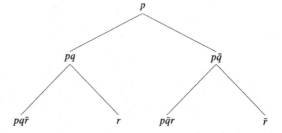

Corollary 4.22 (Soundness) *Let S be a set of clauses. If there is a refutation by resolution for S then S is unsatisfiable.*

Proof Immediate from Theorem 4.21 and Lemma 4.10.   ∎

Theorem 4.23 (Completeness) *If a set of clauses is unsatisfiable then the empty clause □ will be derived by the resolution procedure.*

We have to prove that given an unsatisfiable set of clauses, the resolution procedure will eventually terminate producing □, rather than continuing indefinitely or terminating but failing to produce □. The resolution procedure was defined so that the same pair of clauses is never chosen more than once. Since there are only a finite number of distinct clauses on the finite set of atomic propositions appearing in a set of clauses, the procedure terminates. We need only prove that when the procedure terminates, the empty clause is produced.

Semantic Trees

The proof will use *semantic trees* (which must not be confused with semantic tableaux). A semantic tree is a data structure for recording assignments of T and F to the atomic propositions of a formula in the process of searching for a model (satisfying interpretation). If the formula is unsatisfiable, the search for a model must end in failure. Clauses that are created during a resolution refutation will be associated with nodes of the tree called failure nodes; these nodes represent assignments that falsify the associated clauses. Eventually, the root node (associated with the empty clause □) will be shown to be a failure node.

Definition 4.24 (Semantic tree) Let S be a set of clauses and let $P_S = \{p_1, \ldots, p_n\}$ be the set of atomic propositions appearing in S. $\mathscr{T}$, the *semantic tree* for S, is a complete binary tree of depth n such that for $1 \le i \le n$, every left-branching edge from a node at depth $i - 1$ is labeled p_i and every right-branching edge is labeled by $\bar{p}_i$.

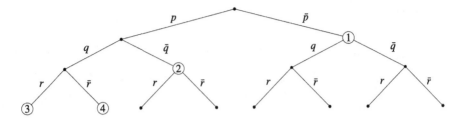

Fig. 4.3 Semantic tree

Every branch b from the root to a leaf in $\mathscr{T}$ is labeled by a sequence of literals $\{l_1, \ldots, l_n\}$, where $l_i = p_i$ or $l_i = \bar{p}_i$. b defines an interpretation by:

$$\mathscr{I}_b(p_i) = T \quad \text{if } l_i = p_i,$$
$$\mathscr{I}_b(p_i) = F \quad \text{if } l_i = \bar{p}_i.$$

A branch b is *closed* if $v_b(S) = F$, otherwise b is *open*. $\mathscr{T}$ is *closed* if all branches are closed, otherwise $\mathscr{T}$ is *open*. ∎

Example 4.25 The semantic tree for $S = \{p, \bar{p}q, \bar{r}, \bar{p}\bar{q}r\}$ is shown in Fig. 4.3 where the numbers on the nodes will be explained later. The branch b ending in the leaf labeled 4 defines the interpretation:

$$\mathscr{I}_b(p) = T, \qquad \mathscr{I}_b(q) = T, \qquad \mathscr{I}_b(r) = F.$$

Since $v_{\mathscr{I}_b}(\bar{p}\bar{q}r) = F$, $v_{\mathscr{I}_b}(S) = F$ (a set of clauses is the conjunction of its members) and the branch b is closed. We leave it to the reader to check that every branch in this tree is closed. ∎

Lemma 4.26 *Let S be a set of clauses and let $\mathscr{T}$ a semantic tree for S. Every interpretation $\mathscr{I}$ for S corresponds to $\mathscr{I}_b$ for some branch b in $\mathscr{T}$, and conversely, every $\mathscr{I}_b$ is an interpretation for S.*

Proof By construction. ∎

Theorem 4.27 *The semantic tree $\mathscr{T}$ for a set of clauses S is closed if and only if the set S is unsatisfiable.*

Proof Suppose that $\mathscr{T}$ is closed and let $\mathscr{I}$ be an arbitrary interpretation for S. By Lemma 4.26, $\mathscr{I}$ is $\mathscr{I}_b$ for some branch in $\mathscr{T}$. Since $\mathscr{T}$ is closed, $v_b(S) = F$. But $\mathscr{I} = \mathscr{I}_b$ was arbitrary so S is unsatisfiable.

Conversely, let S be an unsatisfiable set of clauses, $\mathscr{T}$ the semantic tree for S and b an arbitrary branch in $\mathscr{T}$. Then v_b is an interpretation for S by Lemma 4.26, and $v_b(S) = F$ since S is unsatisfiable. Since b was arbitrary, $\mathscr{T}$ is closed. ∎

Failure Nodes

When traversing a branch of the semantic tree top-down, a (partial) branch from the root to a node represents a partial interpretation (Definition 2.18) defined by the labels of the edges that were traversed. It is possible that this partial interpretation is sufficiently defined to evaluate the truth value of some clauses; in particular, some clause might evaluate to F. Since a set of clauses is an implicit conjunction, if even one clause evaluates to F, the partial interpretation is sufficient to conclude that the entire set of clauses is false. In a *closed* semantic tree, there must be such a node on every branch. However, if a clause contains the literal labeling the edge to a leaf, a (full) interpretation may be necessary to falsify the clause.

Example 4.28 In the semantic tree for $S = \{p, \bar{p}q, \bar{r}, \bar{p}\bar{q}r\}$ (Fig. 4.3), the partial branch $b_{p\bar{q}}$ from the root to the node numbered 2 defines a partial interpretation $\mathscr{I}_{b_{p\bar{q}}}(p) = T$, $\mathscr{I}_{b_{p\bar{q}}}(q) = F$, which falsifies the clause $\bar{p}q$ and thus the entire set of clauses S.

Consider now the partial branches b_p and b_{pq} and the full branch b_{pqr} that are obtained by always taking the child labeled by a positive literal. The partial interpretation $\mathscr{I}_{b_p}(p) = T$ does not falsify any of the clauses, nor does the partial interpretation $\mathscr{I}_{b_{pq}}(p) = T$, $\mathscr{I}b_{pq}(q) = T$. Only the full interpretation $\mathscr{I}_{b_{pqr}}$ that assigns T to r falsifies one of the clauses ($\bar{r}$). ∎

Definition 4.29 Let $\mathscr{T}$ be a closed semantic tree for a set of clauses S and let b be a branch in $\mathscr{T}$. The node in b closest to the root which falsifies S is a *failure node*.

Example 4.30 Referring again to Fig. 4.3, the node numbered 2 is a failure node since neither its parent node (which defines the partial interpretation $\mathscr{I}_{b_p}$) nor the root itself falsifies any of the clauses in the set. We leave it to the read to check that all the numbered nodes are failure nodes. ∎

Since a failure node falsifies S (an implicit conjunction of clauses), it must falsify at least once clause in S.

Definition 4.31 A clause falsified by a failure node is a *clause associated with the node*. ∎

Example 4.32 The failure nodes in Fig. 4.3 are labeled with the number of a clause associated with it; the numbers were given in Examples 4.19. ∎

It is possible that more than one clause is associated with a failure node; for example, if q is added to the set of clauses, then q is another clause associated with failure node numbered 2.

We can characterize the clauses associated with failure nodes. For C to be falsified at a failure node n, *all* the literals in C must be assigned F in the partial interpretation.

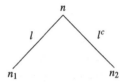

Example 4.33 In Fig. 4.3, $\bar{r}$ is a clause associated with the failure node numbered 3. $\{\bar{r}\}$ is a proper subset of $\{\bar{p}, \bar{q}, \bar{r}\}$, the set of *complements* of the literals assigned to on the branch.   ∎

Lemma 4.34 *A clause C associated with a failure node n is a subset of the complements of the literals appearing on the partial branch b from the root to n.*

Proof Let $C = l_1 \cdots l_k$ and let $E = \{e_1, \ldots, e_m\}$ be the set of literals labeling edges in the branch. Since C is the clause associated with the failure node n, $v_b(C) = F$ for the interpretation $\mathscr{I}_b$ defined by $\mathscr{I}_b(e_j) = T$ for all $e_j \in E$. C is a disjunction so for each $l_i \in C$, $\mathscr{I}_b(l_i)$ must be assigned F. Since $\mathscr{I}_b$ only assigns to the literals in E, it follows that $l_i = e_j^c$ for some $e_j \in E$. Therefore, $C = l_1 \cdots l_k \subseteq \{e_1^c, \ldots, e_m^c\}$.   ∎

Inference Nodes

Definition 4.35 n is an *inference node* iff its children are failure nodes.   ∎

Example 4.36 In Fig. 4.3, the parent of nodes 3 and 4 is an inference node.   ∎

Lemma 4.37 *Let $\mathscr{T}$ be a closed semantic tree for a set of clauses S. If there are at least two failure nodes in $\mathscr{T}$, then there is at least one inference node.*

Proof Suppose that n_1 is a failure node and that its sibling n_2 is not (Fig. 4.4). Then no ancestor of n_2 can be a failure node, because its ancestors are also ancestors of n_1, which is, by assumption, a failure node and thus the node *closest* to the root on its branch which falsifies S.

$\mathscr{T}$ is closed so every branch in $\mathscr{T}$ is closed, in particular, any branch b that includes n_2 is closed. By definition of a closed branch, $\mathscr{I}_b$, the full interpretation associated with the leaf of b, must falsify S. Since neither n_2 nor any ancestor of n_2 is a failure node, some node below n_2 on b (perhaps the leaf itself) must be the highest node which falsifies a clause in S.

We have shown that given an arbitrary failure node n_1, either its sibling n_2 is a failure node (and hence their parent is an inference node), or there is a failure node at a *greater* depth than n_1 and n_2. Therefore, if there is no inference node, there must be an infinite sequence of failure nodes. But this is impossible, since a semantic tree is finite (its depth is the number of different atomic propositions in S).   ∎

Lemma 4.38 *Let $\mathcal{T}$ be closed semantic tree and let n be an inference node whose children n_1 and n_2 of n are (by definition) failure nodes with clauses C_1 and C_2 associated with them, respectively. Then C_1, C_2 clash and the partial interpretation defined by the branch from the root to n falsifies their resolvent.*

Proof of the Notation follows Fig. 4.4. Let b_1 and b_2 be the partial branches from the root to the nodes n_1 and n_2, respectively. Since n_1 and n_2 are failure nodes and since C_1 and C_2 are clauses associated with the nodes, they are *not* falsified by any node higher up in the tree. By Lemma 4.34, the clauses C_1 and C_2 are subsets of the complements of the literals labeling the nodes of b_1 and b_2, respectively. Since b_1 and b_2 are identical except for the edges from n to n_1 and n_2, we must have $\bar{l} \in C_1$ and $\bar{l}^c \in C_2$ so that the clauses are falsified by the assignments to the literals.

Since the nodes n_1 and n_2 are failure nodes, $v_{\mathcal{I}_{b_1}}(C_1) = v_{\mathcal{I}_{b_2}}(C_2) = F$. But clauses are disjunctions so $v_{\mathcal{I}_{b_1}}(C_1 - \{\bar{l}\}) = v_{\mathcal{I}_{b_2}}(C_2 - \{\bar{l}^c\}) = F$ and this also holds for the interpretation $\mathcal{I}_b$. Therefore, their resolvent is also falsified:

$$v_{\mathcal{I}_b}((C_1 - \{\bar{l}\}) \cup (C_2 - \{\bar{l}^c\})) = F.$$

∎

Example 4.39 In Fig. 4.3, $\bar{r}$ and $\bar{p}\bar{q}r$ are clauses associated with failure nodes 3 and 4, respectively. The resolvent $\bar{p}\bar{q}$ is falsified by $\mathcal{I}_{pq}(p) = T, \mathcal{I}_{pq}(q) = T$, the partial interpretation associated with the parent node of 3 and 4. The parent node is now a failure node for the set of clauses $S \cup \{\bar{p}\bar{q}\}$. ∎

There is a technicality that must be dealt with before we can prove completeness. A semantic tree is defined by choosing an ordering for the set of atoms that appear in *all* the clauses in a set; therefore, an inference node may not be a failure node.

Example 4.40 The semantic tree in Fig. 4.3 is also a semantic tree for the set of clauses $\{p, \bar{p}q, \bar{r}, \bar{p}r\}$. Node 3 is a failure node associated with $\bar{r}$ and 4 is a failure node associated with $\bar{p}r$, but their parent is *not* a failure node for their resolvent $\bar{p}$, since it is already falsified by a node higher up in the tree. (Recall that a failure node was defined to be the node *closest* to the root which falsifies the set of clauses.) ∎

Lemma 4.41 *Let n be an inference node, $C_1, C_2 \in S$ clauses associated with the failure nodes that are the children of n, and C their resolvent. Then $S \cup \{C\}$ has a failure node that is either n or an ancestor of n and C is a clause associated with the failure node.*

Proof By Lemma 4.38, $v_{\mathcal{I}_b}(C) = F$, where $\mathcal{I}_b$ is the partial interpretation associated with the partial branch b from the root to the inference node. By Lemma 4.34, $C \subseteq \{l_1^c, \ldots, l_n^c\}$, the set of complements of the literals labeling b. Let j be the smallest index such $C \cap \{l_{j+1}^c, \ldots, l_n^c\} = \emptyset$. Then $C \subseteq \{l_1^c, \ldots, l_j^c\} \subseteq \{l_1^c, \ldots, l_n^c\}$ so $v_{\mathcal{I}_b^j}(C) = v_{\mathcal{I}_b}(C) = F$ where $\mathcal{I}_b^j$ is the partial interpretation defined by the partial branch from the root to node j. It follows that j is a failure node and C is a clause associated with it. ∎

Example 4.42 Returning to the set of clauses $\{p, \bar{p}q, \bar{r}, \bar{p}r\}$ in Example 4.40, the resolvent at the inference node is $C = \{\bar{p}\}$. Now $C = \{\bar{p}\} \subseteq \{\bar{p}, \bar{q}\}$, the complements of the literals on the partial branch from the root to the inference node. Let $j = 1$. Then $\{\bar{p}\} \cap \{\bar{q}\} = \emptyset$, $\{\bar{p}\} \subseteq \{\bar{p}\}$ and $C = \{\bar{p}\}$ is falsified by the partial interpretation $\mathscr{I}_{b_p}(p) = T$.   ∎

We now have all the machinery needed to proof completeness.

Proof of Completeness of resolution If S is an unsatisfiable set of clauses, there is a closed semantic tree $\mathscr{T}$ for S. If S is unsatisfiable and does not already contain □, there must be at least two failure nodes in $\mathscr{T}$ (exercise), so by Lemma 4.37, there is at least one inference node in $\mathscr{T}$.

An application of the resolution rule at the inference node adds the resolvent to the set, creating a failure node by Lemma 4.41 and deleting two failure nodes, thus decreasing the number of failure nodes. When the number of failure nodes has decreased to one, it must be the root which is associated with the derivation of the empty clause by the resolution rule.   ∎

4.5 Hard Examples for Resolution *

If you try the resolution procedure on formulas in clausal form, you will find that is usually quite efficient. However, there are families of formulas on which *any* resolution refutation is necessarily inefficient. We show how an unsatisfiable set of clauses can be associated with an arbitrarily large graph such that a resolution refutation of a set of clauses from this family produces an exponential number of new clauses.

Let G be an undirected graph. Label the nodes with 0 or 1 and the edges with distinct atoms. The following graph will be used as an example throughout this section.

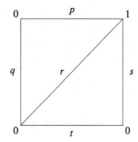

Definition 4.43

- The *parity* of a natural number i is 0 if i is even and 1 if i is odd.
- Let C be a clause. $\Pi(C)$, the *parity of C*, is the parity of the number of complemented literals in C.
- Let $\mathscr{I}$ be an interpretation for a set of atomic propositions $\mathscr{P}$. $\Pi(\mathscr{I})$, the *parity of $\mathscr{I}$*, is the parity of the number of atoms in $\mathscr{P}$ assigned T in $\mathscr{I}$.   ∎

Example 4.44 $\Pi(p\bar{r}\bar{s}) = 2$ and $\Pi(\bar{p}\bar{r}\bar{s}) = 3$. For the interpretation $\mathscr{I}$ defined by $\mathscr{I}(p) = T$, $\mathscr{I}(q) = T$, $\mathscr{I}(r) = F$, the parity $\Pi(\mathscr{I})$ is 2.   ∎

With each graph we associate a set of clauses.

Definition 4.45 Let G be an undirected, connected graph, whose nodes are labeled with 0 or 1 and whose edges are labeled with distinct atomic propositions. Let n be a node of G labeled a_n (0 or 1) and let $\mathscr{P}_n = \{p_1, \ldots, p_k\}$ be the set of atoms labeling edges incident with n.

$C(n)$, the *set of clauses associated with* n, is the set of all clauses C that can be formed as follows: the literals of C are *all* the atoms in $\mathscr{P}_n$, some of which are negated so that $\Pi(C) \neq a_n$.

$C(G)$, the *set of clauses associated with* G, is $\bigcup_{n \in G} C(n)$.

Let $\mathscr{I}$ be an interpretation on all the atomic propositions $\bigcup_n \mathscr{P}_n$ in G. $\mathscr{I}_n$ is the *restriction* of $\mathscr{I}$ to node n which assigns truth values only to the literals in $C(n)$. ∎

Example 4.46 The sets of clauses associated with the four nodes of the graph are (clockwise from the upper-left corner):

$$\{\bar{p}q,\ p\bar{q}\,\}, \qquad \{prs,\ \bar{p}\bar{r}s,\ \bar{p}r\bar{s},\ p\bar{r}\bar{s}\}, \qquad \{\bar{s}t,\ s\bar{t}\,\}, \qquad \{\bar{q}rt,\ q\bar{r}t,\ qr\bar{t},\ \bar{q}\bar{r}\bar{t}\}.$$

By definition, the parity of each clause associated with a node n must be opposite the parity of n. For example:

$$\Pi(\bar{p}\bar{r}s) = 0 \neq 1,$$
$$\Pi(\bar{q}rt) = 1 \neq 0.$$

                                              ∎

Lemma 4.47 $\mathscr{I}_n$ *is a model for* $C(n)$ *if and only if* $\Pi(\mathscr{I}_n) = a_n$.

Proof Suppose that $\Pi(\mathscr{I}_n) \neq a_n$ and consider the clause $C \in C(n)$ defined by:

$$l_i = p_i \qquad \text{if } \mathscr{I}_n(p_i) = F,$$
$$l_i = \bar{p}_i \qquad \text{if } \mathscr{I}_n(p_i) = T.$$

Then:

$$
\begin{aligned}
\Pi(C) \ &= \ \text{parity of negated atoms of } C \quad \text{(by definition)} \\
&= \ \text{parity of literals assigned } T \quad \text{(by construction)} \\
&= \ \Pi(\mathscr{I}_n) \qquad\qquad\qquad\qquad \text{(by definition)} \\
&\neq \ a_n \qquad\qquad\qquad\qquad\quad\ \text{(by assumption)}.
\end{aligned}
$$

But $\mathscr{I}_n(C) = F$ since $\mathscr{I}_n$ assigns F to each literal $l_i \in C$ (T to negated literals and F to atoms). Therefore, $\mathscr{I}_n$ does not satisfy all clauses in $C(n)$.

We leave the proof of the converse as an exercise.   ∎

Example 4.48 Consider an interpretation $\mathscr{I}$ such that $\mathscr{I}_n$ is:

$$\mathscr{I}_n(p) = \mathscr{I}_n(r) = \mathscr{I}_n(s) = T$$

for n the upper right node in the graph. For such interpretations, $\Pi(\mathscr{I}_n) = 1 = a_n$, and it is easy to see that $v_n(prs) = v_n(\bar{p}\bar{r}s) = v_n(\bar{p}q\bar{s}) = v_n(p\bar{r}\bar{s}) = T$ so $\mathscr{I}$ is a model for $C(n)$.

Consider an interpretation $\mathscr{I}$ such that $\mathscr{I}_n$ is:

$$\mathscr{I}_n(p) = \mathscr{I}_n(r) = \mathscr{I}_n(s) = F.$$

$\Pi(\mathscr{I}_n) = 0 \neq a_n$ and $v_n(prs) = F$ so $\mathscr{I}$ is not a model for $C(n)$.   ■

$C(G)$ is the set of clauses obtained by taking the union of the clauses associated with all the nodes in the graph. Compute the sum modulo 2 (denoted $\sum$ in the following lemma) of the labels of the nodes and the sum of the parities of the restrictions of an interpretation to each node. Since each atom appears twice, the sum of the parities of the restricted interpretations must be 0. By Lemma 4.47, for the clauses to be satisfiable, the sum of the node labels must be the same as the sum of the parities of the interpretations, namely zero.

Lemma 4.49 *If $\sum_{n \in G} a_n = 1$ then $C(G)$ is unsatisfiable.*

Proof Suppose that there exists a model $\mathscr{I}$ for $C(G) = \bigcup_{n \in G} C(n)$. By Lemma 4.47, for all n, $\Pi(\mathscr{I}_n) = a_n$, so:

$$\sum_{n \in G} \Pi(\mathscr{I}_n) = \sum_{n \in G} a_n = 1.$$

Let p_e be the atom labeling an arbitrary edge e in G; it is incident with (exactly) two nodes, n_1 and n_2. The sum of the parities of the restricted interpretations can be written:

$$\sum_{n \in G} \Pi(\mathscr{I}_n) = \Pi(\mathscr{I}_{n_1}) + \Pi(\mathscr{I}_{n_2}) + \sum_{n \in (G - \{n_1, n_2\})} \Pi(\mathscr{I}_n).$$

Whatever the value of the assignment of $\mathscr{I}$ to p_e, it appears once in the first term, once in the second term and not at all in the third term above. By modulo 2 arithmetic, the total contribution of the assignment to p_e to $\sum_{n \in G} \Pi(\mathscr{I}_n)$ is 0. Since e was arbitrary, this is true for all atoms, so:

$$\sum_{n \in G} \Pi(\mathscr{I}_n) = 0,$$

contradicting $\sum_{n \in G} \Pi(\mathscr{I}_n) = 1$ obtained above. Therefore, $\mathscr{I}$ cannot be a model for $C(G)$, so $C(G)$ must be unsatisfiable.   ■

Tseitin (1968) defined a family G_n of graphs of arbitrary size n and showed that for a restricted form of resolution the number of distinct clauses that appear a resolution refutation of $C(G_n)$ is *exponential* in n. About twenty years later, the restriction was removed by Urquhart (1987).

4.5.1   Tseitin Encoding

The standard procedure for transforming a formula into CNF (Sect. 4.1) can lead to formulas that are significantly larger than the original formula. In practice, an alternate transformation by Tseitin (1968) yields a more compact set of clauses at the expense of adding new atoms.

Algorithm 4.50 (Tseitin encoding) Let A be a formula in propositional logic. Define a sequence of formulas $A = A_0, A_1, A_2, \ldots$ by repeatedly performing the transformation:

- Let $B_i' \circ B_i''$ be a subformula of A_i, where B_i', B_i'' are literals.
- Let p_i be a new atom that does not appear in A_i. Construct A_{i+1} by replacing the subformula $B_i' \circ B_i''$ by p_i and adding the CNF of:

$$p_i \leftrightarrow B_i' \circ B_i''.$$

- Terminate the transformation when A_n is in CNF.   ∎

Theorem 4.51 *Let A be a formula in propositional logic and apply Algorithm 4.50 to obtain the CNF formula A_n. Then A is satisfiable if and only if A_n is satisfiable.*

Example 4.52 Let n be a node labeled 1 with five incident edges labeled by the atoms p, q, r, s, t. $C(n)$ consists of all clauses of even parity defined on these atoms:

$$pqrst,$$
$$\bar{p}\bar{q}rst, \ \bar{p}q\bar{r}st, \ \ldots, \ pqr\bar{s}\bar{t}, \ pqr\bar{s}\bar{t}$$
$$p\bar{q}\bar{r}\bar{s}\bar{t}, \ \bar{p}q\bar{r}\bar{s}\bar{t}, \ \bar{p}\bar{q}r\bar{s}\bar{t}, \ \bar{p}\bar{q}\bar{r}s\bar{t}, \ \bar{p}\bar{q}\bar{r}\bar{s}t.$$

There are 16 clauses in $C(n)$ since there $2^5 = 32$ clauses on five atoms and half of them have even parity: one clause with parity 0, $\frac{5!}{2! \cdot (5-2)!} = 10$ clauses with parity 2 and five clauses with parity 4. We leave it to the reader to show that this set of clauses is logically equivalent to the formula:

$$(p \leftrightarrow (q \leftrightarrow (r \leftrightarrow (s \leftrightarrow t)))),$$

where we have used parentheses to bring out the structure of subformulas. Applying the Tseitin encoding, we choose four new atoms a, b, c, d and obtain the set of formulas:

$$\{a \leftrightarrow (s \leftrightarrow t), \ b \leftrightarrow (r \leftrightarrow a), \ c \leftrightarrow (q \leftrightarrow b), \ d \leftrightarrow (s \leftrightarrow c)\}.$$

Each of the new formulas is logically equivalent to one in CNF that contains four disjunctions of three literals each; for example:

$$a \leftrightarrow (s \leftrightarrow t) \equiv \{a \vee s \vee t, \ \bar{a} \vee \bar{s} \vee t, \ \bar{a} \vee s \vee \bar{t}, \ a \vee \bar{s} \vee \bar{t}\}.$$

Sixteen clauses of five literals have been replaced by the same number of clauses but each clause has only three literals.   ∎

4.6 Summary

Resolution is a highly efficient refutation procedure that is a decision procedure for unsatisfiability in propositional logic. It works on formulas in clausal form, which is a set representation of conjunctive normal form (a conjunction of disjunctions of literals). Each resolution step takes two clauses that clash on a pair of complementary literals and produces a new clause called the resolvent. If the formula is unsatisfiable, the empty clause will eventually be produced.

4.7 Further Reading

Resolution for propositional logic is presented in the advanced textbooks by Nerode and Shore (1997) and Fitting (1996).

4.8 Exercises

4.1 A formula is in *disjunctive normal form (DNF)* iff it is a disjunction of conjunctions of literals. Show that every formula is equivalent to one in DNF.

4.2 A formula A is in *complete DNF* iff it is in DNF and each propositional letter in A appears in a literal in each conjunction. For example, $(p \wedge q) \vee (\bar{p} \wedge q)$ is in complete DNF. Show that every formula is equivalent to one in complete DNF.

4.3 Simplify the following sets of literals, that is, for each set S find a simpler set S', such that S' is satisfiable if and only if S is satisfiable.

$$\{p\bar{q}, \ q\bar{r}, \ rs, \ p\bar{s}\},$$
$$\{pqr, \ \bar{q}, \ p\bar{r}s, \ qs, \ p\bar{s}\},$$
$$\{pqrs, \ \bar{q}rs, \ \bar{p}rs, \ qs, \ \bar{p}s\},$$
$$\{\bar{p}q, \ qrs, \ \bar{p}\bar{q}rs, \ \bar{r}, \ q\}.$$

4.4 Given the set of clauses $\{\bar{p}\bar{q}r, \ pr, \ qr, \ \bar{r}\}$ construct two refutations: one by resolving the literals in the order $\{p, q, r\}$ and the other in the order $\{r, q, p\}$.

4.5 Transform the set of formulas

$$\{p, \ p \to ((q \lor r) \land \neg (q \land r)), \ p \to ((s \lor t) \land \neg (s \land t)), \ s \to q, \ \neg r \to t, \ t \to s \}$$

into clausal form and refute using resolution.

4.6 * The half-adder of Example 1.2 implements the pair of formulas:

$$s \leftrightarrow \neg (b1 \land b2) \land (b1 \lor b2), \qquad c \leftrightarrow b1 \land b2.$$

Transform the formulas to a set of clauses. Show that the addition of the unit clauses $\{b1, b2, \bar{s}, \bar{c}\}$ gives an unsatisfiable set while the addition of $\{b1, b2, \bar{s}, c\}$ gives a satisfiable set. Explain what this means in terms of the behavior of the circuit.

4.7 Prove that if the set of clauses labeling the leaves of a resolution tree is satisfiable then the clause at the root is satisfiable (Theorem 4.21).

4.8 Construct a resolution refutation for the set of Tseitin clauses given in Example 4.46.

4.9 * Construct the set of Tseitin clauses corresponding to a labeled complete graph on five vertices and give a resolution refutation of the set.

4.10 * Construct the set of Tseitin clauses corresponding to a labeled complete bipartite graph on three vertices on each side and give a resolution refutation of the set.

4.11 * Show that if $\Pi(v_n) = b_n$, then v_n satisfies all clauses in $C(n)$ (the converse direction of Lemma 4.47).

4.12 * Let $\{q_1, \ldots, q_n\}$ be literals on *distinct* atoms. Show that $q_1 \leftrightarrow \cdots \leftrightarrow q_n$ is satisfiable iff $\{p \leftrightarrow q_1, \ldots, p \leftrightarrow q_n\}$ is satisfiable, where p is a new atom. Construct an efficient decision procedure for formulas whose only operators are $\neg$, $\leftrightarrow$ and $\oplus$.

4.13 Prove Theorem 4.13 on the correctness of the CNF-to-3CNF algorithm.

4.14 Carry out the Tseitin encoding on the formula $(a \to (c \land d)) \lor (b \to (c \land e))$.

References

M. Fitting. *First-Order Logic and Automated Theorem Proving (Second Edition)*. Springer, 1996.

A. Nerode and R.A. Shore. *Logic for Applications (Second Edition)*. Springer, 1997.

G.S. Tseitin. On the complexity of derivation in propositional calculus. In A.O. Slisenko, editor, *Structures in Constructive Mathematics and Mathematical Logic, Part II*, pages 115–125. Steklov Mathematical Institute, 1968.

A. Urquhart. Hard examples for resolution. *Journal of the ACM*, 34:209–219, 1987.

Chapter 5
Propositional Logic: Binary Decision Diagrams

The problem of deciding the satisfiability of a formula in propositional logic has turned out to have many important applications in computer science. This chapter and the next one present two widely used approaches for computing with formulas in propositional logic.

A binary decision diagram (BDD) is a data structure for representing the semantics of a formula in propositional logic. A formula is represented by a directed graph and an algorithm is used to *reduce* the graph. Reduced graphs have the property that the graphs for logically equivalent formulas are identical. Clearly, this gives a decision procedure for logical equivalence: transform A_1 and A_2 into BDDs and check that they are identical. A formula is valid iff its BDD is identical to the trivial BDD for *true* and a formula is satisfiable iff its BDD is not identical to the trivial BDD for *false*.

Before defining BDDs formally, the next section motivates the concept by reducing truth tables for formulas.

5.1 Motivation Through Truth Tables

Suppose that we want to decide if two formulas A_1 and A_2 in propositional logic are logically equivalent. Let us construct systematic truth tables, where *systematic* means that the assignments to the atomic propositions are arranged in some consistent order, for example, in lexicographic order by placing T before F and varying the values assigned to the atoms from the right to the left. Now, all we have to do is to check if the truth tables for A_1 and A_2 are identical. Of course, this is very inefficient, because 2^n rows are needed for each formula with n variables. Can we do better?

M. Ben-Ari, *Mathematical Logic for Computer Science*,
DOI 10.1007/978-1-4471-4129-7_5, © Springer-Verlag London 2012

Consider the following truth table for $p \lor (q \land r)$, where we have numbered the rows for convenience in referring to them:

	p	q	r	$p \lor (q \land r)$
1	T	T	T	T
2	T	T	F	T
3	T	F	T	T
4	T	F	F	T
5	F	T	T	T
6	F	T	F	F
7	F	F	T	F
8	F	F	F	F

From rows 1 and 2, we see that when p and q are assigned T, the formula evaluates to T *regardless* of the value of r, and similarly for rows 3 and 4. The first four rows can therefore be condensed into two rows:

	p	q	r	$p \lor (q \land r)$
1	T	T	$*$	T
2	T	F	$*$	T

where $*$ indicates that the value assigned to r is immaterial. We now see that the value assigned to q is immaterial, so these two rows collapse into one:

	p	q	r	$p \lor (q \land r)$
1	T	$*$	$*$	T

After similarly collapsing rows 7 and 8, the truth table has four rows:

	p	q	r	$p \lor (q \land r)$
1	T	$*$	$*$	T
2	F	T	T	T
3	F	T	F	F
4	F	F	$*$	F

Let us try another example, this time for the formula $p \oplus q \oplus r$. It is easy to compute the truth table for a formula whose only operator is $\oplus$, since a row evaluates to T if and only if an odd number of atoms are assigned T:

	p	q	r	$p \oplus q \oplus r$
1	T	T	T	T
2	T	T	F	F
3	T	F	T	F
4	T	F	F	T
5	F	T	T	F
6	F	T	F	T
7	F	F	T	T
8	F	F	F	F

Here, adjacent rows cannot be collapsed, but careful examination reveals that rows 5 and 6 show the same dependence on r as do rows 3 and 4. Rows 7 and 8 are similarly related to rows 1 and 2. Instead of explicitly writing the truth table entries for these rows, we can simply refer to the previous entries:

	p	q	r	$p \oplus q \oplus r$
1	T	T	T	T
2	T	T	F	F
3	T	F	T	F
4	T	F	F	T
5, 6	F	T	$*$	(See rows 3 and 4.)
7, 8	F	F	$*$	(See rows 1 and 2.)

The size of the table has been reduced by removing repetitions of computations of truth values.

5.2 Definition of Binary Decision Diagrams

A binary decision diagram, like a truth table, is a representation of the value of a formula under all possible interpretations. Each node of the tree is labeled with an atom, and solid and dotted edges leaving the node represent the assignment of T and F, respectively, to this atom. Along each branch, there is an edge for every atom in the formula, so there is a one-to-one correspondence between branches and interpretations. The leaf of a branch is labeled with the value of the formula under its interpretation.

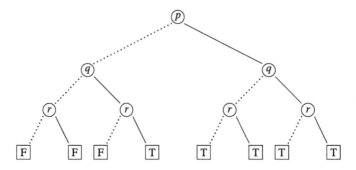

Fig. 5.1 A binary decision diagram for $p \vee (q \wedge r)$

Definition 5.1 A *binary decision diagram (BDD)* for a formula A in propositional logic is a directed acyclic graph. Each leaf is labeled with a truth value T or F. Each interior node is labeled with an atom and has two outgoing edges: one, the *false edge*, is denoted by a dotted line, while the other, the *true edge*, is denoted by a solid line. No atom appears more than once in a branch from the root to an edge.

A full or partial interpretation $\mathscr{I}_b$ for A is associated with each branch b from the root to a leaf. $\mathscr{I}_b(p) = T$ if the true edge was taken at the node labeled p and $\mathscr{I}_b(p) = F$ if the false edge was taken at the node labeled p.

Given a branch b and its associated interpretation $\mathscr{I}_b$, the leaf is labeled with $v_{\mathscr{I}_b}(A)$, the truth value of the formula under $\mathscr{I}_b$. If the interpretation is partial, it must assign to enough atoms so that the truth value is defined.   ∎

Example 5.2 Figure 5.1 is a BDD for $A = p \vee (q \wedge r)$. The interpretation associated with the branch that goes left, right, left is

$$\mathscr{I}(p) = F, \qquad \mathscr{I}(q) = T, \qquad \mathscr{I}(r) = F.$$

The leaf is labeled F so we can conclude that for this interpretation, $v_{\mathscr{I}}(A) = F$. Check that the value of the formula for the interpretation associated with each branch is the same as that given in the first truth table on page 96.   ∎

The BDD in the figure is a special case, where the directed acyclic graph is a tree and a *full* interpretation is associated with each branch.

5.3 Reduced Binary Decision Diagrams

We can modify the structure of a tree such as the one in Fig. 5.1 to obtain a more concise representation without losing the ability to evaluate the formula under all interpretations. The modifications are called *reductions* and they transform the tree into a *directed acyclic graph*, where the direction of an edge is implicitly from a node to its child. When no more reductions can be done, the BDD is *reduced*.

Algorithm 5.3 (Reduce)
Input: A binary decision diagram *bdd*.
Output: A reduced binary decision diagram *bdd'*.

- If *bdd* has more than two distinct leaves (one labeled T and one labeled F), remove duplicate leaves. Direct all edges that pointed to leaves to the remaining two leaves.
- Perform the following steps as long as possible:

  1. If both outgoing edges of a node labeled p_i point to the same node labeled p_j, delete this node for p_i and direct p_i's incoming edges to p_j.
  2. If two nodes labeled p_i are the roots of identical sub-BDDs, delete one sub-BDD and direct its incoming edges to the other node.   ∎

Definition 5.4 A BDD that results from applying the algorithm **Reduce** is a *reduced binary decision diagram*.   ∎

See Bryant (1986) or Baier and Katoen (2008, Sect. 6.7.3) for a proof of the following theorem:

Theorem 5.5 *The reduced BDD bdd' returned by the algorithm* **Reduce** *is logically equivalent to the input BDD bdd.*

Let us apply the algorithm **Reduce** to the two formulas used as motivating examples in Sect. 5.1.

Example 5.6 Figure 5.1 shows a non-reduced BDD for $A = p \vee (q \wedge r)$.
First, merge all leaves into just two, one for T and one for F:

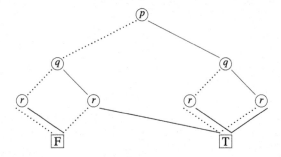

Now we apply Step (1) of the algorithm repeatedly in order to remove nodes that are not needed to evaluate the formula. Once on the left-hand side of the diagram and twice on the right-hand side, the node for r has both outgoing edges leading to the same node. This means that the partial assignment to p and q is sufficient to determine the value of the formula. The three nodes labeled r and their outgoing edges can be deleted and the incoming edges to the r nodes are directed to the joint target nodes:

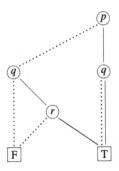

Step (1) can now be applied again to delete the right-hand node for q:

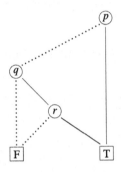

Since neither Step (1) nor Step (2) can be applied, the BDD is reduced.

There are four branches in the reduced BDD for $p \vee (q \wedge r)$. The interpretations associated with the branches are (from left to right):

$$
\begin{aligned}
&\mathscr{I}_{b_1}(p) = F, && \mathscr{I}_{b_1}(q) = F, \\
&\mathscr{I}_{b_2}(p) = F, && \mathscr{I}_{b_2}(q) = T, && \mathscr{I}_{b_2}(r) = F, \\
&\mathscr{I}_{b_3}(p) = F, && \mathscr{I}_{b_3}(q) = T, && \mathscr{I}_{b_3}(r) = T, \\
&\mathscr{I}_{b_4}(p) = T.
\end{aligned}
$$

The interpretations $\mathscr{I}_{b_1}$ and $\mathscr{I}_{b_4}$ are partial interpretations, but they assign truth values to enough atoms for the truth values of the formula to be computed.   ∎

Example 5.7 Consider now the formula $A' = p \oplus q \oplus r$. We start with a tree that defines full interpretations for the formula and delete duplicate leaves. Here is the BDD that results:

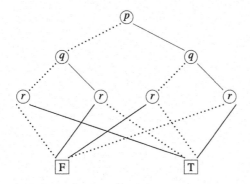

The reduction of Step (1) is not applicable, but examination of the BDD reveals that the subgraphs rooted at the left and right outermost nodes for r have the same structure: their F and T edges point to the same subgraphs, in this case the leaves $\boxed{F}$ and $\boxed{T}$, respectively. Applying Step (2), the T edge from the rightmost node for q can be directed to the leftmost node for r:

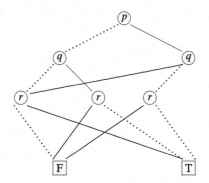

Similarly, the two innermost nodes for r are the roots of identical subgraphs and the F from the rightmost node for q can be directed to the second r node from the left:

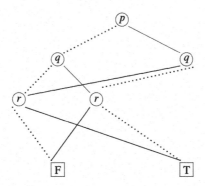

Neither Step (1) nor Step (2) can be applied so the BDD is reduced. By rearranging the nodes, the following symmetric representation of the BDD is obtained:

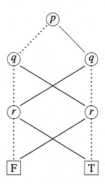

Check that the truth values of A' under the interpretations associated with each branch correspond to those in the reduced truth table on page 97.   ∎

5.4 Ordered Binary Decision Diagrams

The definition of BDDs did not place any requirements on the order in which atoms appear on a branch from the root to the leaves. Since branches can represent partial interpretations, the set of atoms appearing on one branch can be different from the set on another branch. Algorithms on BDDs require that the different orderings do not contradict each other.

Definition 5.8 Let $\mathcal{O} = \{\mathcal{O}_A^1, \ldots, \mathcal{O}_A^n\}$, where for each i, $\mathcal{O}_A^i$ is a sequence of the elements of $\mathcal{P}_A$ (the set of atoms in A) defined by $<^i_{\mathcal{P}_A}$, a total relation that orders $\mathcal{P}_A$. $\mathcal{O}$ is a *compatible set of orderings for* $\mathcal{P}_A$ iff for all $i \neq j$, there are no atoms p, p' such that $p <^i_{\mathcal{P}_A} p'$ in $\mathcal{O}_A^i$ while $p' <^j_{\mathcal{P}_A} p$ in $\mathcal{O}_A^j$.   ∎

Example 5.9 Here is a BDD that is the same as the one in Fig. 5.1, except that the orderings are not compatible because q appears before r on the left branches, while r appears before q on the right branches:

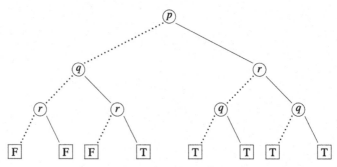

∎

Example 5.10 Consider again the reduced BDD for $p \vee (q \wedge r)$:

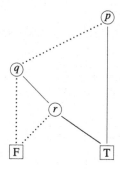

The four branches define three distinct orderings of the atoms:

$$\{(p,q),\ (p,q,r),\ (p)\},$$

but the orderings are compatible.   ■

Definition 5.11 An *ordered binary decision diagram (OBDD)* is a BDD such that the set of orderings of atoms defined by the branches is compatible.   ■

The proofs of the following theorems can be found in Bryant (1986).

Theorem 5.12 *The algorithm **Reduce** constructs an OBDD if the original BDD is ordered. For a given ordering of atoms, the reduced OBDDs for logically equivalent formulas are structurally identical.*

The theorem means that a reduced, ordered BDD is a *canonical* representation of a formula. It immediately provides a set of algorithms for deciding properties of formulas. Let A and B be formulas in propositional logic; construct reduced OBDDs for both formulas using a compatible ordering of $\{\mathscr{P}_A, \mathscr{P}_B\}$. Then:

- A is satisfiable iff $\boxed{T}$ appears in its reduced OBDD.
- A is falsifiable iff $\boxed{F}$ appears in its reduced OBDD.
- A is valid iff its reduced OBDD is the single node $\boxed{T}$.
- A is unsatisfiable iff its reduced OBDD is the single node $\boxed{F}$.
- If the reduced OBDDs for A and B are identical, then $A \equiv B$.

The usefulness of OBDDs depends of course on the efficiency of the algorithm **Reduce** (and others that we will describe), which in turn depends on the size of reduced OBDDs. In many cases the size is quite small, but, unfortunately, the size of the reduced OBDD for a formula depends on the ordering and the difference in sizes among different orderings can be substantial.

Theorem 5.13 *The OBDD for the formula*:

$$(p_1 \wedge p_2) \vee \cdots \vee (p_{2n-1} \wedge p_{2n})$$

has $2n + 2$ nodes under the ordering $p_1, \ldots, p_{2n}$, and 2^{n+1} nodes under the ordering $p_1, p_{n+1}, p_2, p_{n+2}, \ldots, p_n, p_{2n}$.

Fortunately, you can generally use heuristics to choose an efficient ordering, but there are formulas that have large reduced OBDDs under any ordering.

Theorem 5.14 *There is a formula A with n atoms such that the reduced OBDD for any ordering of the atoms has at least 2^{cn} nodes for some $c > 0$.*

5.5 Applying Operators to BDDs

It hardly seems worthwhile to create a BDD if we start from the full binary tree whose size is about the same as the size of the truth table. The power of BDDs comes from the ability to perform operations directly on two reduced BDDs. The algorithm **Apply** recursively constructs the BDD for $A_1 \, op \, A_2$ from the reduced BDDs for A_1 and A_2. It can also be used to construct an initial BDD for an arbitrary formula by building it up from the BDDs for atoms.

*The algorithm **Apply** works only on ordered BDDs.*

Algorithm 5.15 (Apply)
Input: OBDDs bdd_1 for formula A_1 and bdd_2 for formula A_2, using a compatible ordering of $\{\mathscr{P}_{A_1}, \mathscr{P}_{A_2}\}$; an operator op.
Output: An OBDD for the formula $A_1 \, op \, A_2$.

- If bdd_1 and bdd_2 are both leaves labeled w_1 and w_2, respectively, return the leaf labeled by $w_1 \, op \, w_2$.
- If the roots of bdd_1 and bdd_2 are labeled by the same atom p, return the following BDD: (a) the root is labeled by p; (b) the left sub-BDD is obtained by recursively performing this algorithm on the left sub-BDDs of bdd_1 and bdd_2; (c) the right sub-BDD is obtained by recursively performing this algorithm on the right sub-BDDs of bdd_1 and bdd_2.
- If the root of bdd_1 is labeled p_1 and the root of bdd_2 is labeled p_2 such that $p_1 < p_2$ in the ordering, return the following BDD: (a) the root is labeled by p_1; (b) the left sub-BDD is obtained by recursively performing this algorithm on the left sub-BDD of bdd_1 and on (the entire BDD) bdd_2; (c) the right sub-BDD is obtained by recursively performing this algorithm on the right sub-BDD of bdd_1 and on (the entire BDD) bdd_2.
  This construction is also performed if bdd_2 is a leaf, but bdd_1 is not.
- Otherwise, we have a symmetrical case to the previous one. The BDD returned has its root labeled by p_2 and its left (respectively, right) sub-BDD obtained by recursively performing this algorithm on bdd_1 and on the left (respectively, right) sub-BDD of bdd_2.   ∎

We now work out a complete example of the application of the **Apply** algorithm. It is quite lengthy, but each step in the recursive algorithm should not be difficult to follow.

Example 5.16 We construct the BDD for the formula $(p \oplus q) \oplus (p \oplus r)$ from the BDDs for $p \oplus q$ and $p \oplus r$. In the following diagram, we have drawn the two BDDs with the operator $\oplus$ between them:

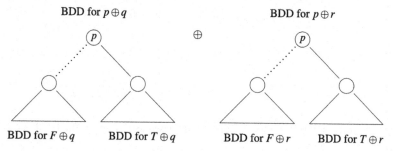

The sub-BDDs will be BDDs for the four subformulas obtained by substituting T and F for p. Notations such as $F \oplus r$ will be used to denote the formula obtained by partially evaluating a formula, in this case, partially evaluating $p \oplus r$ under an interpretation such that $\mathscr{I}(p) = F$.

Since there is only one atom in each sub-BDD, we know what the labels of their roots are:

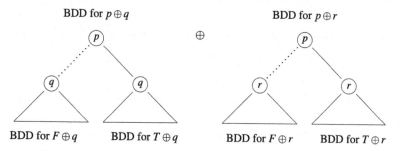

Let us now take the right-hand branch in both BDDs that represent assigning T to p. Evaluating the partial assignment gives $T \oplus q \equiv \neg q$ and $T \oplus r \equiv \neg r$. To obtain the right-hand sub-BDD of the result, we have to compute $\neg q \oplus \neg r$:

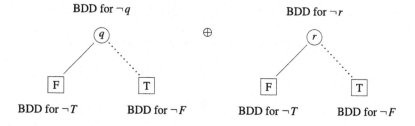

The recursion can be continued by taking the right-hand branch of the BDD for $\neg q$ and assigning F to q. Since the BDD for $\neg r$ does not depend on the assignment to q, it does not split into two recursive subcases. Instead, the algorithm must be applied for each sub-BDD of $\neg q$ together with the *entire* BDD for $\neg r$. The following diagram shows the computation that is done when the right-hand branch of the BDD for $\neg q$ is taken:

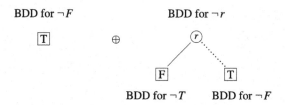

Recursing now on the BDD for $\neg r$ also gives base cases, one for the left-hand (true) branch:

BDD for $\neg F$		BDD for $\neg T$		BDD for $\neg F \oplus \neg T$
T	$\oplus$	F	$\equiv$	T

and one for the right-hand (false) branch:

BDD for $\neg F$		BDD for $\neg F$		BDD for $\neg F \oplus \neg F$
T	$\oplus$	T	$\equiv$	F

When returning from the recursion, these two results are combined:

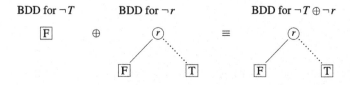

Similarly, taking the left-hand branch of the BDD for $\neg q$ gives:

BDD for $\neg T$ BDD for $\neg r$ BDD for $\neg T \oplus \neg r$

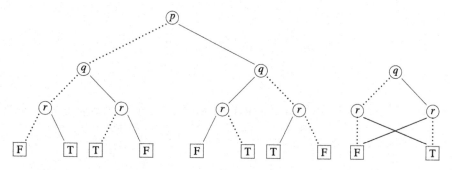

Fig. 5.2 BDD after the **Apply** and **Reduce** algorithms terminate

Returning from the recursion to the BDD for $\neg q$ gives:

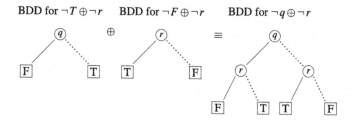

The BDD obtained upon termination of the algorithm is shown in Fig. 5.2 and to its right is the BDD that results from reducing the BDD. Check that this is the reduced BDD for $q \oplus r$:

$$(p \oplus q) \oplus (p \oplus r) \equiv (p \oplus p) \oplus (q \oplus r) \equiv \textit{false} \oplus (q \oplus r) \equiv q \oplus r.$$

■

5.6 Restriction and Quantification *

This section presents additional important algorithms on BDDs.

5.6.1 Restriction

Definition 5.17 The *restriction* operation takes a formula A, an atom p and a truth value $w = T$ or $w = F$. It returns the formula obtained by substituting w for p and partially evaluating A. Notation: $A|_{p=w}$.   ■

Example 5.18 Let $A = p \lor (q \land r)$; its restrictions are:

$$A|_{r=T} \equiv p \lor (q \land T) \equiv p \lor q,$$
$$A|_{r=F} \equiv p \lor (q \land F) \equiv p \lor F \equiv p.$$

∎

The correctness of the algorithm **Reduce** is based upon the following theorem which expresses the application of an operator in terms of its application to restrictions. We leave its proof as an exercise.

Theorem 5.19 (*Shannon expansion*)

$$A_1 \; op \; A_2 \equiv (p \; \land \; (A_1|_{p=T} \; op \; A_2|_{p=T})) \; \lor \; (\neg p \; \land \; (A_1|_{p=F} \; op \; A_2|_{p=F})).$$

Restriction is very easy to implement on OBDDs.

Algorithm 5.20 (Restrict)
Input: An OBDD *bdd* for a formula A; a truth value w.
Output: An OBDD for $A|_{p=w}$.
  Perform a recursive traversal of the OBDD:

- If the root of *bdd* is a leaf, return the leaf.
- If the root of *bdd* is labeled p, return the sub-BDD reached by its true edge if $w = T$ and the sub-BDD reached by its false edge if $w = F$.
- Otherwise (the root is labeled p' for some atom which is not p), apply the algorithm to the left and right sub-BDDs, and return the BDD whose root is p' and whose left and right sub-BDDs are those returned by the recursive calls. ∎

The BDD that results from **Restrict** may not be reduced, so the **Reduce** algorithm is normally applied immediately afterwards.

Example 5.21 The OBDD of $A = p \lor (q \land r)$ is shown in (a) below. (b) is $A|_{r=T}$, (c) is $A|_{r=F}$ and (d) is (c) after reduction.

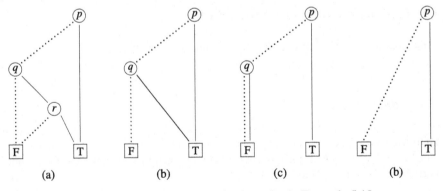

(a)   (b)   (c)   (b)

Compare the OBDDs in (b) and (d) with the formulas in Example 5.18.   ∎

5.6.2 Quantification

Definition 5.22 Let A be a formula and p an atom. The *existential quantification of* A is the formula denoted $\exists p A$ and the *universal quantification of* A is the formula denoted $\forall p A$. $\exists p A$ is true iff A is true for *some* assignment to p, while $\forall p A$ is true iff for *all* assignments to p, A is true. ∎

These formulas are in an extension of propositional logic called *quantified propositional logic*. The proof of the following theorem is left as an exercise.

Theorem 5.23

$$\exists p A \equiv A|_{p=F} \vee A|_{p=T}, \qquad \forall p A \equiv A|_{p=F} \wedge A|_{p=T}.$$

Quantification is easily computed using OBDDs:

$$\exists p A \quad \text{is} \quad \textbf{Apply}(\textbf{Restrict}(A, p, F), or, \textbf{Restrict}(A, p, T)),$$
$$\forall p A \quad \text{is} \quad \textbf{Apply}(\textbf{Restrict}(A, p, F), and, \textbf{Restrict}(A, p, T)).$$

Example 5.24 For the formulas $A = p \vee (q \wedge r)$, we can use $A|_{r=F} \equiv p$ and $A|_{r=T} \equiv p \vee q$ from Example 5.18 to compute its quantifications on r:

$$\exists r \, (p \vee (q \wedge r)) \equiv p \vee (p \vee q) \equiv p \vee q,$$
$$\forall r \, (p \vee (q \wedge r)) \equiv p \wedge (p \vee q) \equiv p.$$

We leave it as an exercise to perform these computations using OBDDs. ∎

5.7 Summary

Binary decision diagrams are a data structure for representing formulas in propositional logic. A BDD is a directed graph that reduces redundancy when compared with a truth table or a semantic tree. Normally, one ensures that all branches of a BDD use compatible orderings of the atomic propositions. An OBDD can be reduced and reduced OBDDs of two formulas are structurally identical if and only if the formulas are logically equivalent. A recursive algorithm can be used to efficiently compute $A \, op \, B$ given the OBDDs for A and B. BDDs have been widely used in model checkers for the verification of computer hardware.

5.8 Further Reading

Bryant's original papers on BDDs (Bryant, 1986, 1992) are relatively easy to read.
There is an extensive presentation of BDDs in Baier and Katoen (2008, Sect. 6.7).

5.9 Exercises

5.1 Construct reduced OBDDs for $p \uparrow (q \uparrow r)$ and $(p \uparrow q) \uparrow r$. What does this
show?

5.2 Construct reduced OBDDs for the formula $(p_1 \wedge p_2) \vee (p_3 \wedge p_4)$ using two
orderings of the variables: p_1, p_2, p_3, p_4 and p_1, p_3, p_2, p_4.

5.3 How can OBDDs be used to check if $A \models B$?

5.4 Compute the Shannon expansion of $(p \rightarrow (q \rightarrow r)) \rightarrow ((p \rightarrow q) \rightarrow (p \rightarrow r))$
with respect to each one of its atomic propositions. Why do you know the answer
even before you start the computation?

5.5 Prove the Shannon expansion (Theorem 5.19) and the formula for propositional
quantification (Theorem 5.23).

5.6 Prove that $\exists r (p \vee (q \wedge r)) = p \vee q$ and $\forall r (p \vee (q \wedge r)) = p$ using BDDs
(Example 5.24).

References

C. Baier and J.-P. Katoen. *Principles of Model Checking*. MIT Press, 2008.
R.E. Bryant. Graph-based algorithms for Boolean function manipulation. *IEEE Transactions on Computers*, C-35:677–691, 1986.
R.E. Bryant. Symbolic Boolean manipulation with ordered binary-decision diagrams. *ACM Computing Surveys*, 24:293–318, 1992.

Chapter 6
Propositional Logic: SAT Solvers

Although it is believed that there is no efficient algorithm for the decidability of satisfiability in propositional logic, many algorithms are efficient in practice. This is particularly true when a formula is satisfiable; for example, when you build a truth table for an unsatisfiable formula of size n you will have to generate all 2^n rows, but if the formula is satisfiable you might get lucky and find a model after generating only a few rows. Even an *incomplete algorithm*—one that can find a model if one exists but may not be able to detect if a formula is unsatisfiable—can be useful in practice.

A computer program that searches for a model for a propositional formula is called a *SAT Solver*. This is a highly developed area of research in computer science because many problems in computer science can be encoded in propositional logic so that a model for a formula is a solution to the problem.

We begin the chapter by proving properties of formulas in clausal form. These properties are the basis of classical algorithms for satisfiability by Davis and Putnam (DP), and Davis, Logemann and Loveland (DLL), which we present next. The joint contribution of these two papers is usually recognized by the use of the acronym DPLL. Then we give an overview of two of the main approaches used in modern SAT solvers that are based upon modifications of the DPLL algorithm. In one approach, algorithms and heuristics are used to guide the search for a model; the other approach uses random search.

6.1 Properties of Clausal Form

This section collects several theorems that describe transformations on sets of clauses that do not affect its satisfiability. These theorems are important because they justify the algorithms presented in the next section.

Definition 6.1 Let S, S' be sets of clauses. $S \approx S'$ denotes that S is satisfiable if and only if S' is satisfiable. ∎

M. Ben-Ari, *Mathematical Logic for Computer Science*,
DOI 10.1007/978-1-4471-4129-7_6, © Springer-Verlag London 2012

It is important to understand that $S \approx S'$ (S is satisfiable if and only if S' is satisfiable) does *not* imply that $S \equiv S'$ (S is logically equivalent to S').

Example 6.2 Consider the two sets of clauses:

$$S = \{pq\bar{r}, p\bar{q}, \bar{p}q\}, \qquad S' = \{p\bar{q}, \bar{p}q\}.$$

S is satisfied by the interpretation:

$$\mathscr{I}(p) = F, \qquad \mathscr{I}(q) = F, \qquad \mathscr{I}(r) = F,$$

while S' is satisfied by the interpretation:

$$\mathscr{I}'(p) = F, \qquad \mathscr{I}'(q) = F.$$

Therefore, $S \approx S'$. However, under the interpretation:

$$\mathscr{I}''(p) = F, \qquad \mathscr{I}''(q) = F, \qquad \mathscr{I}''(r) = T,$$

$v_{\mathscr{I}''}(S) = F$ (since $v_{\mathscr{I}''}(pq\bar{r}) = F$) but $v_{\mathscr{I}''}(S') = T$, so $S \not\equiv S'$.   ∎

Pure Literals

Definition 6.3 Let S be a set of clauses. A *pure literal* in S is a literal l that appears in at least one clause of S, but its complement l^c does not appear in any clause of S.   ∎

Theorem 6.4 *Let S be a set of clauses and let l be a pure literal in S. Let S' be obtained from S by deleting every clause containing l. Then $S \approx S'$.*

Proof If S' is satisfiable, there is a model $\mathscr{I}'$ for S' such that $v_{\mathscr{I}'}(C') = T$ for every $C' \in S'$. Extend $\mathscr{I}'$ to a new interpretation $\mathscr{I}$ by defining $\mathscr{I}(l) = T$ and $\mathscr{I}(p) = \mathscr{I}'(p)$ for all other atoms.

Let us show that $\mathscr{I}$ is a model for S by showing that $v_{\mathscr{I}}(C) = T$ for every $C \in S$. If $C \in S'$, $v_{\mathscr{I}}(C) = v_{\mathscr{I}'}(C)$ since $\mathscr{I}(p) = \mathscr{I}'(p)$ for all atoms p in C. If $C \in S - S'$, $v_{\mathscr{I}}(C) = T$ since $l \in C$ and $\mathscr{I}(l) = T$.

Conversely, if S is satisfiable, S' is obviously satisfiable since $S' \subset S$.   ∎

Example 6.5 For the sets of clauses in Example 6.2, S' was obtained from S by deleting the clause $pq\bar{r}$ containing $\bar{r}$ since $\bar{r}^c = r$ does not appear in S. The interpretation $\mathscr{I}$ was obtained by extending the interpretation $\mathscr{I}'$ by $\mathscr{I}(\bar{r}) = T$ so that $v_{\mathscr{I}}(pq\bar{r}) = T$.   ∎

Unit Clauses

Theorem 6.6 *Let $\{l\} \in S$ be a unit clause and let S' be obtained from S by deleting every clause containing l and by deleting l^c from every (remaining) clause. Then $S \approx S'$.*

Proof Let $\mathscr{I}$ be a model for S and let $\mathscr{I}'$ be the interpretation defined by $\mathscr{I}'(p) = \mathscr{I}(p)$ for all atoms $p \in \mathscr{P}_{S'}$. $\mathscr{I}'$ is the same as $\mathscr{I}$ except no assignment is made to the atom for l which does not occur in S'. Since $\{l\}$ is a unit clause, for $\mathscr{I}$ be a model for S it must be true that $\mathscr{I}(l) = T$ and therefore $\mathscr{I}(l^c) = F$.

Let C' be an arbitrary clause in S'. We must show that $v_{\mathscr{I}'}(C) = T$. There are two cases:

- C' is also a member of S. C is not the unit clause $\{l\}$ (which was deleted); therefore, $\mathscr{I}$ and $\mathscr{I}'$ coincide on the literals of C', so $v_{\mathscr{I}'}(C') = v_{\mathscr{I}}(C) = T$.
- $C' = C - \{l^c\}$ for some $C \in S$. By the first paragraph of the proof, $\mathscr{I}(l^c) = F$, so $v_{\mathscr{I}}(C) = T$ holds only if $\mathscr{I}(l'') = T$ for some other literal $l'' \in C$. But $l'' \in C'$ which implies $v_{\mathscr{I}'}(C') = v_{\mathscr{I}}(C) = T$.

The proof of the converse is similar to the proof of Theorem 6.4.   ∎

Example 6.7 Let:

$$S = \{r, pq\bar{r}, p\bar{q}, \bar{q}p\}, \qquad S' = \{pq, p\bar{q}, \bar{q}p\}.$$

S' was obtained by deleting the unit clause $\{r\}$ from S and the literal $\bar{r}$ from the second clause of S. Since $\mathscr{I}(r) = T$ in any model $\mathscr{I}$ for S, $v_{\mathscr{I}}(pq\bar{r}) = T$ can hold only if either $\mathscr{I}(p) = T$ or $\mathscr{I}(q) = T$ from which we have $v_{\mathscr{I}'}(pq) = T$.   ∎

Here is a proof of the unsatisfiability of the empty clause $\square$ that does not use reasoning about vacuous sets.

Corollary 6.8 $\square$ *is unsatisfiable.*

Proof $\{\{p\}, \{\bar{p}\}\}$ is the clausal form of the unsatisfiable formula $p \wedge \neg p$. Delete the first clause $\{p\}$ from the formula and the literal $\bar{p}$ from the second clause; the result is $\{\{\}\} = \{\square\}$. By Theorem 6.6, $\{\square\} \approx \{\{p\}, \{\bar{p}\}\}$ and therefore $\square$ is unsatisfiable.   ∎

Subsumption

Definition 6.9 Let $C_1 \subseteq C_2$ be two clauses. The clause C_1 *subsumes* the clause C_2 and C_2 *is subsumed by* C_1.   ∎

Theorem 6.10 *Let $C_1, C_2 \in S$ be clauses such that C_1 subsumes C_2, and let $S' = S - \{C_2\}$. Then $S \approx S'$.*

Proof Trivially, if S is satisfiable, so is S' since it is a subset of S.

Conversely, let $\mathscr{I}'$ be an interpretation for S'. If C_2 contains atoms not in S', we might have to extend $\mathscr{I}'$ to an interpretation $\mathscr{I}$ of S, but $C_1 \subseteq C_2$, so $v_{\mathscr{I}'}(C_1) = v_{\mathscr{I}}(C_1) = T$ implies $v_{\mathscr{I}}(C_2) = T$ since a clause is an implicit disjunction. Therefore, $\mathscr{I}$ is a model for S. ∎

The concept of subsumption is somewhat confusing because the smaller clause subsumes (is stronger than) the larger clause. From the proof of the theorem, however, it is easy to see that if C_1 subsumes C_2 then $C_1 \to C_2$.

Example 6.11 Let:

$$S = \{pr, \bar{p}q\bar{r}, q\bar{r}\}, \qquad S' = \{pr, q\bar{r}\},$$

where $\{q\bar{r}\}$, the third clause of S, subsumes $\{\bar{p}q\bar{r}\}$, the second clause of S. Any interpretation which satisfies $\{q\bar{r}\}$ can be extended to an interpretation that satisfies $\{\bar{p}q\bar{r}\}$ because it doesn't matter what is assigned to p. ∎

Renaming

Definition 6.12 Let S be a set of clauses and U a set of atomic propositions. $R_U(S)$, the *renaming of S by U*, is obtained from S by replacing each literal l on an atomic proposition in U by l^c. ∎

Theorem 6.13 $S \approx R_U(S)$.

Proof Let $\mathscr{I}$ be a model for S. Define an interpretation $\mathscr{I}'$ for $R_U(S)$ by:

$$\mathscr{I}'(p) = \mathscr{I}(\bar{p}), \quad \text{if } p \in U,$$
$$\mathscr{I}'(p) = \mathscr{I}(p), \quad \text{if } p \notin U.$$

Let $C \in S$ and $C' = R_U(\{C\})$. Since $\mathscr{I}$ is a model for S, $v_{\mathscr{I}}(C) = T$ and $\mathscr{I}(l) = T$ for some $l \in C$. If the atom p of l is not in U then $l \in C'$ so $\mathscr{I}'(l) = \mathscr{I}(l) = T$ and $v_{\mathscr{I}'}(C') = T$. If $p \in U$ then $l^c \in C'$ so $\mathscr{I}'(l^c) = \mathscr{I}(l) = T$ and $v_{\mathscr{I}'}(C') = T$. ∎

The converse is similar.

Example 6.14 The set of clauses:

$$S = \{pqr, \bar{p}q, \bar{q}\bar{r}, r\}$$

is satisfied by the interpretation:

$$v_{\mathscr{I}}(p) = F, \qquad v_{\mathscr{I}}(q) = F, \qquad v_{\mathscr{I}}(r) = T.$$

The renaming:

$$R_{\{p,q\}}(S) = \{\bar{p}\bar{q}r, p\bar{q}, q\bar{r}, r\}$$

is satisfied by:

$$v_{\mathscr{I}'}(p) = T, \qquad v_{\mathscr{I}'}(q) = T, \qquad v_{\mathscr{I}'}(r) = T.$$

∎

6.2 Davis-Putnam Algorithm

The Davis-Putnam (DP) algorithm was one of the first algorithms proposed for deciding satisfiability. It uses two rules based upon the concepts introduced in the previous section, as well as the resolution rule (Chap. 4).

Algorithm 6.15 (Davis-Putnam algorithm)
Input: A formula A in clausal form.
Output: Report that A is *satisfiable* or *unsatisfiable*.
   Perform the following rules repeatedly, but the third rule is used only if the first two do not apply:

- **Unit-literal rule:** If there is a unit clause $\{l\}$, delete all clauses containing l and delete all occurrences of l^c from all other clauses.
- **Pure-literal rule:** If there is a pure literal l, delete all clauses containing l.
- **Eliminate a variable by resolution:** Choose an atom p and perform all possible resolutions on clauses that clash on p and $\bar{p}$. Add these resolvents to the set of clauses and then delete all clauses containing p or $\bar{p}$.

Terminate the algorithm under the following conditions:

- If empty clause □ is produced, report that the formula is *unsatisfiable*.
- If no more rules are applicable, report that the formula is *satisfiable*.   ∎

   Clearly, the algorithm terminates because the number of atoms in a formula is finite, as is the number of possible clauses that can be produced by resolution. The soundness of the three rules is justified by Theorem 6.6, Theorem 6.4 and Theorem 4.17, respectively.

Example 6.16 Consider the set of clauses:

$$\{p, \bar{p}q, \bar{q}r, \bar{r}st\}.$$

Performing the unit-literal rule on p leads to the creation of a new unit clause q upon which the rule can be applied again. This leads to a new unit clause r and applying the rule results in the singleton set of clauses $\{st\}$. Since no more rules are applicable, the set of clauses is satisfiable.   ∎

Definition 6.17 Repeatedly applying the unit-literal rule until it is no longer applicable is called *unit propagation* or *Boolean constraint propagation*.   ∎

6.3 DPLL Algorithm

Creating *all* possible resolvents on an atom is very inefficient. The DPLL algorithm improves on the DP algorithm by replacing the variable elimination step with a search for a model of the formula.

Definition 6.18 Let A be a set of clauses and let $\mathscr{I}$ be a partial interpretation (Definition 2.18) for A. For $C \in A$, if $v_{\mathscr{I}}(C) = T$, the interpretation $\mathscr{I}$ *satisfies* C, while if $v_{\mathscr{I}}(C) = F$, then C is a *conflict clause* for $\mathscr{I}$.   ∎

Example 6.19 Let $A = \{pqr, \bar{p}q, \bar{q}\bar{r}, r\}$ and let $\mathscr{I}_{\bar{q}r}$ be the partial interpretation defined by:

$$\mathscr{I}_{\bar{q}r}(q) = F, \qquad \mathscr{I}_{\bar{q}r}(r) = T.$$

$\mathscr{I}_{\bar{q}r}$ satisfies all the clauses except for $\bar{p}q$, which cannot be satisfied or falsified without also assigning a truth value to p.

The fourth clause r is a conflict clause for the partial interpretation $\mathscr{I}_r$ defined by $\mathscr{I}_r(r) = F$. Clearly, no interpretation that is an extension of this partial interpretation can satisfy A.   ∎

The DPLL algorithm recursively extends a partial interpretation by adding an assignment to some atom that has not yet been assigned a truth value. The current set of clauses is evaluated using the new partial interpretation and simplified by unit propagation. If the set of clauses contains a conflict clause, there is no need to continue extending this partial interpretation and the search backtracks to try another one.

Algorithm 6.20 (DPLL algorithm)
Input: A formula A in clausal form.
Output: Report that A is *unsatisfiable* or report that A is *satisfiable* and return a partial interpretation that satisfies A.

The algorithm is expressed as the recursive function $DPLL(B, \mathscr{I})$ which takes two parameters: a formula B in clausal form and a partial interpretation $\mathscr{I}$. It is initially called with the formula A and the empty partial interpretation.

$DPLL(B, \mathscr{I})$

- Construct the set of clauses B' by performing unit propagation on B. Construct $\mathscr{I}'$ by adding to $\mathscr{I}$ all the assignments made during propagation.
- Evaluate B' under the partial interpretation $\mathscr{I}'$:
  - If B' contains a conflict clause return 'unsatisfiable';
  - If B' is satisfied return $\mathscr{I}'$;
  - (otherwise, continue).
- *Choose* an atom p in B'; *choose* a truth value *val* as T or F; $\mathscr{I}_1$ is the interpretation $\mathscr{I}'$ together with the assignment of *val* to p.

- *result* ← *DPLL*(*B*′, $\mathscr{I}_1$).
  - If *result* is not 'unsatisfiable' return *result*;
  - (otherwise, continue).
- $\mathscr{I}_2$ is the interpretation $\mathscr{I}'$ together with the assignment of the complement of *val* to *p*.
- *result* ← *DPLL*(*B*′, $\mathscr{I}_2$).
  - Return *result*.   ∎

The DPLL algorithm is highly nondeterministic: it must choose an unassigned atom and then choose which truth value will be assigned to it first.

6.4 An Extended Example of the DPLL Algorithm

We now give an extended example of the DPLL algorithm by solving the 4-queens problem, a smaller instance of the 8-queens problem. Given a 4 × 4 chess board, place four queens so that no one can capture any of the others. Here is a solution:

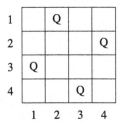

6.4.1 Encoding the Problem in Propositional Logic

First, we have to *encode* this problem as a formula in propositional logic. It should not be too surprising that any finite computational problem can be encoded by binary numbers, which in turn can be represented by truth values. Here we take a more direct approach to the encoding. Suppose that we want to encode the fact that a variable can take one of the values 1, 2, 3. Let us use three atoms p_1, p_2, p_3; the intended meaning is that p_i is true if the variable has the value i. The formula:

$$p_1 \lor p_2 \lor p_3,$$

states that the variable must have *at least* one of these values, while the following formula in CNF states that the variable can have *at most* one of the values:

$$(\overline{p_1} \lor \overline{p_2}) \land (\overline{p_1} \lor \overline{p_3}) \land (\overline{p_2} \lor \overline{p_3}).$$

For example, if p_1 is assigned T, then $\overline{p_1}$ is false, so both $\overline{p_2}$ and $\overline{p_3}$ must be true, that is, p_2 and p_3 must be false for the formula to encode that the variable has the value 1. The conjunction of $p_1 \vee p_2 \vee p_3$ with this formula states that the variable must have exactly one of the values.

In the 4-queens problem, we need 16 atoms: p_{ij}, $1 \le i \le 4$, $1 \le i \le 4$, where p_{ij} is true iff a queen is placed on the square at row i and column j. To simplify notation, instead of $p_{11}, p_{12}, \ldots, p_{43}, p_{44}$, we will use the subscripts $11, 12, \ldots, 43, 44$ alone to denote each atom.

The clauses that claim that there is at least one queen in each row are:

$$11 \vee 12 \vee 13 \vee 14,$$
$$21 \vee 22 \vee 23 \vee 24,$$
$$31 \vee 32 \vee 33 \vee 34,$$
$$41 \vee 42 \vee 43 \vee 44.$$

But no more than one queen may be placed in each row:

$$\overline{11} \vee \overline{12}, \quad \overline{11} \vee \overline{13}, \quad \overline{11} \vee \overline{14}, \quad \overline{12} \vee \overline{13}, \quad \overline{12} \vee \overline{14}, \quad \overline{13} \vee \overline{14},$$
$$\overline{21} \vee \overline{22}, \quad \overline{21} \vee \overline{23}, \quad \overline{21} \vee \overline{24}, \quad \overline{22} \vee \overline{23}, \quad \overline{22} \vee \overline{24}, \quad \overline{23} \vee \overline{24},$$
$$\overline{31} \vee \overline{32}, \quad \overline{31} \vee \overline{33}, \quad \overline{31} \vee \overline{34}, \quad \overline{32} \vee \overline{33}, \quad \overline{32} \vee \overline{34}, \quad \overline{33} \vee \overline{34},$$
$$\overline{41} \vee \overline{42}, \quad \overline{41} \vee \overline{43}, \quad \overline{41} \vee \overline{44}, \quad \overline{42} \vee \overline{43}, \quad \overline{42} \vee \overline{44}, \quad \overline{43} \vee \overline{44},$$

and no more than one queen in each column:

$$\overline{11} \vee \overline{21}, \quad \overline{11} \vee \overline{31}, \quad \overline{11} \vee \overline{41}, \quad \overline{21} \vee \overline{31}, \quad \overline{21} \vee \overline{41}, \quad \overline{31} \vee \overline{41},$$
$$\overline{12} \vee \overline{22}, \quad \overline{12} \vee \overline{32}, \quad \overline{12} \vee \overline{42}, \quad \overline{22} \vee \overline{32}, \quad \overline{22} \vee \overline{42}, \quad \overline{32} \vee \overline{42},$$
$$\overline{13} \vee \overline{23}, \quad \overline{13} \vee \overline{33}, \quad \overline{13} \vee \overline{43}, \quad \overline{23} \vee \overline{33}, \quad \overline{23} \vee \overline{43}, \quad \overline{33} \vee \overline{43},$$
$$\overline{14} \vee \overline{24}, \quad \overline{14} \vee \overline{34}, \quad \overline{14} \vee \overline{44}, \quad \overline{24} \vee \overline{34}, \quad \overline{24} \vee \overline{44}, \quad \overline{34} \vee \overline{44}.$$

We also have to ensure that no more than one queen is placed in each diagonal. To do this systematically, we check each square (i, j), starting at the top left and enumerate the squares that are diagonally *below* it, which are $(i - 1, j + 1)$, $(i + 1, j + 1)$, $(i - 2, j + 2)$, $(i + 2, j + 2)$, as long as both numbers are within the range from 1 to 4. By commutativity, $\overline{12} \vee \overline{21} \equiv \overline{21} \vee \overline{12}$, we do not have to check the squares above. Here are the clauses:

$$\overline{11} \vee \overline{22}, \quad \overline{11} \vee \overline{33}, \quad \overline{11} \vee \overline{44}, \qquad \overline{12} \vee \overline{21}, \quad \overline{12} \vee \overline{23}, \quad \overline{12} \vee \overline{34},$$
$$\overline{13} \vee \overline{22}, \quad \overline{13} \vee \overline{31}, \quad \overline{13} \vee \overline{24}, \qquad \overline{14} \vee \overline{23}, \quad \overline{14} \vee \overline{32}, \quad \overline{14} \vee \overline{41},$$
$$\overline{21} \vee \overline{32}, \quad \overline{21} \vee \overline{43}, \qquad\qquad\qquad\quad \overline{22} \vee \overline{31}, \quad \overline{22} \vee \overline{33}, \quad \overline{22} \vee \overline{44},$$
$$\overline{23} \vee \overline{32}, \quad \overline{23} \vee \overline{41}, \quad \overline{23} \vee \overline{34}, \qquad \overline{24} \vee \overline{33}, \quad \overline{24} \vee \overline{42},$$
$$\overline{31} \vee \overline{42}, \qquad\qquad\qquad\qquad\qquad\quad \overline{32} \vee \overline{41}, \quad \overline{32} \vee \overline{43},$$
$$\overline{33} \vee \overline{42}, \quad \overline{33} \vee \overline{44}, \qquad\qquad\qquad\quad \overline{34} \vee \overline{43}.$$

Check this by drawing a 4×4 chess board on a piece of paper and tracing each of the diagonals.

The total number of clauses is:

$$
\begin{array}{lr}
4 & + \\
(4 \times 6) & + \\
(4 \times 6) & + \\
(3 + 3 + 3 + 3 + 2 + 3 + 3 + 2 + 1 + 2 + 2 + 1) & = \\
4 + 24 + 24 + 28 & = \quad 80.
\end{array}
$$

The 4-queens problem has a solution if and only if this 80-clause formula is satisfiable. If an algorithm not only decides that the formula is satisfiable, but also returns a model, the atoms assigned T in the model will tell us where to place the queens.

6.4.2 Solving the Problem with the DP Algorithm?

Let us try to use the DP algorithm to solve the 4-queens problem. There are no unit clauses, so we much choose an atom to eliminate. In the absence of any other information, let us start with the first atom 11. The atom appears as a positive literal only in the first clause $11 \lor 12 \lor 13 \lor 14$, so that clause must participate in the resolution rule. Negative literals appear in all three sets that exclude two queens in a row, column or diagonal. However, the row exclusion clauses $\overline{11} \lor \overline{12}$, $\overline{11} \lor \overline{13}$, $\overline{11} \lor \overline{14}$, cannot be resolved with $11 \lor 12 \lor 13 \lor 14$ because they clash on more than one literal, so resolving them would result in trivial clauses (Lemma 4.16). This leaves six clashing clauses—three for column exclusion and three for diagonal exclusion—and the resolvents are:

$$
\begin{array}{l}
\overline{21} \lor 12 \lor 13 \lor 14, \quad \overline{31} \lor 12 \lor 13 \lor 14, \quad \overline{41} \lor 12 \lor 13 \lor 14, \\
\overline{22} \lor 12 \lor 13 \lor 14, \quad \overline{33} \lor 12 \lor 13 \lor 14, \quad \overline{44} \lor 12 \lor 13 \lor 14,
\end{array}
$$

The ten original clauses with 11 or $\overline{11}$ are now removed from the set. We don't seem to be making much progress, so let us turn to the DPLL algorithm.

6.4.3 Solving the Problem with the DPLL Algorithm

In this section we will carry out the DPLL algorithm in a purely formal manner as a computer would. We suggest, however, that you 'cheat' by referring to the 4×4 chessboard, which will clarify what happens at each step.

We start by assigning T to 11 (and F to $\overline{11}$). The clause $11 \lor 12 \lor 13 \lor 14$ becomes true and can be deleted, while $\overline{11}$ can be deleted from all other clauses. This results in nine new (unit) clauses:

$$\overline{12}, \ \overline{13}, \ \overline{14}, \ \overline{21}, \ \overline{31}, \ \overline{41}, \ \overline{22}, \ \overline{33}, \ \overline{44}.$$

The next step is to carry out unit propagation for each of these literals. Negated atoms like $\overline{12}$ are assigned T so they are erased from clauses with positive literals and all clauses contain the negated atoms are deleted:

$\geq$ one / row	$23 \lor 24, \ 32 \lor 34, \ 42 \lor 43,$
$\leq$ one / row	$\overline{23} \lor \overline{24}, \ \overline{32} \lor \overline{34}, \ \overline{42} \lor \overline{43},$
$\leq$ one / column	$\overline{32} \lor \overline{42}, \ \overline{23} \lor \overline{43}, \ \overline{24} \lor \overline{34},$
$\leq$ one / diagonal	$\overline{24} \lor \overline{42}, \ \overline{23} \lor \overline{32}, \ \overline{23} \lor \overline{34}, \ \overline{32} \lor \overline{43}, \ \overline{34} \lor \overline{43}.$

By choosing the value of only one literal and propagating units, 80 clauses have been reduced to only 14 clauses!

Let us now assign T to 23 (and F to $\overline{23}$); this creates four new unit clauses:

$$\overline{24}, \ \overline{43}, \ \overline{32}, \ \overline{34}$$

and the other clauses are:

$$32 \lor 34, \ 42 \lor 43,$$
$$\overline{32} \lor \overline{34}, \ \overline{42} \lor \overline{43},$$
$$\overline{32} \lor \overline{42}, \ \overline{24} \lor \overline{34},$$
$$\overline{24} \lor \overline{42}, \ \overline{32} \lor \overline{43}, \ \overline{34} \lor \overline{43}.$$

Propagating the unit $\overline{24}$ gives:

$$32 \lor 34, \ 42 \lor 43,$$
$$\overline{32} \lor \overline{34}, \ \overline{42} \lor \overline{43},$$
$$\overline{32} \lor \overline{42},$$
$$\overline{32} \lor \overline{43}, \ \overline{34} \lor \overline{43}$$

and then propagating the unit $\overline{43}$ gives:

$$32 \lor 34, \ 42,$$
$$\overline{32} \lor \overline{34},$$
$$\overline{32} \lor \overline{42}.$$

The next unit to propagate is $\overline{32}$; the result is the pair of clauses 34, 42, which we can written more formally as $\{\{34\}, \{42\}\}$. The remaining unit to propagate is $l = \overline{34}$, but erasing the literal $l^c = 34$ from the clause $\{34\}$ produces the empty clause $\Box$, which is unsatisfiable. Just by choosing values for the two literals 11 and 23, unit propagation has caused the entire set of 80 clauses to collapse into the empty clause. We have ruled out 2^{14} of the 2^{16} possible interpretations, because *any* interpretation which assigns T to 11 and 23 cannot satisfy the set of clauses.

We should now backtrack and assign F to 23. But wait, let us 'cheat' and notice that there are no solutions with a queen placed on the top left square. Instead, we backtrack to the very start of the algorithm and assign T to 12. The first clause is deleted and unit propagation produces new unit clauses:

$$\overline{11}, \ \overline{13}, \ \overline{14}, \ \overline{22}, \ \overline{32}, \ \overline{42}, \ \overline{21}, \ \overline{23}, \ \overline{34}.$$

Clearly, propagating the unit clauses: $\overline{11}, \ \overline{13}, \ \overline{14}, \ \overline{21}, \ \overline{22}, \ \overline{23}$ removes all clauses with literals from the first two rows except those with 24 or $\overline{24}$:

$\geq$ one / row	$24, \ 31 \vee 32 \vee 33 \vee 34, \ 41 \vee 42 \vee 43 \vee 44,$
$\leq$ one / row	$\overline{31} \vee \overline{32}, \ \overline{31} \vee \overline{33}, \ \overline{31} \vee \overline{34}, \ \overline{32} \vee \overline{33}, \ \overline{32} \vee \overline{34}, \ \overline{33} \vee \overline{34},$
	$\overline{41} \vee \overline{42}, \ \overline{41} \vee \overline{43}, \ \overline{41} \vee \overline{44}, \ \overline{42} \vee \overline{43}, \ \overline{42} \vee \overline{44}, \ \overline{43} \vee \overline{44},$
$\leq$ one / column	$\overline{24} \vee \overline{34}, \ \overline{24} \vee \overline{44}, \ \overline{31} \vee \overline{41}, \ \overline{32} \vee \overline{42}, \ \overline{33} \vee \overline{43}, \ \overline{34} \vee \overline{44},$
$\leq$ one / diagonal	$\overline{24} \vee \overline{33}, \ \overline{24} \vee \overline{42}, \ \overline{31} \vee \overline{42}, \ \overline{32} \vee \overline{41}, \ \overline{32} \vee \overline{43}, \ \overline{33} \vee \overline{42},$
	$\overline{33} \vee \overline{44}, \ \overline{34} \vee \overline{43}.$

We can now propagate $\overline{32}, \ \overline{34}, \ \overline{42}$ to obtain:

$$24, \ 31 \vee 33, \ 41 \vee 43 \vee 44,$$
$$\overline{31} \vee \overline{33}, \ \overline{41} \vee \overline{43}, \ \overline{41} \vee \overline{44}, \ \overline{43} \vee \overline{44},$$
$$\overline{24} \vee \overline{44}, \ \overline{31} \vee \overline{41}, \ \overline{33} \vee \overline{43},$$
$$\overline{24} \vee \overline{33}, \ \overline{33} \vee \overline{44}.$$

There is now a new unit clause 24 which can be propagated:

$$31 \vee 33, \ 41 \vee 43 \vee 44,$$
$$\overline{31} \vee \overline{33}, \ \overline{41} \vee \overline{43}, \ \overline{41} \vee \overline{44}, \ \overline{43} \vee \overline{44},$$
$$\overline{44}, \ \overline{31} \vee \overline{41}, \ \overline{33} \vee \overline{43},$$
$$\overline{33}, \ \overline{33} \vee \overline{44}.$$

Propagating $\overline{33}$ gives:

$$31, \;\; 41 \vee 43 \vee 44,$$
$$\overline{41} \vee \overline{43}, \;\; \overline{41} \vee \overline{44}, \;\; \overline{43} \vee \overline{44},$$
$$\overline{44}, \;\; \overline{31} \vee \overline{41}$$

and then propagating $\overline{44}$ gives:

$$31, \;\; 41 \vee 43,$$
$$\overline{41} \vee \overline{43}, \;\; \overline{31} \vee \overline{41}.$$

The new unit clause is 31 can be propagated:

$$41 \vee 43, \;\; \overline{41} \vee \overline{43}, \;\; \overline{41}.$$

Finally, propagating $\overline{41}$ leaves one last clause 43. We conclude that the set of clauses is satisfiable. If you check which atomic propositions are assigned T, you will find that they are 12, 24, 31, 43, which is precisely the placement of queens shown in the diagram at the beginning of this section!

6.5 Improving the DPLL Algorithm

An efficient implementation of the DPLL algorithm must use data structures designed so that operations like unit propagation are efficient. Furthermore, an iterative algorithm must replace the recursive one. Beyond such issues of implementation, the DPLL algorithm has become a practical approach for SAT solving because of optimizations to the algorithm itself. We will survey several of these: heuristics to resolve the nondeterministic choice of which assignments to make, learning from conflicts and non-chronological backtracking.

6.5.1 Branching Heuristics

The DPLL algorithm is nondeterministic since when branching occurs we have to choose an atom and an assignment of a truth value. As we saw in the formula for the 4-queens problem, choosing 12 as the first literal to branch on was more efficient than choosing 11.

Various heuristics have been developed for choosing literals to branch on. The choice of a literal is based upon some measurable characteristic of the formula, such as the size of the clauses and number of the literals in a clause. The 4-queens problem is symmetric so measure-based heuristics are unlikely to help. Consider, instead, the following set of clauses:

$$\bar{p}qs, \ p\bar{q}s, \ pq\bar{s}, \ \bar{p}\bar{q}\bar{s},$$
$$\bar{p}tv, \ p\bar{t}v, \ pt\bar{v}, \ \bar{p}\bar{t}\bar{v},$$
$$t\bar{u}, \ \bar{t}u, \ u\bar{v}, \ \bar{u}v,$$
$$q\bar{r}, \ \bar{q}r, \ r\bar{s}, \ \bar{r}s.$$

These are the Tseitin clauses (Sect. 4.5) associated with two triangles qrs and tuv connected by an edge labeled p. However, the parity on each node is zero so that clauses are satisfiable.

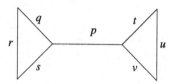

Let us try the heuristic: branch on the literal that occurs most often in the set of clauses. This is p (or $\bar{p}$) which occur four times. Deleting all clauses with p and all occurrences of $\bar{p}$ gives:

$$qs, \ \bar{q}\bar{s}, \ tv, \ \bar{t}\bar{v}$$
$$t\bar{u}, \ \bar{t}u, \ u\bar{v}, \ \bar{u}v,$$
$$q\bar{r}, \ \bar{q}r, \ r\bar{s}, \ \bar{r}s.$$

We haven't progressed very far.

Let us try, instead, the heuristic: branch on a literal whose atom occurs most often in a clause of shortest length. The intuition behind this heuristic is that assigning to literals in short clauses will bring us rapidly to a unit clause that can be propagated or even to an unsatisfiable clause. In this set of clauses, we can choose r or u; suppose that we choose r and assign T to r. Immediately, we obtain two unit clauses q and s. Propagating these units leads to another unit $\bar{p}$ and propagating that unit results in:

$$t\bar{v}, \ \bar{t}v, \ t\bar{u}, \ \bar{t}u, \ u\bar{v}, \ \bar{u}v.$$

This heuristic leads to fewer clauses than the previous one.

6.5.2 Non-chronological Backtracking

Consider now the set of clauses:

$$pq, \ qr, \ \bar{p}\bar{s}t, \ \bar{p}su, \ \bar{p}\bar{t}u, \ \bar{p}s\bar{u}, \ \bar{p}\bar{s}\bar{u}.$$

Let us set the atom p to T. The first clause is deleted and $\bar{p}$ is deleted from the other clauses, resulting in:

$$qr, \ \bar{s}t, \ su, \ \bar{t}u, \ s\bar{u}, \ \bar{s}\bar{u}.$$

Obviously, it is possible to assign values to q and r to satisfy the clause qr without in any way affecting the satisfiability of the rest of the formula, but the DPLL algorithm doesn't know that. It might just choose to branch on the these atoms, first assigning T to q:

$$\bar{s}t, \quad su, \quad \bar{t}u, \quad s\bar{u}, \quad \bar{s}\bar{u}.$$

The algorithms next choice might be to assign T to s, resulting in:

$$t, \quad \bar{t}u, \quad \bar{u}.$$

Unit propagation immediately produces the empty clause showing that this set of assignments does not satisfy the formula. Backtracking and assigning F to s leads to:

$$u, \quad \bar{t}u, \quad \bar{u}$$

and then to the empty clause.

The algorithm returns from the recursion and tries the assignment F to q even though that assignment will also lead to an unsatisfiable clause. The DPLL algorithm can be modified to analyze the sequence of assignments and to discover that the assignments of T to p and T or F to s are sufficient by themselves to show that the formula is unsatisfiable. Therefore, once the two calls on s have returned, the algorithm can directly return all the way up to try the assignment of F to p without checking the other assignment to q.

An algorithm which returns up the tree of assignments to an ancestor that is not its parent is said to engage in *non-chronological backtracking*. These algorithms are significantly more efficient than the DPLL algorithm that performs chronological backtracking.

6.5.3 Learning Conflict Clauses

We showed that the assignment of T to both p and s necessarily falsifies the formula in the previous section. Unfortunately, backtracking causes this information to be lost. In a large set of clauses, the algorithm might again try a sequence of assignments that includes assignments known lead to interpretations that falsify the formula. The DPLL algorithm can be modified to prevent this by *adding* clauses to the formula, such as $\bar{p}\bar{s}$ in this case; these clauses will immediately force the set to be unsatisfiable on an interpretation that contains the known assignments.

A clause like $\bar{p}\bar{s}$ is called a *conflict clause*, because it is obtained from an analysis of the assignments that led to the detection of a conflict—a partial assignment that falsifies the formula. An algorithm that performs conflict analysis *learns* (adds) conflict clauses in the hope of improving performance. Since memory is limited, an algorithm must also include a policy for deleting clauses that have been learned.

6.6 Stochastic Algorithms

On the surface, nothing appears to be less random than algorithms because they formally specify the steps taken to solve a problem. It may come as a surprise that algorithms that use randomness can be very effective. A random algorithm will not be *complete*—it may not return an answer—but many random algorithms can be shown to return the correct answer with high probability. In practice, an efficient incomplete algorithm can be more useful than an inefficient complete algorithm.

Many SAT solvers employ *stochastic algorithms* that use randomness. Of course, they can only be used when we are looking for a model, because an incomplete algorithm can never declare that a formula is unsatisfiable.

The basic form of a stochastic algorithm for SAT is very simple:

Algorithm 6.21 (Stochastic algorithm for SAT)
Input: A formula A in clausal form.
Output: A model for A (or failure to return any answer).

- Choose a random interpretation $\mathscr{I}$ for A.
- Repeat indefinitely:
  - If $v_{\mathscr{I}}(A) = T$ return $\mathscr{I}$;
  - Otherwise:
    · Choose an atom p in A;
    · Modify $\mathscr{I}$ by flipping p (changing its assignment to the complementary assignment). ■

In practice, Algorithm 6.21 is modified to limit the number of attempts to flip an atom in the interpretation; when the limit is reached, the loop restarts after choosing a new random interpretation. Of course, you might want to limit the number of restarts so that the algorithm does not run indefinitely.

Stochastic algorithms for SAT differ in the strategy used to choose an atom to flip. A simple strategy is to choose to flip the atom that will cause the largest number of currently unsatisfied clauses to become satisfied. An algorithm can add randomness to avoid getting stuck in a *local minimum*: a partial interpretation where no flip of an atom can improve the chance of obtaining a model.

6.6.1 Solving the 4-Queens Problem with a Stochastic Algorithm

The n-queens problem is quite unsuited for stochastic algorithms, because the number of models (solutions) is very small compared to the number of interpretations so a random algorithm has a low probability of finding one. For the 4-queens problem, there are only two solutions, but there are $2^{16} = 65536$ interpretations! Nevertheless, we will use this problem to an example of how the algorithm works.

Consider the random assignment associated with the configuration:

	1	2	3	4
1	Q	Q		
2				Q
3	Q			
4			Q	

The atoms 11, 12, 24, 31, 43 are assigned T and the rest are assigned F. There are two unsatisfied clauses:

$$\overline{11} \lor \overline{12}, \quad \overline{11} \lor \overline{31},$$

corresponding to having two queens in the first row and two in the first column. Obviously, we want to flip the assignment to 11 from T to F because that reduces the number of unsatisfied clauses from two to zero, but let us see what other choices give.

Flipping 12 will satisfy $\overline{11} \lor \overline{12}$ but leave $\overline{11} \lor \overline{31}$ unsatisfied, reducing the number of unsatisfied clauses from two to one. Flipping 31 will also satisfy one of the unsatisfied clauses, but it will make the previously satisfied clause $31 \lor 32 \lor 33 \lor 34$ (at least one queen in a row) unsatisfied; therefore, the number of unsatisfied clauses in unchanged. Flipping 24 or 43 will satisfy no unsatisfied clause and make the corresponding row clause unsatisfied, *increasing* the number of unsatisfied clauses. Flipping any of the atoms that have been assigned F is even worse, because several clauses will become unsatisfied. For example, flipping 22 will falsify $\overline{22} \lor \overline{24}$, $\overline{12} \lor \overline{22}$, $\overline{11} \lor \overline{22}$ and $\overline{22} \lor \overline{31}$.

For this example, the heuristic of flipping the atom that causes the largest reduction in the number of unsatisfied clauses works very well and it leads immediately to a solution.

6.7 Complexity of SAT *

The problems of deciding satisfiability and validity in propositional logic are central to complexity theory. In this section we survey some of the basic results. It assumes that you are familiar with fundamental concepts of computational complexity: deterministic and nondeterministic algorithms, polynomial and exponential time and space, the complexity classes $\mathscr{P}$, $\mathscr{NP}$, $co\text{-}\mathscr{NP}$.

The method of truth tables is a deterministic algorithm for deciding both satisfiability and validity in propositional logic. The algorithm is exponential, because the size of a formula is polynomial in n, the number of variables, while the truth table has 2^n rows.

The method of semantic tableaux is a nondeterministic algorithm for both satisfiability and validity, because at any stage of the construction, we can choose a leaf to expand and choose a formula in the label of the leaf to which a rule will be applied. Nevertheless, it can be shown that there are families of formulas for which the method of semantic tableaux is exponential, as are the David-Putnam procedure and resolution (Sect. 4.5).

There is a very simple nondeterministic algorithm for deciding the *satisfiability* of a formula A in propositional logic:

Choose an interpretation $\mathscr{I}$ for A.
Compute the truth value $v_{\mathscr{I}}(A)$.
If $v_{\mathscr{I}}(A) = T$ then A is satisfiable.

If A is satisfiable, for *some* computation (choice of $\mathscr{I}$), the algorithm returns with the answer that A is satisfiable. Of course, other choices may not give the correct answer, but that does not affect the correctness of the *nondeterministic* algorithm. Furthermore, the algorithm is very efficient, since choosing an interpretation and computing the truth value of a formula are linear in the size of the formula. This shows that the problem of satisfiability in propositional logic is in the class $\mathscr{N}\mathscr{P}$ of problems solvable by a *N*ondeterministic algorithm in *P*olynomial time.

In the context of deciding satisfiability, the difference between a deterministic and a nondeterministic algorithm seems to be that guessing and checking is efficient whereas searching is inefficient. One can conjecture that satisfiability is *not* in the class $\mathscr{P}$ of problems solvable in *P*olynomial time by a deterministic algorithm.

A famous theorem by Cook and Levin from 1971 showed that if satisfiability *is* in $\mathscr{P}$, then for *every* problem in $\mathscr{N}\mathscr{P}$, there is a deterministic polynomial algorithm! A problem with this property is called an $\mathscr{N}\mathscr{P}$-*complete problem*. The theorem on $\mathscr{N}\mathscr{P}$-completeness is proved by showing how to transform an arbitrary nondeterministic Turing machine into a formula in propositional logic such that the Turing machine produces an answer if and only if the corresponding formula is satisfiable. (It must also be shown that the size of the formula is a polynomial of the size of the Turing machine.) Satisfiability was the first problem shown to be $\mathscr{N}\mathscr{P}$-complete, although since then thousands of problems have been proven to be in this class.

A major open theoretical question in computer science is called $\mathscr{P} = \mathscr{N}\mathscr{P}$?: Are the two classes the same or are nondeterministic algorithms more efficient? The problem can be settled by demonstrating a polynomial algorithm for one of these problems like satisfiability or by proving that no such algorithm exists.

Unsatisfiability (validity) in propositional logic is in the class *co-*$\mathscr{N}\mathscr{P}$ of problems whose complement (here, satisfiability) is in $\mathscr{N}\mathscr{P}$. It can be shown that *co-*$\mathscr{N}\mathscr{P} = \mathscr{N}\mathscr{P}$ if and only if unsatisfiability is in $\mathscr{N}\mathscr{P}$, but it is not known if there is a nondeterministic polynomial decision procedure for unsatisfiability.

6.8 Summary

The problems of deciding satisfiability and validity in propositional logic are almost certainly intractable: the former is in $\mathcal{NP}$ and the latter in $co\text{-}\mathcal{NP}$. However, algorithms and data structures like the ones described in this chapter have proved themselves to be highly efficient in many practical applications. The DPLL algorithm uses elementary properties of clauses to search for a model. SAT solvers based upon the DPLL algorithm employ heuristics and randomness to make the search for a model more efficient.

6.9 Further Reading

The original papers on the DP and DLL algorithms are: Davis and Putnam (1960) and Davis et al. (1962). Malik and Zhang (2009) and Zhang (2003) contain good introductions to SAT solvers. For the state of the art on SAT algorithms, see the handbook by Biere et al. (2009). The presentation in Sect. 6.5.2 was adapted from Sect. 3.6.4.1 of this book.

See http://www.satlive.org/ for links to software for SAT solvers.

The encoding of the 8-queens problem is taken from Biere et al. (2009, Sect. 2.3.1), which is based on Nadel (1990).

There are many textbooks on computational models: Gopalakrishnan (2006), Sipser (2005), Hopcroft et al. (2006).

6.10 Exercises

6.1 Are there other solutions to the 4-queens problem? If so, compute these solutions using the DPLL algorithm and an appropriate choice of assignments.

6.2 A variant of the n-queens problem is the *n-rooks problem*. A rook can only capture horizontally and vertically, not diagonally. Solve the 4-rooks problem using DPLL.

6.3 The pigeon-hole problem is to place $n + 1$ pigeons into n holes such that each hole contains at most one pigeon. There is no solution, of course!

1. Encode the pigeon-hole problem for 3 holes and 4 pigeons as a formula in clausal form.
2. Use the DPLL algorithm to show that the formula is unsatisfiable for one assignment.
3. Develop an expression for the number of clauses in the formula for the pigeon-hole problem with n holes.

6.4 Let G be a connected undirected graph. The *graph coloring problem* is to decide if one of k colors $\{c_1, \ldots, c_k\}$ can be assigned to each vertex $color(v_i) = c_j$ such that $color(v_{i_1}) \neq color(v_{i_2})$ if (v_{i_1}, v_{i_1}) is an edge in E. Show how to translate the graph coloring problem for any G into SAT. Use the DPLL algorithm to show that $K_{2,2}$ is 2-colorable and that the triangle is 3-colorable.

6.5 What is the relation between the DP algorithm and resolution?

6.6 * Let 3SAT be the problem of deciding satisfiability of formulas in CNF such that there are *three* literals in each clause. The proof that SAT is $\mathcal{NP}$-complete actually shows that 3SAT is $\mathcal{NP}$-complete. Let 2SAT be the problem of deciding satisfiability of formulas in CNF such that there are *two* literals in each clause. Show that there is an efficient algorithm for 2SAT.

6.7 * Show that there is an efficient algorithm for Horn-SAT, deciding if a set of Horn clauses is satisfiable.

References

A. Biere, M. Heule, H. Van Maaren, and T. Walsh, editors. *Handbook of Satisfiability*, volume 185 of *Frontiers in Artificial Intelligence and Applications*. IOS Press, 2009.

M. Davis and H. Putnam. A computing procedure for quantification theory. *Journal of the ACM*, 7:201–215, 1960.

M. Davis, G. Logemann, and D. Loveland. A machine program for theorem-proving. *Communications of the ACM*, 5:394–397, 1962.

G. Gopalakrishnan. *Computational Engineering: Applied Automata Theory and Logic*. Springer, 2006.

J.E. Hopcroft, R. Motwani, and J.D. Ullman. *Introduction to Automata Theory, Languages and Computation (Third Edition)*. Addison-Wesley, 2006.

S. Malik and L. Zhang. Boolean satisfiability: From theoretical hardness to practical success. *Communications of the ACM*, 52(8):76–82, 2009.

B.A. Nadel. Representation selection for constraint satisfaction: A case study using n-queens. *IEEE Expert: Intelligent Systems and Their Applications*, 5:16–23, June 1990.

M. Sipser. *Introduction to the Theory of Computation (Second Edition)*. Course Technology, 2005.

L. Zhang. *Searching for truth: Techniques for satisfiability of Boolean formulas*. PhD thesis, Princeton University, 2003. http://research.microsoft.com/en-us/people/lintaoz/thesis_lintao_zhang.pdf.

Chapter 7
First-Order Logic: Formulas, Models, Tableaux

7.1 Relations and Predicates

The axioms and theorems of mathematics are defined on sets such as the set of integers $\mathcal{Z}$. We need to be able to write and manipulate logical formulas that contain relations on values from arbitrary sets. *First-order logic* is an extension of propositional logic that includes predicates interpreted as relations on a domain.

Before continuing, you may wish to review Appendix on set theory.

Example 7.1 $\mathcal{P}(x) \subset \mathcal{N}$ is the unary relation that is the subset of natural numbers that are prime: $\{2, 3, 5, 7, 11, \ldots\}$. ∎

Example 7.2 $\mathcal{S}(x, y) \subset \mathcal{N}^2$ is the binary relation that is the subset of pairs (x, y) of natural numbers such that $y = x^2$: $\{(0, 0), (1, 1), (2, 4), (3, 9), \ldots\}$. ∎

It would be more usual in mathematics to define a unary function $f(x) = x^2$ which maps a natural number x into its square. As shown in the example, functions are special cases of relations. For simplicity, we limit ourselves to relations in this chapter and the next; the extension of first-order logic to include functions is introduced in Sect. 9.1.

Definition 7.3 Let $\mathcal{R}$ be an n-ary relation on a domain D, that is, $\mathcal{R}$ is a subset of D^n. The relation $\mathcal{R}$ can be *represented* by the Boolean-valued function $P_{\mathcal{R}} : D^n \mapsto \{T, F\}$ that maps an n-tuple to T if and only if the n-tuple is an element of the relation:

$$P_{\mathcal{R}}(d_1, \ldots, d_n) = T \text{ iff } (d_1, \ldots, d_n) \in \mathcal{R},$$
$$P_{\mathcal{R}}(d_1, \ldots, d_n) = F \text{ iff } (d_1, \ldots, d_n) \notin \mathcal{R}.$$

∎

Example 7.4 The set of primes $\mathcal{P}$ is represented by the function $P_{\mathcal{P}}$:

$$P_{\mathcal{P}}(0) = F, \ P_{\mathcal{P}}(1) = F, \ P_{\mathcal{P}}(2) = T,$$
$$P_{\mathcal{P}}(3) = T, \ P_{\mathcal{P}}(4) = F, \ P_{\mathcal{P}}(5) = T,$$
$$P_{\mathcal{P}}(6) = F, \ P_{\mathcal{P}}(7) = T, \ P_{\mathcal{P}}(8) = F, \ \ldots$$

∎

M. Ben-Ari, *Mathematical Logic for Computer Science*,
DOI 10.1007/978-1-4471-4129-7_7, © Springer-Verlag London 2012

Example 7.5 The set of squares $\mathscr{S}$ is represented by the function $P_{\mathscr{S}}$:

$$P_{\mathscr{S}}(0,0) = P_{\mathscr{S}}(1,1) = P_{\mathscr{S}}(2,4) = P_{\mathscr{S}}(3,9) = \cdots = T,$$

$$P_{\mathscr{S}}(0,1) = P_{\mathscr{S}}(1,0) = P_{\mathscr{S}}(0,2) = P_{\mathscr{S}}(2,0) =$$
$$P_{\mathscr{S}}(1,2) = P_{\mathscr{S}}(2,1) = P_{\mathscr{S}}(0,3) = P_{\mathscr{S}}(2,2) = \cdots = F.$$

■

This correspondence provides the link necessary for a logical formalization of mathematics. All the logical machinery—formulas, interpretations, proofs—that we developed for propositional logic can be applied to predicates. The presence of a domain upon which predicates are interpreted considerably complicates the technical details but not the basic concepts.

Here is an overview of our development of first-order logic:

- Syntax (Sect. 7.2): Predicates are used to represent functions from a domain to truth values. Quantifiers allow a purely syntactical expression of the statement that the relation represented by a predicate is true for *some* or *all* elements of the domain.
- Semantics (Sect. 7.3): An interpretation consists of a domain and an assignment of relations to the predicates. The semantics of the Boolean operators remains unchanged, but the evaluation of the truth value of the formula must take the quantifiers into account.
- Semantic tableaux (Sect. 7.5): The construction of a tableau is potentially infinite because a formula can be interpreted in an infinite domain. It follows that the method of semantic tableaux is not decision procedure for satisfiability in first-order logic. However, if the construction of a tableau for a formula A terminates in a closed tableau, then A is unsatisfiable (soundness); conversely, a systematic tableau for an unsatisfiable formula will close (completeness).
- Deduction (Sects. 8.1, 8.2): There are Gentzen and Hilbert deductive systems which are sound and complete. A valid formula is provable and we can construct a proof of the formula using tableaux, but given an *arbitrary* formula we cannot decide if it is valid and hence provable.
- Functions (Sect. 9.1): The syntax of first-order logic can be extended with function symbols that are interpreted as functions on the domain. With functions we can reason about mathematical operations, for example:

$$((x > 0 \wedge y > 0) \vee (x < 0 \wedge y < 0)) \rightarrow (x \cdot y > 0).$$

- Herbrand interpretations (Sect. 9.3): There are canonical interpretations called Herbrand interpretations. If a formula in *clausal form* has a model, it has a model which is an Herbrand interpretation, so to check satisfiability, it is sufficient to check if there is an Herbrand model for a formula.
- Resolution (Chap. 10): Resolution can be generalized to first-order logic with functions.

7.2 Formulas in First-Order Logic

7.2.1 Syntax

Definition 7.6 Let $\mathscr{P}$, $\mathscr{A}$ and $\mathscr{V}$ be countable sets of *predicate symbols*, *constant symbols* and *variables*. Each predicate symbol $p^n \in \mathscr{P}$ is associated with an *arity*, the number $n \geq 1$ of *arguments* that it takes. p^n is called an *n*-ary predicate. For $n = 1, 2$, the terms *unary* and *binary*, respectively, are also used. ∎

Notation

- We will drop the word 'symbol' and use the words 'predicate' and 'constant' by themselves for the syntactical symbols.
- By convention, the following lower-case letters, possibly with subscripts, will denote these sets: $\mathscr{P} = \{p, q, r\}$, $\mathscr{A} = \{a, b, c\}$, $\mathscr{V} = \{x, y, z\}$.
- The superscript denoting the arity of the predicate will not be written since the arity can be inferred from the number of arguments.

Definition 7.7
∀ is the *universal quantifier* and is read *for all*.
∃ is the *existential quantifier* and is read *there exists*. ∎

Definition 7.8 An *atomic formula* is an *n*-ary predicate followed by a list of n arguments in parentheses: $p(t_1, t_2, \ldots, t_n)$, where each argument t_i is either a variable or a constant. A *formula* in first-order logic is a tree defined recursively as follows:

- A formula is a leaf labeled by an atomic formula.
- A formula is a node labeled by ¬ with a single child that is a formula.
- A formula is a node labeled by ∀x or ∃x (for some variable x) with a single child that is a formula.
- A formula is a node labeled by a binary Boolean operator with two children both of which are formulas.

A formula of the form ∀$x A$ is a *universally quantified formula* or, simply, a *universal formula*. Similarly, a formula of the form ∃$x A$ is an *existentially quantified formula* or an *existential formula*. ∎

The definition of derivation and formation trees, and the concept of induction on the structure of a formula are taken over unchanged from propositional logic. When writing a formula as a string, the quantifiers are considered to have the same precedence as negation and a higher precedence than the binary operators.

Example 7.9 Figure 7.1 shows the tree representation of the formula:

$$\forall x (\neg \exists y p(x, y) \vee \neg \exists y p(y, x)).$$

The parentheses in $p(x, y)$ are part of the syntax of the atomic formula. ∎

Fig. 7.1 Tree for
$\forall x(\neg \exists y p(x, y) \vee \neg \exists y p(y, x))$

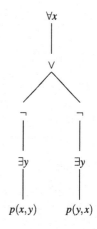

Example 7.10 Here are some examples of formulas in first-order logic:

$$\forall x \forall y(p(x, y) \rightarrow p(y, x)),$$
$$\forall x \exists y p(x, y),$$
$$\exists x \exists y(p(x) \wedge \neg p(y)),$$
$$\forall x p(a, x),$$
$$\forall x(p(x) \wedge q(x)) \leftrightarrow (\forall x p(x) \wedge \forall x q(x)),$$
$$\exists x(p(x) \vee q(x)) \leftrightarrow (\exists x p(x) \vee \exists x q(x)),$$
$$\forall x(p(x) \rightarrow q(x)) \rightarrow (\forall x p(x) \rightarrow \forall x q(x)),$$
$$(\forall x p(x) \rightarrow \forall x q(x)) \rightarrow \forall x(p(x) \rightarrow q(x)).$$

For now, they are just given as examples of the syntax of formulas in first-order logic; their meaning will be discussed in Sect. 7.3.2.   ∎

7.2.2 The Scope of Variables

Definition 7.11 A universal or existential formula $\forall x\, A$ or $\exists x\, A$ is a *quantified formula*. x is the *quantified variable* and its *scope* is the formula A. It is not required that x actually appear in the scope of its quantification.   ∎

The concept of the scope of variables in formulas of first-order logic is similar to the concept of the scope of variables in block-structured programming languages. Consider the program in Fig. 7.2. The variable x is declared twice, once globally and once locally in method p. The scope of the global declaration includes p, but the local declaration *hides* the global one. Within p, the value printed will be 1, the

Fig. 7.2 Global and local
variables

```
class MyClass {
  int x;

  void p() {
    int x;
    x = 1;
    // Print the value of x
  }

  void q() {
    // Print the value of x
  }

  ... void main(...) {
  x = 5;
  p;
  q;
  }
```

value of the local variable. Within the method q, the global variable x is in scope but not hidden and the value 5 will be printed. As in programming, hiding a quantified variable within its scope is confusing and should be avoided by giving different names to each quantified variable.

Definition 7.12 Let A be a formula. An occurrence of a variable x in A is a *free variable* of A iff x is not within the scope of a quantified variable x. A variable which is not free is *bound*.

If a formula has no free variables, it is *closed*. If $\{x_1, \ldots, x_n\}$ are all the free variables of A, the *universal closure* of A is $\forall x_1 \cdots \forall x_n A$ and the *existential closure* is $\exists x_1 \cdots \exists x_n A$.

$A(x_1, \ldots, x_n)$ indicates that the set of free variables of the formula A is a subset of $\{x_1, \ldots, x_n\}$. ∎

Example 7.13 $p(x, y)$ has two free variables x and y, $\exists y p(x, y)$ has one free variable x and $\forall x \exists y p(x, y)$ is closed. The universal closure of $p(x, y)$ is $\forall x \forall y p(x, y)$ and its existential closure is $\exists x \exists y p(x, y)$. ∎

Example 7.14 In $\forall x p(x) \wedge q(x)$, the occurrence of x in $p(x)$ is bound and the occurrence in $q(x)$ is free. The universal closure is $\forall x (\forall x p(x) \wedge q(x))$. Obviously, it would have been better to write the formula as $\forall x p(x) \wedge q(y)$ with y as the free variable; its universal closure is $\forall y (\forall x p(x) \wedge q(y))$. ∎

7.2.3 A Formal Grammar for Formulas *

As with propositional logic (Sect. 2.1.6), formulas in first-order logic can be defined as the strings generated by a context-free grammar.

Definition 7.15 The following grammar defines *atomic formulas* and *formulas* in first-order logic:

argument	::=	x	for any $x \in \mathcal{V}$
argument	::=	a	for any $a \in \mathcal{A}$
argument_list	::=	*argument*	
argument_list	::=	*argument, argument_list*	
atomic_formula	::=	$p\,(argument\_list)$	for any n-ary $p \in \mathcal{P}, n \geq 1$
formula	::=	*atomic_formula*	
formula	::=	$\neg$ *formula*	
formula	::=	*formula* $\vee$ *formula*	similarly for $\wedge, \cdots$
formula	::=	$\forall\, x$ *formula*	for any $x \in \mathcal{V}$
formula	::=	$\exists\, x$ *formula*	for any $x \in \mathcal{V}$

An n-ary predicate p must have an argument list of length n.   ∎

7.3 Interpretations

In propositional logic, an interpretation is a mapping from atomic propositions to truth values. In first-order logic, the analogous concept is a mapping from atomic formulas to truth values. However, atomic formulas contain variables and constants that must be assigned elements of some domain; once that is done, the predicates are interpreted as relations over the domain.

Definition 7.16 Let A be a formula where $\{p_1, \ldots, p_m\}$ are all the predicates appearing in A and $\{a_1, \ldots, a_k\}$ are all the constants appearing in A. An *interpretation* $\mathcal{I}_A$ for A is a triple:

$$(D, \{R_1, \ldots, R_m\}, \{d_1, \ldots, d_k\}),$$

where D is a *non-empty* set called the *domain*, R_i is an n_i-ary relation on D that is assigned to the n_i-ary predicate p_i and $d_i \in D$ is assigned to the constant a_i.   ∎

Example 7.17 Here are three interpretations for the formula $\forall x p(a, x)$:

$$\mathcal{I}_1 = (\mathcal{N}, \{\leq\}, \{0\}), \qquad \mathcal{I}_2 = (\mathcal{N}, \{\leq\}, \{1\}), \qquad \mathcal{I}_3 = (\mathcal{Z}, \{\leq\}, \{0\}).$$

The domain is either the $\mathcal{N}$, the set of natural numbers, or $\mathcal{Z}$, the set of integers. The binary relation $\leq$ (*less-than*) is assigned to the binary predicate p and either 0 or 1 is assigned to the constant a.

The formula can also be interpreted over strings:

$$\mathcal{I}_4 = (\mathcal{S}, \{substr\}, \{"\ "\}).$$

The domain $\mathcal{S}$ is a set of strings, *substr* is the binary relation such that $(s_1, s_2) \in$ *substr* iff s_1 is a substring of s_2, and $"\ "$ is the null string. ∎

A formula might have free variables and its truth value depends on the assignment of domain elements to the variables. For example, it doesn't make sense to ask if the formula $p(x, a)$ is true in the interpretation $(\mathcal{N}, \{>\}, \{10\})$. If x is assigned 15 the truth value of the formula is T, while if x is assigned 6 the truth value of the formula is F.

Definition 7.18 Let $\mathcal{I}_A$ be an interpretation for a formula A. An *assignment* $\sigma_{\mathcal{I}_A}$: $\mathcal{V} \mapsto D$ is a function which maps every free variable $v \in \mathcal{V}$ to an element $d \in D$, the domain of $\mathcal{I}_A$.

$\sigma_{\mathcal{I}_A}[x_i \leftarrow d_i]$ is an assignment that is the same as $\sigma_{\mathcal{I}_A}$ except that x_i is mapped to d_i. ∎

We can now define the truth value of a formula of first-order logic.

Definition 7.19 Let A be a formula, $\mathcal{I}_A$ an interpretation and $\sigma_{\mathcal{I}_A}$ an assignment. $v_{\sigma_{\mathcal{I}_A}}(A)$, the *truth value of A under $\mathcal{I}_A$ and $\sigma_{\mathcal{I}_A}$*, is defined by induction on the structure of A (where we have simplified the notation by writing v_σ for $v_{\sigma_{\mathcal{I}_A}}$):

- Let $A = p_k(c_1, \ldots, c_n)$ be an atomic formula where each c_i is either a variable x_i or a constant a_i. $v_\sigma(A) = T$ iff $(d_1, \ldots, d_n) \in R_k$ where R_k is the relation assigned by $\mathcal{I}_A$ to p_k, and d_i is the domain element assigned to c_i, either by $\mathcal{I}_A$ if c_i is a constant or by $\sigma_{\mathcal{I}_A}$ if c_i is a variable.
- $v_\sigma(\neg A_1) = T$ iff $v_\sigma(A_1) = F$.
- $v_\sigma(A_1 \vee A_2) = T$ iff $v_\sigma(A_1) = T$ or $v_\sigma(A_2) = T$, and similarly for the other Boolean operators.
- $v_\sigma(\forall x A_1) = T$ iff $v_{\sigma[x \leftarrow d]}(A_1) = T$ for *all* $d \in D$.
- $v_\sigma(\exists x A_1) = T$ iff $v_{\sigma[x \leftarrow d]}(A_1) = T$ for *some* $d \in D$. ∎

7.3.1 Closed Formulas

We define satisfiability and validity only on closed formulas. The reason is both convenience (not having to deal with assignments in addition to interpretations) and simplicity (because we can use the closures of formulas).

Theorem 7.20 *Let A be a* closed *formula and let $\mathscr{I}_A$ be an interpretation for A. Then $v_{\sigma_{\mathscr{I}_A}}(A)$ does not depend on $\sigma_{\mathscr{I}_A}$.*

Proof Call a formula independent of $\sigma_{\mathscr{I}_A}$ if its value does not depend on $\sigma_{\mathscr{I}_A}$. Let $A' = \forall x A_1(x)$ be a (not necessarily proper) subformula of A, where A' is *not* contained in the scope of any other quantifier. Then $v_{\sigma_{\mathscr{I}_A}}(A') = T$ iff $v_{\sigma_{\mathscr{I}_A}[x \leftarrow d]}(A_1)$ for all $d \in D$. But x is the only free variable in A_1, so A_1 is independent of $\sigma_{\mathscr{I}_A}$ since what is assigned to x is replaced by the assignment $[x \leftarrow d]$. A similar results holds for an existential formula $\exists x A_1(x)$.

The theorem can now be proved by induction on the depth of the quantifiers and by structural induction, using the fact that a formula constructed using Boolean operators on independent formulas is also independent.   ∎

By the theorem, if A is a closed formula we can use the notation $v_{\mathscr{I}}(A)$ without mentioning an assignment.

Example 7.21 Let us check the truth values of the formula $A = \forall x p(a, x)$ under the interpretations given in Example 7.17:

- $v_{\mathscr{I}_1}(A) = T$: For *all* $n \in \mathscr{N}$, $0 \le n$.
- $v_{\mathscr{I}_2}(A) = F$: It is not true that for *all* $n \in \mathscr{N}$, $1 \le n$. If $n = 0$ then $1 \not\le 0$.
- $v_{\mathscr{I}_3}(A) = F$: There is no smallest integer.
- $v_{\mathscr{I}_4}(A) = T$: By definition, the null string is a substring of every string.

The proof of the following theorem is left as an exercise.

Theorem 7.22 *Let $A' = A(x_1, \ldots, x_n)$ be a (non-closed) formula with free variables $x_1, \ldots, x_n$, and let $\mathscr{I}$ be an interpretation. Then:*

- $v_{\sigma_{\mathscr{I}_A}}(A') = T$ *for* some *assignment $\sigma_{\mathscr{I}_A}$ iff $v_{\mathscr{I}}(\exists x_1 \cdots \exists x_n A') = T$.*
- $v_{\sigma_{\mathscr{I}_A}}(A') = T$ *for* all *assignments $\sigma_{\mathscr{I}_A}$ iff $v_{\mathscr{I}}(\forall x_1 \cdots \forall x_n A') = T$.*

7.3.2 Validity and Satisfiability

Definition 7.23 Let A be a closed formula of first-order logic.

- A is *true in $\mathscr{I}$* or $\mathscr{I}$ is a *model* for A iff $v_{\mathscr{I}}(A) = T$. Notation: $\mathscr{I} \models A$.
- A is *valid* if for *all* interpretations $\mathscr{I}$, $\mathscr{I} \models A$. Notation: $\models A$.
- A is *satisfiable* if for *some* interpretation $\mathscr{I}$, $\mathscr{I} \models A$.
- A is *unsatisfiable* if it is not satisfiable.
- A is *falsifiable* if it is not valid.   ∎

Example 7.24 The closed formula $\forall x p(x) \rightarrow p(a)$ is valid. If it were not, there would be an interpretation $\mathscr{I} = (D, \{R\}, \{d\})$ such that $v_{\mathscr{I}}(\forall x p(x)) = T$ and $v_{\mathscr{I}}(p(a)) = F$. By Theorem 7.22, $v_{\sigma_{\mathscr{I}}}(p(x)) = T$ for all assignments $\sigma_{\mathscr{I}}$, in particular for the assignment $\sigma'_{\mathscr{I}}$ that assigns d to x. But $p(a)$ is closed, so $v_{\sigma'_{\mathscr{I}}}(p(a)) = v_{\mathscr{I}}(p(a)) = F$, a contradiction.   ∎

Let us now analyze the semantics of the formulas in Example 7.10.

Example 7.25

- $\forall x \forall y (p(x, y) \rightarrow p(y, x))$
  The formula is satisfiable in an interpretation where p is assigned a symmetric relation like $=$. It is not valid because the formula is falsified in an interpretation that assigns to p a non-symmetric relation like $<$.
- $\forall x \exists y p(x, y)$
  The formula is satisfiable in an interpretation where p is assigned a relation that is a total function, for example, $(x, y) \in R$ iff $y = x + 1$ for $x, y \in \mathscr{Z}$. The formula is falsified if the domain is changed to the negative numbers because there is no negative number y such that $y = -1 + 1$.
- $\exists x \exists y (p(x) \wedge \neg p(y))$
  This formula is satisfiable only in a domain with at least two elements.
- $\forall x p(a, x)$
  This expresses the existence of an element with special properties. For example, if p is interpreted by the relation $\leq$ on the domain $\mathscr{N}$, then the formula is true for $a = 0$. If we change the domain to $\mathscr{Z}$ the formula is false for the same assignment of $\leq$ to p.
- $\forall x (p(x) \wedge q(x)) \leftrightarrow (\forall x p(x) \wedge \forall x q(x))$
  The formula is valid. We prove the forward direction and leave the converse as an exercise. Let $\mathscr{I} = (D, \{R_1, R_2\}, \{\})$ be an arbitrary interpretation. By Theorem 7.22, $v_{\sigma_{\mathscr{I}}}(p(x) \wedge q(x)) = T$ for all assignments $\sigma_{\mathscr{I}}$, and by the inductive definition of an interpretation, $v_{\sigma_{\mathscr{I}}}(p(x)) = T$ and $v_{\sigma_{\mathscr{I}}}(q(x)) = T$ for all assignments $\sigma_{\mathscr{I}}$. Again by Theorem 7.22, $v_{\mathscr{I}}(\forall x p(x)) = T$ and $v_{\mathscr{I}}(\forall x q(x)) = T$, and by the definition of an interpretation $v_{\mathscr{I}}(\forall x p(x) \wedge \forall x q(x)) = T$.
  Show that $\forall$ does not distribute over disjunction by constructing a falsifying interpretation for $\forall x (p(x) \vee q(x)) \leftrightarrow (\forall x p(x) \vee \forall x q(x))$.
- $\forall x (p(x) \rightarrow q(x)) \rightarrow (\forall x p(x) \rightarrow \forall x q(x))$
  We leave it as an exercise to show that this is a valid formula, but its converse $(\forall x p(x) \rightarrow \forall x q(x)) \rightarrow \forall x (p(x) \rightarrow q(x))$ is not. ∎

7.3.3 An Interpretation for a Set of Formulas

In propositional logic, the concept of interpretation and the definition of properties such as satisfiability can be extended to sets of formulas (Sect. 2.2.4). The same holds for first-order logic.

Definition 7.26 Let $U = \{A_1, \ldots\}$ be a set of formulas where $\{p_1, \ldots, p_m\}$ are all the predicates appearing in all $A_i \in S$ and $\{a_1, \ldots, a_k\}$ are all the constants appearing in all $A_i \in S$. An *interpretation* $\mathscr{I}_U$ for S is a triple:

$$(D, \{R_1, \ldots, R_m\}, \{d_1, \ldots, d_k\}),$$

where D is a *non-empty* set called the *domain*, R_i is an n_i-ary relation on D that is assigned to the n_i-ary predicate p_i and $d_i \in D$ is an element of D that is assigned to the constant a_i. ∎

Similarly, an assignment needs to assign elements of the domain to the free variables (if any) in all formulas in U. For simplicity, the following definition is given only for closed formulas.

Definition 7.27 A set of closed formulas $U = \{A_1, \ldots\}$ is *(simultaneously) satisfiable* iff there exists an interpretation $\mathscr{I}_U$ such that $v_{\mathscr{I}_{\mathscr{U}}}(A_i) = T$ for all i. The satisfying interpretation is a *model* of U. U is *valid* iff for every interpretation $\mathscr{I}_U$, $v_{\mathscr{I}_{\mathscr{U}}}(A_i) = T$ for all i. ∎

The definitions of unsatisfiable and falsifiable are similar.

7.4 Logical Equivalence

Definition 7.28

- Let $U = \{A_1, A_2\}$ be a pair of closed formulas. A_1 is *logically equivalent* to A_2 iff $v_{\mathscr{I}_U}(A_1) = v_{\mathscr{I}_U}(A_2)$ for all interpretations $\mathscr{I}_U$. Notation: $A_1 \equiv A_2$.
- Let A be a closed formula and U a set of closed formulas. A is a *logical consequence* of U iff for all interpretations $\mathscr{I}_{U \cup \{A\}}$, $v_{\mathscr{I}_{U \cup \{A\}}}(A_i) = T$ for all $A_i \in U$ implies $v_{\mathscr{I}_{U \cup \{A\}}}(A) = T$. Notation: $U \models A$. ∎

As in propositional logic, the metamathematical concept $A \equiv B$ is not the same as the formula $A \leftrightarrow B$ in the logic, and similarly for logical consequence and implication. The relations between the concepts is given by the following theorem whose proof is similar to the proofs of Theorems 2.29, 2.50.

Theorem 7.29 *Let A, B be closed formulas and $U = \{A_1, \ldots, A_n\}$ be a set of closed formulas. Then:*

$$A \equiv B \quad \text{iff} \quad \models A \leftrightarrow B,$$
$$U \models A \quad \text{iff} \quad \models (A_1 \wedge \cdots \wedge A_n) \rightarrow A.$$

7.4.1 Logical Equivalences in First-Order Logic

Duality

The two quantifiers are duals:

$$\models \forall x\, A(x) \leftrightarrow \neg\exists x\,\neg A(x),$$
$$\models \exists x\, A(x) \leftrightarrow \neg\forall x\,\neg A(x).$$

In many presentations of first-order logic, $\forall$ is defined in the logic and $\exists$ is considered to be an abbreviation of $\neg\forall\neg$.

Commutativity and Distributivity

Quantifiers of the same type commute:

$$\models \forall x\forall y\, A(x, y) \leftrightarrow \forall y\forall x\, A(x, y),$$
$$\models \exists x\exists y\, A(x, y) \leftrightarrow \exists y\exists x\, A(x, y),$$

but $\forall$ and $\exists$ commute only in one direction:

$$\models \exists x\forall y\, A(x, y) \rightarrow \forall y\exists x\, A(x, y).$$

Universal quantifiers distribute over conjunction, and existential quantifiers distribute over disjunction:

$$\models \exists x(A(x) \vee B(x)) \leftrightarrow \exists x\, A(x) \vee \exists x\, B(x),$$
$$\models \forall x(A(x) \wedge B(x)) \leftrightarrow \forall x\, A(x) \wedge \forall x\, B(x),$$

but only one direction holds when distributing universal quantifiers over disjunction and existential quantifiers over conjunction:

$$\models \forall x\, A(x) \vee \forall x\, B(x) \rightarrow \forall x(A(x) \vee B(x)),$$
$$\models \exists x(A(x) \wedge B(x)) \rightarrow \exists x\, A(x) \wedge \exists x\, B(x).$$

To see that the converse direction of the second formula is falsifiable, let $D = \{d_1, d_2\}$ be a domain with two elements and consider an interpretation such that:

$$v(A(d_1)) = T, \qquad v(A(d_2)) = F, \qquad v(B(d_1)) = F, \qquad v(B(d_2)) = T.$$

Then $v(\exists x\, A(x) \wedge \exists x\, B(x)) = T$ but $v(\exists x(A(x) \wedge B(x))) = F$. A similar counterexample can be found for the first formula with the universal quantifiers and disjunction.

In the formulas with more than one quantifier, the scope rules ensure that each quantified variable is distinct. You may wish to write the formulas in the equivalent form with distinct variables names:

$$\models \forall x(A(x) \wedge B(x)) \leftrightarrow \forall y\, A(y) \wedge \forall z\, B(z).$$

Quantification Without the Free Variable in Its Scope

When quantifying over a disjunction or conjunction, if one subformula does not contain the quantified variable as a free variable, then distribution may be freely performed. If x is not free in B then:

$$\models \exists x A(x) \lor B \;\leftrightarrow\; \exists x (A(x) \lor B), \qquad \models \forall x A(x) \lor B \;\leftrightarrow\; \forall x (A(x) \lor B),$$
$$\models B \lor \exists x A(x) \;\leftrightarrow\; \exists x (B \lor A(x)), \qquad \models B \lor \forall x A(x) \;\leftrightarrow\; \forall x (B \lor A(x)),$$
$$\models \exists x A(x) \land B \;\leftrightarrow\; \exists x (A(x) \land B), \qquad \models \forall x A(x) \land B \;\leftrightarrow\; \forall x (A(x) \land B),$$
$$\models B \land \exists x A(x) \;\leftrightarrow\; \exists x (B \land A(x)), \qquad \models B \land \forall x A(x) \;\leftrightarrow\; \forall x (B \land A(x)).$$

Quantification over Implication and Equivalence

Distributing a quantifier over an equivalence or an implication is not trivial.

As with the other operators, if the quantified variable does not appear in one of the subformulas there is no problem:

$$\models \forall x (A \to B(x)) \;\leftrightarrow\; (A \to \forall x B(x)),$$
$$\models \forall x (A(x) \to B) \;\leftrightarrow\; (\exists x A(x) \to B).$$

Distribution of universal quantification over equivalence works in one direction:

$$\models \forall x (A(x) \leftrightarrow B(x)) \;\to\; (\forall x A(x) \leftrightarrow \forall x B(x)),$$

while for existential quantification, we have the formula:

$$\models \forall x (A(x) \leftrightarrow B(x)) \;\to\; (\exists x A(x) \leftrightarrow \exists x B(x)).$$

For distribution over an implication, the following formulas hold:

$$\models \exists x (A(x) \to B(x)) \;\leftrightarrow\; (\forall x A(x) \to \exists x B(x)),$$
$$\models (\exists x A(x) \to \forall x B(x)) \;\to\; \forall x (A(x) \to B(x)),$$
$$\models \forall x (A(x) \to B(x)) \;\to\; (\exists x A(x) \to \exists x B(x)),$$
$$\models \forall x (A(x) \to B(x)) \;\to\; (\forall x A(x) \to \exists x B(x)).$$

To derive these formulas, replace the implication or equivalence by the equivalent disjunction and conjunction and use the previous equivalences.

Example 7.30

$$\begin{aligned}
\exists x (A(x) \to B(x)) \;&\equiv\; \exists x (\neg A(x) \lor B(x)) \\
&\equiv\; \exists x \neg A(x) \lor \exists x B(x) \\
&\equiv\; \neg \exists x \neg A(x) \to \exists x B(x) \\
&\equiv\; \forall x A(x) \to \exists x B(x).
\end{aligned}$$

∎

The formulas for conjunction and disjunction can be proved directly using the semantic definitions.

Example 7.31 Prove: $\models \forall x (A(x) \vee B(x)) \to \forall x A(x) \vee \exists x B(x)$.

Use logical equivalences of propositional logic (considering each atomic formula as an atomic proposition) to transform the formula:

$$\begin{aligned}
\forall x (A(x) \vee B(x)) \to (\forall x A(x) \vee \exists x B(x)) &\equiv \\
\forall x (A(x) \vee B(x)) \to (\neg \forall x A(x) \to \exists x B(x)) &\equiv \\
\neg \forall x A(x) \to (\forall x (A(x) \vee B(x)) \to \exists x B(x)).
\end{aligned}$$

By duality of the quantifiers, we have:

$$\exists x \neg A(x) \to (\forall x (A(x) \vee B(x)) \to \exists x B(x))).$$

For the formula to be valid, it must be true under all interpretations. Clearly, if $v_{\mathscr{I}}(\exists x \neg A(x)) = F$ or $v_{\mathscr{I}}(\forall x (A(x) \vee B(x))) = F$, the formula is true, so we need only show $v_{\mathscr{I}}(\exists x B(x)) = T$ for interpretations $v_{\mathscr{I}}$ under which these subformulas are true. By Theorem 7.22, for some assignment $\sigma'_{\mathscr{I}}$, $v_{\sigma'_{\mathscr{I}}}(\neg A(x)) = T$ and thus $v_{\sigma'_{\mathscr{I}}}(A(x)) = F$. Using Theorem 7.22 again, $v_{\sigma_{\mathscr{I}}}(A(x) \vee B(x)) = T$ under all assignments, in particular under $\sigma'_{\mathscr{I}}$. By definition of an interpretation for disjunction, $v_{\sigma'_{\mathscr{I}}}(B(x)) = T$, and using Theorem 7.22 yet again, $v_{\mathscr{I}}(\exists x B(x)) = T$.   ∎

7.5 Semantic Tableaux

Before presenting the formal construction of semantic tableaux for first-order logic, we informally construct several tableaux in order to demonstrate the difficulties that must be dealt with and to motivate their solutions.

First, we need to clarify the concept of constant symbols. Recall from Definition 7.6 that formulas of first-order are constructed from countable sets of predicate, variable and constant symbols, although a particular formula such as $\exists x p(a, x)$ will only use a finite subset of these symbols. To build semantic tableaux in first-order logic, we will need to use the entire set of constant symbols $\mathscr{A} = \{a_0, a_1, \ldots\}$. If a formula like $\exists x p(a, x)$ contains a constant symbol, we assume that it is one of the a_i.

Definition 7.32 Let A be a quantified formula $\forall x A_1(x)$ or $\exists x A_1(x)$ and let a be a constant symbol. An *instantiation of A by a* is the formula $A_1(a)$, where all free occurrences of x are replaced by the constant a.   ∎

7.5.1 Examples for Semantic Tableaux

Instantiate Universal Formulas with all Constants

Example 7.33 Consider the valid formula:

$$A = \forall x(p(x) \to q(x)) \to (\forall x p(x) \to \forall x q(x)),$$

and let us build a semantic tableau for its negation. Applying the rule for the α-formula $\neg (A_1 \to A_2)$ twice, we get:

$$\neg (\forall x(p(x) \to q(x)) \to (\forall x p(x) \to \forall x q(x)))$$
$$\downarrow$$
$$\forall x(p(x) \to q(x)), \ \neg (\forall x p(x) \to \forall x q(x))$$
$$\downarrow$$
$$\forall x(p(x) \to q(x)), \ \forall x p(x), \ \neg \forall x q(x)$$
$$\downarrow$$
$$\forall x(p(x) \to q(x)), \ \forall x p(x), \ \exists \neg x q(x)$$

where the last node is obtained by the duality of $\forall$ and $\exists$.

The third formula will be true in an interpretation only if there exists a domain element c such that $c \in R_q$, where R_q is the relation assigned to the predicate q. Let us use the first constant a_1 to represent this element and instantiate the formula with it:

$$\forall x(p(x) \to q(x)), \ \forall x p(x), \ \exists \neg x q(x)$$
$$\downarrow$$
$$\forall x(p(x) \to q(x)), \ \forall x p(x), \ \neg q(a_1).$$

The first two formulas are universally quantified, so they can be true only if they hold for *every* element of the domain of an interpretation. Since any interpretation must include the domain element that is assigned to the constant a_1, we instantiate the universally quantified formulas with this constant:

$$\forall x(p(x) \to q(x)), \ \forall x p(x), \ \neg q(a_1)$$
$$\downarrow$$
$$\forall x(p(x) \to q(x)), \ p(a_1), \ \neg q(a_1)$$
$$\downarrow$$
$$p(a_1) \to q(a_1), \ p(a_1), \ \neg q(a_1).$$

Applying the rule to the β-formula $p(a_1) \to q(a_1)$ immediately gives a closed tableau, which to be expected for the negation of the valid formula A. ∎

From this example we learn that existentially quantified formulas must be instantiated with a constant the represents the domain element that must exist. Once a constant is introduced, instantiations of all universally quantified formulas must be done for that constant.

$$\neg (\forall x(p(x) \vee q(x)) \rightarrow (\forall xp(x) \vee \forall xq(x)))$$
$$\downarrow$$
$$\forall x(p(x) \vee q(x)), \; \neg (\forall xp(x) \vee \forall xq(x))$$
$$\downarrow$$
$$\forall x(p(x) \vee q(x)), \; \neg \forall xp(x), \; \neg \forall xq(x)$$
$$\downarrow$$
$$\forall x(p(x) \vee q(x)), \; \exists \neg xp(x), \; \exists \neg xq(x)$$
$$\downarrow$$
$$\forall x(p(x) \vee q(x)), \; \exists \neg xp(x), \; \neg q(a_1)$$
$$\downarrow$$
$$\forall x(p(x) \vee q(x)), \; \neg p(a_1), \; \neg q(a_1)$$
$$\downarrow$$
$$p(a_1) \vee q(a_1), \; \neg p(a_1), \; \neg q(a_1)$$

$$p(a_1), \; \neg p(a_1), \; \neg q(a_1) \qquad\qquad q(a_1), \; \neg p(a_1), \; \neg q(a_1)$$
$$\times \qquad\qquad\qquad\qquad\qquad \times$$

Fig. 7.3 Semantic tableau for the negation of a satisfiable, but not valid, formula

Don't Use the Same Constant Twice to Instantiate Existential Formulas

Example 7.34 Figure 7.3 shows an attempt to construct a tableau for the negation of the formula:

$$A = \forall x(p(x) \vee q(x)) \rightarrow (\forall xp(x) \vee \forall xq(x)),$$

which is satisfiable but not valid. As a falsifiable formula, its negation $\neg A$ is satisfiable, but the tableau in the figure is closed. What went wrong?

The answer is that instantiation of $\exists x \neg p(x))$ should not have used the constant a_1 once it had already been chosen for the instantiation of $\exists \neg xq(x)$. Choosing the same constant means that the interpretation will assign the same domain element to both occurrences of the constant. In fact, the formula A true (and $\neg A$ is false) in all interpretations over domains of a single element, but the formula might be satisfiable in interpretations with larger domains.

To avoid unnecessary constraints on the domain of a possible interpretation, a *new* constant must be chosen for every instantiation of an existentially quantified formula:

$$\forall x(p(x) \vee q(x)), \; \exists \neg xp(x), \; \exists \neg xq(x)$$
$$\downarrow$$
$$\forall x(p(x) \vee q(x)), \; \exists \neg xp(x), \; \neg q(a_1)$$
$$\downarrow$$
$$\forall x(p(x) \vee q(x)), \; \neg p(a_2), \; \neg q(a_1).$$

Instantiating the universally quantified formula with a_1 gives:

$$p(a_1) \vee q(a_1), \; \neg p(a_2), \; \neg q(a_1).$$

■

Don't Use Up Universal Formulas

Example 7.35 Continuing the tableau from the previous example:

$$\forall x(p(x) \vee q(x)), \; \neg p(a_2), \; \neg q(a_1)$$
$$\downarrow$$
$$p(a_1) \vee q(a_1), \; \neg p(a_2), \; \neg q(a_1)$$

we should now instantiate the universal formula $\forall x(p(x) \vee q(x))$ *again* with a_2, since it must be true for *all* domain elements, but, unfortunately, the formula has been used up by the tableau construction. To prevent this, universal formulas will never be deleted from the label of a node. They remain in the labels of all descendant nodes so as to constrain the possible interpretations of every new constant that is introduced:

$$\forall x(p(x) \vee q(x)), \; \neg p(a_2), \; \neg q(a_1)$$
$$\downarrow$$
$$\forall x(p(x) \vee q(x)), \; p(a_1) \vee q(a_1), \; \neg p(a_2), \; \neg q(a_1)$$
$$\downarrow$$
$$\forall x(p(x) \vee q(x)), \; p(a_2) \vee q(a_2), \; p(a_1) \vee q(a_1), \; \neg p(a_2), \; \neg q(a_1).$$

We leave it to the reader to continue the construction the tableau using the rule for β-formulas. Exactly one branch of the tableau will be open. A model can be defined by specifying a domain with two elements, say, 1 and 2. These elements are assigned to the constants a_1 and a_2, respectively, and the relations R_p and R_q assigned to p and q, respectively, hold for exactly one of the domain elements:

$$\mathscr{I} = (\{1, 2\}, \{R_p = \{1\}, \; R_q = \{2\}\}, \{a_1 = 1, a_2 = 2\}).$$

As expected, this model satisfies $\neg A$, so A is falsifiable.   ∎

A Branch May not Terminate

Example 7.36 Let us construct a semantic tableau to see if the formula $A = \forall x \exists y p(x, y)$ is satisfiable. Apparently, no rules apply since the formula is universally quantified and we only required that they had to be instantiated for constants already appearing in the formulas labeling a node. The constants are those that appear in the original formula and those that were introduced by instantiating existentially quantified formulas.

However, recall from Definition 7.16 that an interpretation is required to have a *non-empty* domain; therefore, we can arbitrarily choose the constant a_1 to represent that element. The tableau construction begins by instantiating A and then instantiating the existential formula with a new constant:

$$\forall x \exists y p(x, y)$$
$$\downarrow$$
$$\forall x \exists y p(x, y), \ \exists y p(a_1, y)$$
$$\downarrow$$
$$\forall x \exists y p(x, y), \ p(a_1, a_2).$$

Since $A = \forall x \exists y p(x, y)$ is universally quantified, it is not used up.

The new constant a_2 is used to instantiate the universal formula A again; this results in an existential formula which must be instantiated with a new constant a_3:

$$\forall x \exists y p(x, y), \ p(a_1, a_2)$$
$$\downarrow$$
$$\forall x \exists y p(x, y), \ \exists y p(a_2, y), \ p(a_1, a_2)$$
$$\downarrow$$
$$\forall x \exists y p(x, y), \ p(a_2, a_3), \ p(a_1, a_2).$$

The construction of this semantic tableau will not terminate and an infinite branch results. It is easy to see that there are models for A with infinite domains, for example, $(\mathcal{N}, \{<\}, \{\})$. ∎

The method of semantic tableaux is *not* a decision procedure for satisfiability in first-order logic, because we can never know if a branch that does not close defines an infinite model or if it will eventually close, say, after one million further applications of the tableau rules.

Example 7.36 is not very satisfactory because the formula $\forall x \exists y p(x, y)$ is satisfiable in a finite model, in fact, even in a model whose domain contains a single element. We were being on the safe side in always choosing new constants to instantiate existentially quantified formulas. Nevertheless, it is easy to find formulas that have no finite models, for example:

$$\forall x \exists y p(x, y) \ \wedge \ \forall x \neg p(x, x) \ \wedge \ \forall x \forall y \forall z (p(x, y) \wedge p(y, z) \rightarrow p(x, z)).$$

Check that $(\mathcal{N}, \{<\}, \{\})$ is an infinite model for this formula; we leave it as an exercise to show that the formula has no finite models.

An Open Branch with Universal Formulas May Terminate

Example 7.37 The first two steps of the tableau for $\{\forall x p(a, x)\}$ are:

$$\{\forall x p(a, x)\}$$
$$\downarrow$$
$$\{p(a, a), \forall x p(a, x)\}$$
$$\downarrow$$
$$\{p(a, a), \forall x p(a, x)\}.$$

There is no point in creating the same node again and again, so we specify that this branch is finite and open. Clearly, $(\{a\}, \{P = (a, a)\}, \{a\})$ is a model for the formula. ∎

$$\forall x \exists y p(x, y) \, \wedge \, \forall x (q(x) \wedge \neg q(x))$$
$$\downarrow$$
$$\forall x \exists y p(x, y), \, \forall x (q(x) \wedge \neg q(x))$$
$$\downarrow$$
$$\forall x \exists y p(x, y), \, \exists y p(a_1, y), \, \forall x (q(x) \wedge \neg q(x))$$
$$\downarrow$$
$$\forall x \exists y p(x, y), \, p(a_1, a_2), \, \forall x (q(x) \wedge \neg q(x))$$
$$\downarrow$$
$$\forall x \exists y p(x, y), \, \exists y p(a_2, y), \, p(a_1, a_2), \, \forall x (q(x) \wedge \neg q(x))$$
$$\downarrow$$
$$\forall x \exists y p(x, y), \, p(a_2, a_3), \, p(a_1, a_2), \, \forall x (q(x) \wedge \neg q(x))$$

Fig. 7.4 A tableau that should close, but doesn't

The Tableau Construction Must Be Systematic

Example 7.38 The tableau in Fig. 7.4 is for the formula which is the conjunction of $\forall x \exists y p(x, y)$, which we already know to be satisfiable, together with the formula $\forall x (q(x) \wedge \neg q(x))$, which is clearly unsatisfiable. However, the branch can be continued indefinitely, because we are, in effect, choosing to apply rules only to subformulas of $\forall x \exists y p(x, y)$, as we did in Example 7.36. This branch will never close although the formula is unsatisfiable. A *systematic* construction is needed to make sure that rules are eventually applied to all the formulas labeling a node. ∎

7.5.2 The Algorithm for Semantic Tableaux

The following definition extends a familiar concept from propositional logic:

Definition 7.39 A *literal* is a closed atomic formula $p(a_1, \ldots, a_k)$, an atomic formula all of whose arguments are constants, or the negation of a closed atomic formula $\neg p(a_1, \ldots, a_k)$. If A is $p(a_1, \ldots, a_k)$ then $A^c = \neg p(a_1, \ldots, a_k)$, while if A is $\neg p(a_1, \ldots, a_k)$ then $A^c = p(a_1, \ldots, a_k)$. ∎

The classification of formulas in propositional logic as α and β formulas (Sect. 2.6.2) is retained and we extend the classification to formulas with quantifiers. γ-formulas are universally quantified formulas $\forall x A(x)$ and the negations of existentially quantified formulas $\neg \exists x A(x)$, while δ-formulas are existentially quantified formulas $\exists x A(x)$ and the negations of universally quantified formulas $\neg \forall x A(x)$. The rules for these formulas are simply instantiation with a constant:

γ	$\gamma(a)$		δ	$\delta(a)$
$\forall x A(x)$	$A(a)$		$\exists x A(x)$	$A(a)$
$\neg \exists x A(x)$	$\neg A(a)$		$\neg \forall x A(x)$	$\neg A(a)$

The algorithm for the construction of a semantic tableau in first-order logic is similar to that for propositional logic with the addition of rules for quantified formulas, together with various constraints designed to avoid the problems were saw in the examples.

Algorithm 7.40 (Construction of a semantic tableau)
Input: A formula ϕ of first-order logic.
Output: A semantic tableau $\mathcal{T}$ for ϕ: each branch may be infinite, finite and marked open, or finite and marked closed.

A semantic tableau is a tree $\mathcal{T}$ where each node is labeled by a pair $W(n) = (U(n), C(n))$, where:

$$U(n) = \{A_{n_1}, \ldots, A_{n_k}\}$$

is a set of formulas and:

$$C(n) = \{c_{n_1}, \ldots, c_{n_m}\}$$

is a set of constants. $C(n)$ contains the list of constants that appear in the formulas in $U(n)$. Of course, the sets $C(n)$ could be created on-the-fly from $U(n)$, but the algorithm in easier to understand if they explicitly label the nodes.

Initially, $\mathcal{T}$ consists of a single node n_0, the root, labeled with

$$(\{\phi\}, \{a_{0_1}, \ldots, a_{0_k}\}),$$

where $\{a_{0_1}, \ldots, a_{0_k}\}$ is the set of constants that appear in ϕ. If ϕ has no constants, take the first constant a_0 in the set $\mathcal{A}$ and label the node with $(\{\phi\}, \{a_0\})$.

The tableau is built inductively by repeatedly *choosing* an unmarked leaf l labeled with $W(l) = (U(l), C(l))$, and applying the *first applicable rule* in the following list:

- If $U(l)$ contains a complementary pair of literals, mark the leaf *closed* $\times$.
- If $U(l)$ is not a set of literals, *choose* a formula A in $U(l)$ that is an α-, β- or δ-formula.
  - If A is an α-formula, create a new node l' as a child of l. Label l' with:

$$W(l') = ((U(l) - \{A\}) \cup \{\alpha_1, \alpha_2\}, C(l)).$$

   (In the case that A is $\neg\neg A_1$, there is no α_2.)
  - If A is a β-formula, create two new nodes l' and l'' as children of l. Label l' and l'' with:

$$\begin{aligned} W(l') &= ((U(l) - \{A\}) \cup \{\beta_1\}, C(l)), \\ W(l'') &= ((U(l) - \{A\}) \cup \{\beta_2\}, C(l)). \end{aligned}$$

  - If A is a δ-formula, create a new node l' as a child of l and label l' with:

$$W(l') = ((U(l) - \{A\}) \cup \{\delta(a')\}, C(l) \cup \{a'\}),$$

   where a' is some constant that *does not appear in* $U(l)$.

- Let $\{\gamma_{l_1}, \ldots, \gamma_{l_m}\} \subseteq U(l)$ be all the γ-formulas in $U(l)$ and let $C(l) = \{c_{l_1}, \ldots, c_{l_k}\}$. Create a new node l' as a child of l and label l' with

$$W(l') = \left(U(l) \cup \left\{\bigcup_{i=1}^{m}\bigcup_{j=1}^{k}\gamma_{l_i}(c_{l_j})\right\}, C(l)\right).$$

However, if $U(l)$ consists only of literals and γ-formulas and if $U(l')$ as constructed would be the same as $U(l)$, do not create node l'; instead, mark the leaf l as *open* ⊙.   ■

Compare the algorithm with the examples in Sect. 7.5.1. The phrase *first applicable rule* ensures that the construction is systematic. For δ-formulas, we added the condition that a *new* constant be used in the instantiation. For γ-formulas, the formula to which the rule is applied is not removed from the set $U(l)$ when $W(l')$ is created. The sentence beginning *however* in the rule for γ-formulas is intended to take care of the case where no new formulas are produced by the application of the rule.

Definition 7.41 A branch in a tableau is *closed* iff it terminates in a leaf marked closed; otherwise (it is infinite or it terminates in a leaf marked open), the branch is *open*.

A tableau is *closed* if all of its branches are closed; otherwise (it has a finite or infinite open branch), the tableau is open.   ■

Algorithm 7.40 is *not* a search procedure for a satisfying interpretation, because it may choose to infinitely expand one branch. Semantic tableaux in first-order logic can only be used to prove the validity of a formula by showing that a tableau for its negation closes. Since all branches close in a closed tableau, the nondeterminism in the application of the rules (choosing a leaf and choosing an α-, β- or γ-formula) doesn't matter.

7.6 Soundness and Completion of Semantic Tableaux

7.6.1 Soundness

The proof of the soundness of the algorithm for constructing semantic tableaux in first-order logic is a straightforward generalization of the one for propositional logic (Sect. 2.7.2).

Theorem 7.42 (Soundness) *Let ϕ be a formula in first-order logic and let $\mathcal{T}$ be a tableau for ϕ. If $\mathcal{T}$ closes, then ϕ is unsatisfiable.*

Proof The theorem is a special case of the following statement: if a subtree rooted at a node n of $\mathcal{T}$ closes, the set of formulas $U(n)$ is unsatisfiable.

The proof is by induction on the height h of n. The proofs of the base case for $h = 0$ and the inductive cases 1 and 2 for α- and β-rules are the same as in propositional logic (Sect. 2.6).

Case 3: The γ-rule was used. Then:

$$U(n) = U_0 \cup \{\forall x A(x)\} \quad \text{and} \quad U(n') = U_0 \cup \{\forall x A(x), \, A(a)\},$$

for some set of formulas U_0, where we have simplified the notation and explicitly considered only one formula.

The inductive hypothesis is that $U(n')$ is unsatisfiable and we want to prove that $U(n)$ is also unsatisfiable. Assume to the contrary that $U(n)$ is satisfiable and let $\mathscr{I}$ be a model for $U(n)$. Then $v_{\mathscr{I}}(A_i) = T$ for all $A_i \in U_0$ and also $v_{\mathscr{I}}(\forall x A(x)) = T$. But $U(n') = U(n) \cup \{A(a)\}$, so if we can show that $v_{\mathscr{I}}(A(a)) = T$, this will contradict the inductive hypothesis that $U(n')$ is unsatisfiable.

Now $v_{\mathscr{I}}(\forall x A(x)) = T$ iff $v_{\sigma_{\mathscr{I}}}(A(x)) = T$ for all assignments $\sigma_{\mathscr{I}}$, in particular for any assignment that assigns the same domain element to x that $\mathscr{I}$ does to a, so $v_{\mathscr{I}}(A(a)) = T$. By the tableau construction, $a \in C(n)$ and it appears in some formula of $U(n)$; therefore, $\mathscr{I}$, a model of $U(n)$, does, in fact, assign a domain element to a.

Case 4: The δ-rule was used. Then:

$$U(n) = U_0 \cup \{\exists x A(x)\} \quad \text{and} \quad U(n') = U_0 \cup \{A(a)\},$$

for some set of formulas U_0 and for some constant a that does not occur in any formula of $U(n)$.

The inductive hypothesis is that $U(n')$ is unsatisfiable and we want to prove that $U(n)$ is also unsatisfiable. Assume to the contrary that $U(n)$ is satisfiable and let:

$$\mathscr{I} = (D, \{R_1, \ldots, R_n\}, \{d_1, \ldots, d_k\})$$

be a model for $U(n)$.

Now $v_{\mathscr{I}}(\exists x A(x)) = T$ iff $v_{\sigma_{\mathscr{I}}}(A(x)) = T$ for some assignment $\sigma_{\mathscr{I}}$, that is, $\sigma_{\mathscr{I}}(x) = d$ for some $d \in D$. Extend $\mathscr{I}$ to the interpretation:

$$\mathscr{I}' = (D, \{R_1, \ldots, R_n\}, \{d_1, \ldots, d_k, d\})$$

by assigning d to the constant a. $\mathscr{I}'$ is well-defined: since a does not occur in $U(n)$, it is not among the constants $\{a_1, \ldots, a_k\}$ already assigned $\{d_1, \ldots, d_k\}$ in $\mathscr{I}$. Since $v_{\mathscr{I}'}(U_0) = v_{\mathscr{I}}(U_0) = T$, $v_{\mathscr{I}'}(A(a)) = T$ contradicts the inductive hypothesis that $U(n')$ is unsatisfiable.   ∎

7.6.2 Completeness

To prove the completeness of the algorithm for semantic tableaux we define a Hintikka set, show that a (possibly infinite) branch in a tableau is a Hintikka set and then prove Hintikka's Lemma that a Hintikka set can be extended to a model. We begin with a technical lemma whose proof is left as an exercise.

Lemma 7.43 *Let b be an open branch of a semantic tableau, n a node on b, and A a formula in $U(n)$. Then some rule is applied to A at node n or at a node m that is a descendant of n on b. Furthermore, if A is a γ-formula and $a \in C(n)$, then $\gamma(a) \in U(m')$, where m' is the child node created from m by applying a rule.*

Definition 7.44 Let U be a set of closed formulas in first-order logic. U is a *Hintikka set* iff the following conditions hold for all formulas $A \in U$:

1. If A is a literal, then either $A \notin U$ or $A^c \notin U$.
2. If A is an α-formula, then $\alpha_1 \in U$ and $\alpha_2 \in U$.
3. If A is a β-formula, then $\beta_1 \in U$ or $\beta_2 \in U$.
4. If A is a γ-formula, then $\gamma(c) \in U$ for *all* constants c in formulas in U.
5. If A is a δ-formula, then $\delta(c) \in U$ for *some* constant c.   ∎

Theorem 7.45 *Let b be a (finite or infinite) open branch of a semantic tableau and let $U = \bigcup_{n \in b} U(n)$. Then U is a Hintikka set.*

Proof Let $A \in U$. We show that the conditions for a Hintikka set hold.

Suppose that A is a literal. By the construction of the tableau, once a literal appears in a branch, it is never deleted. Therefore, if A appears in a node n and A^c appears in a node m which is a descendant of n, then A must also appear in m. By assumption, b is open, so either $A \notin U$ or $A^c \notin U$ and condition 1 holds.

If A is not atomic and not a γ-formula, by Lemma 7.43 eventually a rule is applied to A, and conditions 2, 3 and 5 hold.

Let A be a γ-formula that first appears in $U(n)$, let c be a constant that first appears in $C(m)$ and let $k = \max(n, m)$. By the construction of the tableau, the set of γ-formulas and the set of constants are non-decreasing along a branch, so $A \in U(k)$ and $c \in C(k)$. By Lemma 7.43, $\gamma(c) \in U(k') \subseteq U$, for some $k' > k$.   ∎

Theorem 7.46 (Hintikka's Lemma) *Let U be a Hintikka set. Then there is a (finite or infinite) model for U.*

Proof Let $\mathscr{C} = \{c_1, c_2, \ldots\}$ be the set of constants in formulas of U. Define an interpretation $\mathscr{I}$ as follows. The domain is the same set of symbols $\{c_1, c_2, \ldots\}$. Assign to each constant c_i in U the symbol c_i in the domain. For each n-ary predicate p_i in U, define an n-ary relation R_i by:

$$
\begin{aligned}
(a_{i_1}, \ldots, a_{i_n}) \in R_i \quad &\text{if} \quad p(a_{i_1}, \ldots, a_{i_n}) \in U, \\
(a_{i_1}, \ldots, a_{i_n}) \notin R_i \quad &\text{if} \quad \neg\, p(a_{i_1}, \ldots, a_{i_n}) \in U, \\
(a_{i_1}, \ldots, a_{i_n}) \in R_i \quad &\text{otherwise.}
\end{aligned}
$$

The relations are well-defined by condition 1 in the definition of Hintikka sets. We leave as an exercise to show that $\mathscr{I} \models A$ for all $A \in U$ by induction on the structure of A using the conditions defining a Hintikka set.   ∎

Theorem 7.47 (Completeness) *Let A be a valid formula. Then the semantic tableau for $\neg A$ closes.*

Proof Let A be a valid formula and suppose that the semantic tableau for $\neg A$ does not close. By Definition 7.41, the tableau must contain a (finite or infinite) open branch b. By Theorem 7.45, $U = \bigcup_{n \in b} U(n)$ is a Hintikka set and by Theorem 7.46, there is a model $\mathscr{I}$ for U. But $\neg A \in U$ so $\mathscr{I} \models \neg A$ contradicting the assumption that A is valid. ∎

7.7 Summary

First-order logic adds variables and constants to propositional logic, together with the quantifiers $\forall$ (for all) and $\exists$ (there exists). An interpretation includes a domain; the predicates are interpreted as relations over elements of the domain, while constants are interpreted as domain elements and variables in non-closed formulas are assigned domain elements.

The method of semantic tableaux is sound and complete for showing that a formula is unsatisfiable, but it is not a decision procedure for satisfiability, since branches of a tableau may be infinite. When a tableau is constructed, a universal quantifier followed by an existential quantifier can result in an infinite branch: the existential formula is instantiated with a new constant and then the instantiation of the universal formula results in a new occurrence of the existentially quantified formula, and so on indefinitely. There are formulas that are satisfiable only in an infinite domain.

7.8 Further Reading

The presentation of semantic tableaux follows that of Smullyan (1968) although he uses analytic tableaux. Advanced textbooks that also use tableaux are Nerode and Shore (1997) and Fitting (1996).

7.9 Exercises

7.1 Find an interpretation which falsifies $\exists x p(x) \rightarrow p(a)$.

7.2 Prove the statements left as exercises in Example 7.25:

- $\forall x p(x) \wedge \forall x q(x) \rightarrow \forall x (p(x) \wedge q(x))$ is valid.
- $\forall x (p(x) \rightarrow q(x)) \rightarrow (\forall x p(x) \rightarrow \forall x q(x))$ is a valid formula, but its converse $(\forall x p(x) \rightarrow \forall x q(x)) \rightarrow \forall x (p(x) \rightarrow q(x))$ is not.

7.3 Prove that the following formulas are valid:

$$\exists x(A(x) \rightarrow B(x)) \leftrightarrow (\forall x A(x) \rightarrow \exists x B(x)),$$
$$(\exists x A(x) \rightarrow \forall x B(x)) \rightarrow \forall x(A(x) \rightarrow B(x)),$$
$$\forall x(A(x) \vee B(x)) \rightarrow (\forall x A(x) \vee \exists x B(x)),$$
$$\forall x(A(x) \rightarrow B(x)) \rightarrow (\exists x A(x) \rightarrow \exists x B(x)).$$

7.4 For each formula in the previous exercise that is an implication, prove that the converse is not valid by giving a falsifying interpretation.

7.5 For each of the following formulas, either prove that it is valid or give a falsifying interpretation.

$$\exists x \forall y((p(x, y) \wedge \neg p(y, x)) \rightarrow (p(x, x) \leftrightarrow p(y, y))),$$
$$\forall x \forall y \forall z(p(x, x) \wedge (p(x, z) \rightarrow (p(x, y) \vee p(y, z)))) \rightarrow \exists y \forall z p(y, z).$$

7.6 Suppose that we allowed the domain of an interpretation to be *empty*. What would this mean for the equivalence:

$$\forall y p(y, y) \vee \exists x q(x, x) \equiv \exists x(\forall y p(y, y) \vee q(x, x)).$$

7.7 Prove Theorem 7.22 on the relationship between a non-closed formula and its closure.

7.8 Complete the semantic tableau construction for the negation of

$$\forall x(p(x) \vee q(x)) \rightarrow (\forall x p(x) \vee \forall x q(x)).$$

7.9 Prove that the formula $(\forall x p(x) \rightarrow \forall x q(x)) \rightarrow \forall x(p(x) \rightarrow q(x))$ is not valid by constructing a semantic tableau for its negation.

7.10 Prove that the following formula has no finite models:

$$\forall x \exists y p(x, y) \ \wedge \ \forall x \neg p(x, x) \ \wedge \ \forall x \forall y \forall z(p(x, y) \wedge p(y, z) \rightarrow p(x, z)).$$

7.11 Prove Lemma 7.43, the technical lemma used in the proof of the completeness of the method of semantic tableaux.

7.12 Complete the proof of Lemma 7.46 that every Hintikka set has a model.

References

M. Fitting. *First-Order Logic and Automated Theorem Proving (Second Edition)*. Springer, 1996.
A. Nerode and R.A. Shore. *Logic for Applications (Second Edition)*. Springer, 1997.
R.M. Smullyan. *First-Order Logic*. Springer-Verlag, 1968. Reprinted by Dover, 1995.

Chapter 8
First-Order Logic: Deductive Systems

We extend the deductive systems $\mathcal{G}$ and $\mathcal{H}$ from propositional logic to first-order logic by adding axioms and rules of inference for the universal quantifier. (The existential quantifier is defined as the dual of the universal quantifier.) The construction of semantic tableaux for first-order logic included restrictions on the use of constants and similar restrictions will be needed here.

8.1 Gentzen System $\mathcal{G}$

Figure 8.1 is a closed semantic tableau for the negation of the valid formula

$$\forall x p(x) \vee \forall x q(x) \rightarrow \forall x (p(x) \vee q(x)).$$

The formulas to which rules are applied are underlined, while the sets of constants $C(n)$ in the labels of each node are implicit.

Let us turn the tree upside down and in every node n replace $U(n)$, the set of formulas labeling the node n, by $\bar{U}(n)$, the set of complements of the formulas in $U(n)$. The result (Fig. 8.2) is a Gentzen proof for the formula.

Here is the classification of quantified formulas into γ- and δ-formulas:

γ	$\gamma(a)$	δ	$\delta(a)$
$\exists x A(x)$	$A(a)$	$\forall x A(x)$	$A(a)$
$\neg \forall x A(x)$	$\neg A(a)$	$\neg \exists x A(x)$	$\neg A(a)$

Definition 8.1 The *Gentzen system* $\mathcal{G}$ is a deductive system. Its *axioms* are sets of formulas U containing a complementary pair of literals. The rules of inference are the rules given for α- and β-formulas in Sect. 3.2, together with the following rules for γ- and δ-formulas:

M. Ben-Ari, *Mathematical Logic for Computer Science*,
DOI 10.1007/978-1-4471-4129-7_8, © Springer-Verlag London 2012

$$\neg\,(\forall x p(x) \lor \forall x q(x) \;\rightarrow\; \forall x(p(x) \lor q(x)))$$
$$\downarrow$$
$$\forall x p(x) \lor \forall x q(x),\; \neg\forall x(p(x) \lor q(x))$$

$\forall x p(x),\; \underline{\neg\forall x(p(x) \lor q(x))}$ $\forall x q(x),\; \underline{\neg\forall x(p(x) \lor q(x))}$
$$\downarrow \qquad\qquad\qquad\qquad\qquad\qquad\qquad \downarrow$$
$\forall x p(x),\; \neg\,(p(a) \lor q(a))$ $\forall x q(x),\; \neg\,(p(a) \lor q(a))$
$$\downarrow \qquad\qquad\qquad\qquad\qquad\qquad\qquad \downarrow$$
$\underline{\forall x p(x)},\; \neg p(a),\; \neg q(a)$ $\underline{\forall x q(x)},\; \neg p(a),\; \neg q(a)$
$$\downarrow \qquad\qquad\qquad\qquad\qquad\qquad\qquad \downarrow$$
$\forall x p(x),\; p(a),\; \neg p(a),\; \neg q(a)$ $\forall x q(x),\; q(a),\; \neg p(a),\; \neg q(a)$
$$\times \qquad\qquad\qquad\qquad\qquad\qquad\qquad \times$$

Fig. 8.1 Semantic tableau in first-order logic

$$\frac{U \cup \{\gamma, \gamma(a)\}}{U \cup \{\gamma\}}, \qquad \frac{U \cup \{\delta(a)\}}{U \cup \{\delta\}}.$$

The rule for δ-formulas can be applied only if the constant a does not occur in any formula of U.   ∎

The γ-rule can be read: if an existential formula and some instantiation of it are true, then the instantiation is redundant.

The δ-rules formalizes the following frequently used method of mathematical reasoning: Let a be an *arbitrary* constant. Suppose that $A(a)$ can be proved. Since a was arbitrary, the proof holds for $\forall x A(x)$. In order to generalize from a specific constant to *for all*, it is essential that a be an arbitrary constant and not one of the constants that is constrained by another subformula.

$\underline{\neg\forall x p(x),\; \neg\,p(a)},\; p(a),\; q(a)$ $\underline{\neg\forall x q(x),\; \neg q(a)},\; p(a),\; q(a)$
$$\downarrow \qquad\qquad\qquad\qquad\qquad\qquad\qquad \downarrow$$
$\neg\forall x p(x),\; \underline{p(a),\; q(a)}$ $\neg\forall x q(x),\; \underline{p(a),\; q(a)}$
$$\downarrow \qquad\qquad\qquad\qquad\qquad\qquad\qquad \downarrow$$
$\neg\forall x p(x),\; \underline{p(a) \lor q(a)}$ $\neg\forall x q(x),\; \underline{p(a) \lor q(a)}$
$$\downarrow \qquad\qquad\qquad\qquad\qquad\qquad\qquad \downarrow$$
$\underline{\neg\forall x p(x)},\; \forall x(p(x) \lor q(x))$ $\underline{\neg\forall x q(x)},\; \forall x(p(x) \lor q(x))$
$$\searrow \qquad\qquad\qquad\qquad\qquad\qquad \swarrow$$
$$\underline{\neg\,(\forall x p(x) \lor \forall x q(x)),\; \forall x(p(x) \lor q(x))}$$
$$\downarrow$$
$$\forall x p(x) \lor \forall x q(x) \;\rightarrow\; \forall x(p(x) \lor q(x))$$

Fig. 8.2 Gentzen proof tree in first-order logic

$$\frac{\neg \forall y p(a, y), \; \neg \, p(a, b), \; \exists x p(x, b), \; p(a, b)}{}$$
$$\downarrow$$
$$\frac{\neg \forall y p(a, y), \; \exists x p(x, b), \; p(a, b)}{}$$
$$\downarrow$$
$$\frac{\neg \forall y p(a, y), \; \exists x p(x, b)}{}$$
$$\downarrow$$
$$\frac{\neg \forall y p(a, y), \; \forall y \exists x p(x, y)}{}$$
$$\downarrow$$
$$\frac{\neg \exists x \forall y p(x, y), \; \forall y \exists x p(x, y)}{}$$
$$\downarrow$$
$$\exists x \forall y p(x, y) \to \forall y \exists x p(x, y)$$

Fig. 8.3 Gentzen proof: use rules for γ-formulas followed by rules for δ-formulas

Example 8.2 The proof of $\exists x \forall y p(x, y) \to \forall y \exists x p(x, y)$ in Fig. 8.3 begins with the axiom obtained from the complementary literals $\neg \, p(a, b)$ and $p(a, b)$. Then the rule for the γ-formulas is used twice:

$$\frac{U, \neg \forall y p(a, y), \neg \, p(a, b)}{U, \neg \forall y p(a, y)} \, , \qquad \frac{U, \exists x p(x, b), p(a, b)}{U, \exists x p(x, b)} \, .$$

Once this is done, it is easy to apply rules for the δ-formulas because the constants a and b appear only once so that the condition in the rule is satisfied:

$$\frac{U, \exists x p(x, b)}{U, \forall y \exists x p(x, y)} \, , \qquad \frac{U, \neg \forall y p(a, y)}{U, \neg \exists x \forall y \exists x p(x, y)} \, .$$

A final application of the rule for the α-formula completes the proof.   ∎

We leave the proof of the soundness and completeness of $\mathscr{G}$ as an exercise.

Theorem 8.3 (Soundness and completeness) *Let U be a set of formulas in first-order logic. There is a Gentzen proof for U if and only if there is a closed semantic tableau for $\bar{U}$.*

8.2 Hilbert System $\mathscr{H}$

The Hilbert system $\mathscr{H}$ for propositional logic is extended to first-order logic by adding two axioms and a rule of inference.

Definition 8.4 The axioms of the Hilbert system $\mathscr{H}$ for first-order logic are:

> **Axiom 1** $\vdash (A \to (B \to A))$,
>
> **Axiom 2** $\vdash (A \to (B \to C)) \to ((A \to B) \to (A \to C))$,
>
> **Axiom 3** $\vdash (\neg B \to \neg A) \to (A \to B)$,
>
> **Axiom 4** $\vdash \forall x A(x) \to A(a)$,
>
> **Axiom 5** $\vdash \forall x (A \to B(x)) \to (A \to \forall x B(x))$.

- In Axioms 1, 2 and 3, A, B and C are any formulas of first-order logic.
- In Axiom 4, $A(x)$ is a formula with a free variable x.
- In Axiom 5, $B(x)$ is a formula with a free variable x, while x is *not* a free variable of the formula A.

The rules of inference are **modus ponens** and **generalization**:

$$\frac{\vdash A \to B \quad \vdash A}{\vdash B}, \qquad \frac{\vdash A(a)}{\vdash \forall x A(x)}.$$

■

Propositional Reasoning in First-Order Logic

Axioms 1, 2, 3 and the rule of inference *MP* are generalized to any formulas in first-order logic so all of the theorems and derived rules of inference that we proved in Chap. 3 can be used in first-order logic.

Example 8.5

$$\vdash \forall x p(x) \to (\exists y \forall x q(x, y) \to \forall x p(x))$$

is an instance of Axiom 1 in first-order logic and:

$$\frac{\vdash \forall x p(x) \to (\exists y \forall x q(x, y) \to \forall x p(x)) \quad \vdash \forall x p(x)}{\vdash \exists y \forall x q(x, y) \to \forall x p(x)}$$

uses the rule of inference *modus ponens*.   ■

In the proofs in this chapter, we will not bother to give the details of deductions that use propositional reasoning because these are easy to understand. The notation *PC* will be used for propositional deductions.

Specialization and Generalization

Axiom 4 can also be used as a rule of inference:

Rule 8.6 (Axiom 4)

$$\frac{U \vdash \forall x\, A(x)}{U \vdash A(a)}.$$

Any occurrence of $\forall x\, A(x)$ can be replaced by $A(a)$ for *any* a. If $A(x)$ is true whatever the assignment of a domain element of an interpretation $\mathscr{I}$ to x, then $A(a)$ is true for the domain element that $\mathscr{I}$ assigns to a.

The generalization rule of inference states that if a occurs in a formula, we may bind *all* occurrences of a with the quantifier. Since a is arbitrary, that is the same as saying that $A(x)$ is true for *all* assignments to x.

There is a reason that the generalization rule was given only for formulas that can be proved *without* a set of assumptions U:

$$\frac{\vdash A(a)}{\vdash \forall x\, A(x)}.$$

Example 8.7 Suppose that we were allowed to apply generalization to $A(a) \vdash A(a)$ to obtain $A(a) \vdash \forall x\, A(x)$ and consider the interpretation:

$$(\mathscr{L}, \{even(x)\}, \{2\}).$$

The assumption $A(a)$ is true but $\forall x\, A(x)$ is not, which means that generalization is not sound as it transforms $A(a) \models A(a)$ into $A(a) \not\models \forall x\, A(x)$. ∎

Since proofs invariably have assumptions, a constraint must be placed on the generalization rule to make it useful:

Rule 8.8 (Generalization)

$$\frac{U \vdash A(a)}{U \vdash \forall x\, A(x)},$$

provided that a does not appear in U.

The Deduction Rule

The Deduction rule is essential for proving theorems from assumptions.

Rule 8.9 (Deduction rule)

$$\frac{U \cup \{A\} \vdash B}{U \vdash A \to B}.$$

Theorem 8.10 (Deduction Theorem) *The deduction rule is sound.*

Proof The proof is by induction on the length of the proof of $U \cup \{A\} \vdash B$. We must show how to obtain a proof of $U \vdash A \to B$ that does not use the deduction rule. The proof for propositional logic (Theorem 3.14) is modified to take into account the new axioms and generalization.

The modification for the additional axioms is trivial.

Consider now an application of the generalization rule, where, without loss of generality, we assume that the generalization rule is applied to the immediately preceding formula in the proof:

$$\begin{array}{ll} i & U \cup \{A\} \vdash B(a) \\ i+1 & U \cup \{A\} \vdash \forall x\, B(x) \qquad\qquad\qquad\qquad \text{Generalization} \end{array}$$

By the condition on the generalization rule in the presence of assumptions, a does not appear in either U or A.

The proof that the deduction rule is sound is as follows:

$$\begin{array}{lll} i & U \cup \{A\} \vdash B(a) & \\ i' & U \vdash A \to B(a) & \text{Inductive hypothesis, } i \\ i'+1 & U \vdash \forall x(A \to B) & \text{Generalization, } i' \\ i'+2 & U \vdash \forall x(A \to B) \to (A \to \forall x\, B) & \text{Axiom 5} \\ i'+3 & U \vdash A \to \forall x\, B & \text{MP, } i'+1, i'+2 \end{array}$$

The fact that a does not appear in U is used in line $i'+1$ and the fact that a does not appear in A is used in line $i'+2$. ■

8.3 Equivalence of $\mathcal{H}$ and $\mathcal{G}$

We prove that any theorem that can be proved in $\mathcal{G}$ can also be proved in $\mathcal{H}$. We already know how to transform propositional proofs in $\mathcal{G}$ to proofs in $\mathcal{H}$; what remains is to show that any application of the γ- and δ-rules in $\mathcal{G}$ can be transformed into a proof in $\mathcal{H}$.

Theorem 8.11 *The rule for a γ-formula can be simulated in $\mathcal{H}$.*

Proof Suppose that the rule:

$$\frac{U \vee \neg \forall x\, A(x) \vee \neg A(a)}{U \vee \neg \forall x\, A(x)}$$

was used. This can be simulated in $\mathcal{H}$ as follows:

$$\begin{array}{lll} 1. & \vdash \forall x\, A(x) \to A(a) & \text{Axiom 4} \\ 2. & \vdash \neg \forall x\, A(x) \vee A(a) & \text{PC 1} \\ 3. & \vdash U \vee \neg \forall x\, A(x) \vee A(a) & \text{PC 2} \\ 4. & \vdash U \vee \neg \forall x\, A(x) \vee \neg A(a) & \text{Assumption} \\ 5. & \vdash U \vee \neg \forall x\, A(x) & \text{PC 3, 4} \end{array}$$

  ■

Theorem 8.12 *The rule for a δ-formula can be simulated in $\mathcal{H}$.*

Proof Suppose that the rule:

$$\frac{U \vee A(a)}{U \vee \forall x A(x)}$$

was used. This can be simulated in $\mathcal{H}$ as follows:

1.	$\vdash U \vee A(a)$	Assumption
2.	$\vdash \neg U \to A(a)$	PC 1
3.	$\vdash \forall x(\neg U \to A(x))$	Gen. 2
4.	$\vdash \forall x(\neg U \to A(x)) \to (\neg U \to \forall x A(x))$	Axiom 5
5.	$\vdash \neg U \to \forall x A(x)$	MP 3, 4
6.	$\vdash U \vee \forall x A(x)$	PC 5

The use of Axiom 5 requires that a not occur in U, but we know that this holds by the corresponding condition on the rule for the δ-formula. ∎

Simulations in $\mathcal{G}$ of proofs in $\mathcal{H}$ are left as an exercise. From this follows:

Theorem 8.13 (Soundness and completeness) *The Hilbert system $\mathcal{H}$ is sound and complete.*

8.4   Proofs of Theorems in $\mathcal{H}$

We now give a series of theorems and proofs in $\mathcal{H}$.

The first two are elementary theorems using existential quantifiers.

Theorem 8.14 $\vdash A(a) \to \exists x A(x)$.

Proof

1.	$\vdash \forall x \neg A(x) \to \neg A(a)$	Axiom 4
2.	$\vdash A(a) \to \neg \forall x \neg A(x)$	PC 1
3.	$\vdash A(a) \to \exists x A(x)$	Definition ∃

∎

Theorem 8.15 $\vdash \forall x A(x) \to \exists x A(x)$.

Proof

1.	$\forall x A(x) \vdash \forall x A(x)$	Assumption
2.	$\forall x A(x) \vdash A(a)$	Axiom 4
3.	$\forall x A(x) \vdash A(a) \to \exists x A(x)$	Theorem 8.14
4.	$\forall x A(x) \vdash \exists x A(x)$	MP 2, 3
5.	$\vdash \forall x A(x) \to \exists x A(x)$	Deduction

∎

Theorem 8.16 $\vdash \forall x(A(x) \to B(x)) \to (\forall x A(x) \to \forall x B(x))$.

Proof

1.	$\forall x(A(x) \to B(x)), \forall x A(x) \vdash \forall x A(x)$	Assumption
2.	$\forall x(A(x) \to B(x)), \forall x A(x) \vdash A(a)$	Axiom 4
3.	$\forall x(A(x) \to B(x)), \forall x A(x) \vdash \forall x(A(x) \to B(x))$	Assumption
4.	$\forall x(A(x) \to B(x)), \forall x A(x) \vdash A(a) \to B(a)$	Axiom 4
5.	$\forall x(A(x) \to B(x)), \forall x A(x) \vdash B(a)$	PC 2, 4
6.	$\forall x(A(x) \to B(x)), \forall x A(x) \vdash \forall x B(x)$	Gen. 5
7.	$\forall x(A(x) \to B(x)) \vdash \forall x A(x) \to \forall x B(x)$	Deduction
8.	$\vdash \forall x(A(x) \to B(x)) \to (\forall x A(x) \to \forall x B(x))$	Deduction

∎

Rule 8.17 (Generalization)

$$\frac{\vdash A(a) \to B(a)}{\vdash \forall x A(x) \to \forall x B(x)}.$$

The next theorem was previously proved in the Gentzen system. Make sure that you understand why Axiom 5 can be used.

Theorem 8.18 $\vdash \exists x \forall y A(x, y) \to \forall y \exists x A(x, y)$.

Proof

1.	$\vdash A(a, b) \to \exists x A(x, b)$	Theorem 8.14
2.	$\vdash \forall y A(a, y) \to \forall y \exists x A(x, y)$	Gen 1
3.	$\vdash \neg \forall y \exists x A(x, y) \to \neg \forall y A(a, y)$	PC 2
4.	$\vdash \forall x(\neg \forall y \exists x A(x, y) \to \neg \forall y A(x, y))$	Gen. 3
5.	$\vdash (\forall x(\neg \forall y \exists x A(x, y) \to \neg \forall y A(x, y))) \to$	
	$\quad (\neg \forall y \exists x A(x, y) \to \forall x \neg \forall y A(x, y))$	Axiom 5
6.	$\vdash \neg \forall y \exists x A(x, y) \to \forall x \neg \forall y A(x, y)$	MP 4, 5
7.	$\vdash \neg \forall x \neg \forall y A(x, y) \to \forall y \exists x A(x, y)$	PC 6
8.	$\vdash \exists x \forall y A(x, y) \to \forall y \exists x A(x, y)$	Definition of $\exists$

∎

The proof of the following theorem is left as an exercise:

Theorem 8.19 *Let A be a formula that does not have x as a free variable.*

$$\vdash \forall x(A \to B(x)) \leftrightarrow (A \to \forall x B(x)),$$
$$\vdash \exists x(A \to B(x)) \leftrightarrow (A \to \exists x B(x)).$$

The name of a bound variable can be changed if necessary:

Theorem 8.20 $\vdash \forall x A(x) \leftrightarrow \forall y A(y)$.

Proof

1.	$\vdash \forall x A(x) \rightarrow A(a)$	Axiom 4
2.	$\vdash \forall y (\forall x A(x) \rightarrow A(y))$	Gen. 1
3.	$\vdash \forall x A(x) \rightarrow \forall y A(y)$	Axiom 5
4.	$\vdash \forall y A(y) \rightarrow \forall x A(x)$	Similarly
5.	$\vdash \forall x A(x) \leftrightarrow \forall y A(y)$	PC 3, 4

∎

The next theorem shows a non-obvious relation between the quantifiers.

Theorem 8.21 *Let B be a formula that does not have x as a free variable.*

$$\vdash \forall x (A(x) \rightarrow B) \leftrightarrow (\exists x A(x) \rightarrow B).$$

Proof

1.	$\forall x (A(x) \rightarrow B) \vdash \forall x (A(x) \rightarrow B)$	Assumption
2.	$\forall x (A(x) \rightarrow B) \vdash \forall x (\neg B \rightarrow \neg A(x))$	Exercise
3.	$\forall x (A(x) \rightarrow B) \vdash \neg B \rightarrow \forall x \neg A(x)$	Axiom 5
4.	$\forall x (A(x) \rightarrow B) \vdash \neg \forall x \neg A(x) \rightarrow B$	PC 3
5.	$\forall x (A(x) \rightarrow B) \vdash \exists x A(x) \rightarrow B$	Definition of ∃
6.	$\vdash \forall x (A(x) \rightarrow B) \rightarrow (\exists x A(x) \rightarrow B)$	Deduction
7.	$\exists x A(x) \rightarrow B \vdash \exists x A(x) \rightarrow B$	Assumption
8.	$\exists x A(x) \rightarrow B \vdash \neg \forall x \neg A(x) \rightarrow B$	Definition of ∃
9.	$\exists x A(x) \rightarrow B \vdash \neg B \rightarrow \forall x \neg A(x)$	PC 8
10.	$\exists x A(x) \rightarrow B \vdash \forall x (\neg B \rightarrow \neg A(x))$	Theorem 8.19
11.	$\exists x A(x) \rightarrow B \vdash \forall x (A(x) \rightarrow B)$	Exercise
12.	$\vdash \forall x (A(x) \rightarrow B) \leftrightarrow (\exists x A(x) \rightarrow B)$	PC 6, 11

∎

8.5 The C-Rule *

The C-rule is a rule of inference that is useful in proofs of existentially quantified formulas. The rule is the formalization of the argument: if there exists an object satisfying a certain property, let a be that object.

Definition 8.22 (C-Rule) The following rule may be used in a proof:

i	$U \vdash \exists x A(x)$	(an existentially quantified formula)
$i+1$	$U \vdash A(a)$	C-rule

provided that

- The constant a is new and does not appear in steps $1, \ldots, i$ of the proof.
- Generalization is never applied to a free variable or constant in the formula to which the C-rule is applied:

i	$U \vdash \exists x A(x, y)$	(an existentially quantified formula)
$i+1$	$U \vdash A(a, y)$	C-rule
	$\ldots$	
j	$U \vdash \forall y A(a, y)$	Illegal! ∎

For a proof that the rule is sound, see Mendelson (2009, Proposition 2.10). We use the C-Rule to give a more intuitive proof of Theorem 8.18.

Theorem 8.23 $\vdash \exists x \forall y A(x, y) \rightarrow \forall y \exists x A(x, y)$

Proof

1.	$\exists x \forall y A(x, y) \vdash \exists x \forall y A(x, y)$	Assumption
2.	$\exists x \forall y A(x, y) \vdash \forall y A(a, y)$	C-Rule
3.	$\exists x \forall y A(x, y) \vdash A(a, b)$	Axiom 4
4.	$\exists x \forall y A(x, y) \vdash \exists x A(x, b)$	Theorem 8.14
5.	$\exists x \forall y A(x, y) \vdash \forall y \exists x A(x, y)$	Gen. 4
6.	$\vdash \exists x \forall y A(x, y) \rightarrow \forall y \exists x A(x, y)$	Deduction ∎

The conditions in the C-rule are necessary. The first condition is similar to the condition on the deduction rule. The second condition is needed so that a formula that is true for one specific constant is not generalized for all values of a variable. Without the condition, we could prove the converse of Theorem 8.18, which is not a valid formula:

1.	$\forall x \exists y A(x, y) \vdash \forall x \exists y A(x, y)$	Assumption
2.	$\forall x \exists y A(x, y) \vdash \exists y A(a, y)$	Axiom 4
3.	$\forall x \exists y A(x, y) \vdash A(a, b)$	C-rule
4.	$\forall x \exists y A(x, y) \vdash \forall x A(x, b)$	Generalization (illegal!)
5.	$\forall x \exists y A(x, y) \vdash \exists y \forall x A(x, y)$	Theorem 8.14
6.	$\vdash \forall x \exists y A(x, y) \rightarrow \exists y \forall x A(x, y)$	Deduction

8.6 Summary

Gentzen and Hilbert deductive systems were defined for first-order logic. They are sound and complete. Be careful to distinguish between completeness and decidability. Completeness means that every valid formula has a proof. We can discover the proof by constructing a semantic tableau for its negation. However, we cannot decide if an arbitrary formula is valid and provable.

8.7 Further Reading

Our presentation is adapted from Smullyan (1968) and Mendelson (2009). Chapter X of (Smullyan, 1968) compares various proofs of completeness.

8.8 Exercises

8.1 Prove in $\mathscr{G}$:

$$\vdash \forall x(p(x) \to q(x)) \to (\exists x p(x) \to \exists x q(x)),$$
$$\vdash \exists x(p(x) \to q(x)) \leftrightarrow (\forall x p(x) \to \exists x q(x)).$$

8.2 Prove the soundness and completeness of $\mathscr{G}$ (Theorem 8.3).

8.3 Prove that Axioms 4 and 5 are valid.

8.4 Show that a proof in $\mathscr{H}$ can be simulated in $\mathscr{G}$.

8.5 Prove in $\mathscr{H}$: $\vdash \forall x(p(x) \to q) \leftrightarrow \forall x(\neg q \to \neg p(x))$.

8.6 Prove in $\mathscr{H}$: $\vdash \forall x(p(x) \leftrightarrow q(x)) \to (\forall x p(x) \leftrightarrow \forall x q(x))$.

8.7 Prove the theorems of Exercise 8.1 in $\mathscr{H}$.

8.8 Prove Theorem 8.19 in $\mathscr{H}$. Let A be a formula that does not have x as a free variable.

$$\vdash \forall x(A \to B(x)) \leftrightarrow (A \to \forall x B(x)),$$
$$\vdash \exists x(A \to B(x)) \leftrightarrow (A \to \exists x B(x)).$$

8.9 Let A be a formula built from the quantifiers and the Boolean operators $\neg, \vee, \wedge$ only. A', the dual of A is obtained by exchanging $\forall$ and $\exists$ and exchanging $\vee$ and $\wedge$. Prove that $\vdash A$ iff $\vdash \neg A'$.

References

E. Mendelson. *Introduction to Mathematical Logic (Fifth Edition)*. Chapman & Hall/CRC, 2009.
R.M. Smullyan. *First-Order Logic*. Springer-Verlag, 1968. Reprinted by Dover, 1995.

Chapter 9
First-Order Logic: Terms and Normal Forms

The formulas in first-order logic that we have defined are sufficient to express many interesting properties. Consider, for example, the formula:

$$\forall x \forall y \forall z\, (p\,(x, y) \land p\,(y, z) \rightarrow p\,(x, z)).$$

Under the interpretation:

$$\{\mathscr{Z}, \{<\}, \{\}\},$$

it expresses the true statement that the relation *less-than* is transitive in the domain of the integers. Suppose, now, that we want to express the following statement which is also true in the domain of integers:

$$\textit{for all } x, y, z : (x < y) \rightarrow (x + z < y + z).$$

The difference between this statement and the previous one is that it uses the *function* $+$.

Section 9.1 presents the extension of first-order logic to include functions. In Sect. 9.2, we describe a canonical form of formulas called *prenex conjunctive normal form*, which extends CNF to first-order logic. It enables us to define formulas as sets of clauses and to perform resolution on the clauses. In Sects. 9.3, 9.4, we show that canonical interpretations can be defined from *syntactical* objects like predicate and function letters.

9.1 First-Order Logic with Functions

9.1.1 Functions and Terms

Recall (Definition 7.8) that atomic formulas consist of an n-ary predicate followed by a list of n arguments that are variables and constants. We now generalize the arguments to include terms built from functions.

M. Ben-Ari, *Mathematical Logic for Computer Science*,
DOI 10.1007/978-1-4471-4129-7_9, © Springer-Verlag London 2012

Definition 9.1 Let $\mathcal{F}$ be a countable set of *function symbols*, where each symbol has an *arity* denoted by a superscript. *Terms* are defined recursively as follows:

- A variable, constant or 0-ary function symbol is a term.
- If f^n is an n-ary function symbol ($n > 0$) and $\{t_1, t_2, \ldots, t_n\}$ are terms, then $f^n(t_1, t_2, \ldots, t_n)$ is a term.

An *atomic formula* is an n-ary predicate followed by a list of n *arguments* where each argument t_i is a term: $p(t_1, t_2, \ldots, t_n)$.   ■

Notation

- We drop the word 'symbol' and use the word 'function' alone with the understanding that these are syntactical symbols only.
- By convention, functions are denoted by $\{f, g, h\}$ possibly with subscripts.
- The superscript denoting the arity of the function will not be written since the arity can be inferred from the number of arguments.
- Constant symbols are no longer needed since they are the same as 0-ary functions; nevertheless, we retain them since it is more natural to write $p(a, b)$ than to write $p(f_1, f_2)$.

Example 9.2 Examples of terms are

$$a, \quad x, \quad f(a, x), \quad f(g(x), y), \quad g(f(a, g(b))),$$

and examples of atomic formulas are

$$p(a, b), \quad p(x, f(a, x)), \quad q(f(a, a), f(g(x), g(x))).$$

    ■

9.1.2 Formal Grammar *

The following grammar defines terms and a new rule for atomic formulas:

term	$::= x$	for any $x \in \mathcal{V}$
term	$::= a$	for any $a \in \mathcal{A}$
term	$::= f^0$	for any $f^0 \in \mathcal{F}$
term	$::= f^n(\textit{term\_list})$	for any $f^n \in \mathcal{F}$
term_list	$::= \textit{term}$	
term_list	$::= \textit{term, term\_list}$	
atomic_formula	$::= p\,(\textit{term\_list})$	for any $p \in \mathcal{P}$.

It is required that the number of elements in a *term_list* be equal to the arity of the function or predicate symbol that is applied to the list.

9.1.3 Interpretations

The definition of interpretation in first-order logic is extended so that function symbols are interpreted by functions over the domain.

Definition 9.3 Let U be a set of formulas such that $\{p_1, \ldots, p_k\}$ are all the predicate symbols, $\{f_1^{n_1}, \ldots, f_l^{n_l}\}$ are all the function symbols and $\{a_1, \ldots, a_m\}$ are all the constant symbols appearing in U. An *interpretation* $\mathscr{I}$ is a 4-tuple:

$$\mathscr{I} = (D, \{R_1, \ldots, R_k\}, \{F_1^{n_1}, \ldots, F_l^{n_l}\}, \{d_1, \ldots, d_m\}),$$

consisting of a *non-empty* domain D, an assignment of an n_i-ary relation R_i on D to the n_i-ary predicate symbols p_i for $1 \leq i \leq k$, an assignment of an n_j-ary function $F_j^{n_j}$ on D to the function symbol $f_j^{n_j}$ for $1 \leq j \leq l$, and an assignment of an element $d_n \in D$ to the constant symbol a_n for $1 \leq n \leq m$. ∎

The rest of the semantical definitions in Sect. 7.3 go through unchanged, except for the meaning of an atomic formula. We give an outline and we leave the details as an exercise. In an interpretation $\mathscr{I}$, let $\mathscr{D}_\mathscr{I}$ be a map from terms to domain elements that satisfies:

$$\mathscr{D}_\mathscr{I}(f_i(t_1, \ldots, t_n)) = F_i(\mathscr{D}_\mathscr{I}(t_1), \ldots, \mathscr{D}_\mathscr{I}(t_n)).$$

Given an atomic formula $A = p_k(t_1, \ldots, t_n)$, $v_{\sigma_\mathscr{I}}(A) = T$ iff

$$(\mathscr{D}_\mathscr{I}(t_1), \ldots, \mathscr{D}_\mathscr{I}(d_n)) \in R_k.$$

Example 9.4 Consider the formula:

$$A = \forall x \forall y (p(x, y) \rightarrow p(f(x, a), f(y, a))).$$

We claim that the formula is true in the interpretation:

$$(\mathscr{Z}, \{\leq\}, \{+\}, \{1\}).$$

For arbitrary $m, n \in \mathscr{Z}$ assigned to x, y:

$$\mathscr{D}_\mathscr{I}(f(x, a)) = +(\mathscr{D}_\mathscr{I}(x), \mathscr{D}_\mathscr{I}(a)) = +(m, 1) = m + 1,$$
$$\mathscr{D}_\mathscr{I}(f(y, a)) = +(\mathscr{D}_\mathscr{I}(y), \mathscr{D}_\mathscr{I}(a)) = +(n, 1) = n + 1,$$

where we have changed to infix notation. p is assigned to the relation $\leq$ by $\mathscr{I}$ and $m \leq n$ implies $m + 1 \leq n + 1$ in $\mathscr{Z}$, so the formula is true for this assignment. Since m and n were arbitrary, the quantified formula A is true in this interpretation.

Here is another interpretation for the same formula A:

$$(\{\mathscr{S}^*\}, \{suffix\}, \{\cdot\}, \{\texttt{tuv}\}),$$

where $\mathcal{S}^*$ is the set of strings over some alphabet $\mathcal{S}$, *suffix* is the relation such that $(s_1, s_2) \in$ *suffix* iff s_1 is a suffix of s_2, $\cdot$ is the function that concatenates its arguments, and `tuv` is a string. The formula A is true for arbitrary s_1 and s_2 assigned to x and y. For example, if x is assigned `def` and y is assigned `abcdef`, then `deftuv` is a suffix of `abcdeftuv`.

A is not valid since it is falsified by the interpretation:

$$(\mathscr{Z}, \{>\}, \{\cdot\}, \{-1\}).$$

Obviously, $5 > 4$ does not imply $5 \cdot (-1) > 4 \cdot (-1)$.   ∎

9.1.4 Semantic Tableaux

The algorithm for building semantic tableaux for formulas of first-order logic with function symbols is almost the same as Algorithm 7.40 for first-order logic with constant symbols only. The difference is that any term, not just a constant, can be substituted for a variable. Definition 7.39 of a literal also needs to be generalized.

Definition 9.5

- A *ground term* is a term which does not contain any variables.
- A *ground atomic formula* is an atomic formula, all of whose terms are ground.
- A *ground literal* is a ground atomic formula or the negation of one.
- A *ground formula* is a quantifier-free formula, all of whose atomic formula are ground.
- A is a *ground instance* of a quantifier-free formula A' iff it can be obtained from A' by substituting ground terms for the (free) variables in A'.   ∎

Example 9.6 The terms a, $f(a, b)$, $g(b, f(a, b))$ are ground. $p(f(a, b), a)$ is a ground atomic formula and $\neg\, p(f(a, b), a)$ is a ground literal. $p(f(x, y), a)$ is not a ground atomic formula because of the variables x, y.   ∎

The construction of the semantic tableaux can be modified for formulas with functions. The rule for δ-formulas, which required that a set of formulas be instantiated with a new *constant*, must be replaced with a requirement that the instantiation be done with a new *ground term*. Therefore, we need to ensure that there exists an enumeration of ground terms. By definition, the sets of constant symbols and function symbols were assumed to be countable, but we must show that the set of ground terms constructed from them are also countable. The proof will be familiar to readers who have seen a proof that the set of rational is countable.

Theorem 9.7 *The set of ground terms is countable.*

Proof To simplify the notation, identify the constant symbols with the 0-ary function symbols. By definition, the set of function symbols is countable:

$$\{f_0, \ f_1, \ f_2, \ f_3, \ \ldots\}.$$

Clearly, for every n, there is a finite number k_n of ground terms of height at most n that can be constructed from the first n function symbols $\{f_0, \ldots, f_n\}$, where by the height of a formula we mean the height of its tree representation. For each n, place these terms in a sequence $T^n = (t_1^n, t_2^n, \ldots, t_{k_n}^n)$. The countable enumeration of all ground terms is obtained by concatenating these sequences:

$$t_1^0, \ldots, t_{k_0}^0, \quad t_1^1, \ldots, t_{k_1}^1, \quad t_1^2, \ldots, t_{k_2}^2, \quad \ldots$$

∎

Example 9.8 Let the first four function symbols be $\{a, b, f, g, \ldots\}$, where f is unary and g is binary. Figure 9.1 shows the first four sequences of ground terms (without duplicates). The point is not that one would actually carry out this construction; we only need the theoretical result that such an enumeration is possible.   ∎

$n = 1$ a

$n = 2$ b

$n = 3$ $f(a), \ f(b), \ f(f(a)), \ f(f(b))$

$n = 4$ $f(f(f(a))), \ f(f(f(b))),$

$g(a, a), \ g(a, b), \ g(a, f(a)), \ g(a, f(b)), \ g(a, f(f(a))), \ g(a, f(f(b))),$
six similar terms with b as the first argument of g,

$g(f(a), a), \ g(f(a), b), \ g(f(a), f(a)), \ g(f(a), f(b)),$
$g(f(a), f(f(a))), \ g(f(a), f(f(b))),$
six similar terms with $f(b)$ as the first argument of g,

$g(f(f(a)), a), \ g(f(f(a)), b), \ g(f(f(a)), f(a)), \ g(f(f(a)), f(b)),$
$g(f(f(a)), f(f(a))), \ g(f(f(a)), f(f(b))),$
six similar terms with $f(f(b))$ as the first argument of g,

$f(g(a, a)), \ f(g(a, b)), \ f(g(a, f(a))), \ f(g(a, f(b))),$
twelve similar terms with $b, \ f(a), \ f(b)$ as the first argument of g,

$f(f(g(a, a))), \ f(f(g(a, b))), \ f(f(g(b, a))), \ f(f(g(b, b))).$

Fig. 9.1 Finite sequences of terms

9.2 PCNF and Clausal Form

Recall that a formula of propositional logic is in conjunctive normal form (CNF) iff it is a conjunction of disjunctions of literals. A notational variant of CNF is clausal form: the formula is represented as a set of clauses, where each clause is a set of literals. We now proceed to generalize CNF to first-order logic by defining a normal form that takes the quantifiers into account.

Definition 9.9 A formula is in *prenex conjunctive normal form (PCNF)* iff it is of the form:

$$Q_1 x_1 \cdots Q_n x_n M$$

where the Q_i are quantifiers and M is a quantifier-free formula in CNF. The sequence $Q_1 x_1 \cdots Q_n x_n$ is the *prefix* and M is the *matrix*.   ∎

Example 9.10 The following formula is in PCNF:

$$\forall y \forall z ([p(f(y)) \vee \neg p(g(z)) \vee q(z)] \wedge [\neg q(z) \vee \neg p(g(z)) \vee q(y)]).$$

Definition 9.11 Let A be a *closed* formula in PCNF whose prefix consists only of *universal* quantifiers. The *clausal form* of A consists of the matrix of A written as a set of clauses.   ∎

Example 9.12 The formula in Example 9.10 is closed and has only universal quantifiers, so it can be written in clausal form as:

$$\{\{p(f(y)), \neg p(g(z)), q(z)\}, \{\neg q(z), \neg p(g(z)), q(y)\}\}.$$

                       ∎

9.2.1 Skolem's Theorem

In propositional logic, every formula is equivalent to one in CNF, but this is not true in first-order logic. However, a formula in first-order logic can be transformed into one in clausal form without modifying its satisfiability.

Theorem 9.13 (Skolem) *Let A be a closed formula. Then there exists a formula A' in clausal form such that $A \approx A'$.*

Recall that $A \approx A'$ means that A is satisfiable if and only if A' is satisfiable; that is, there *exists* a model for A if and only if there *exists* a model for A'. This is not the same as logically equivalence $A \equiv A'$, which means that for *all* models $\mathscr{I}$, $\mathscr{I}$ is a model for A if and only if it is a model for A'.

It is straightforward to transform A into a logically equivalent formula in PCNF. It is the removal of the existential quantifiers that causes the new formula not to be equivalent to the old one. The removal is accomplished by defining new function symbols. In $A = \forall x \exists y p(x, y)$, the quantifiers can be read: for all x, *produce* a value y associated with that x such that the predicate p is true. But our intuitive concept of a function is the same: $y = f(x)$ means that given x, f produces a value y associated with x. The existential quantifier can be removed giving $A' = \forall x p(x, f(x))$.

Example 9.14 Consider the interpretation:

$$\mathscr{I} = (\mathscr{Z}, \{>\}, \{\})$$

for the PCNF formula $A = \forall x \exists y p(x, y)$. Obviously, $\mathscr{I} \models A$.

The formula $A' = \forall x p(x, f(x))$ is obtained from A by removing the existential quantifier and replacing it with a function. Consider the following interpretation:

$$\mathscr{I}' = (\mathscr{Z}, \{>\}, \{F(x) = x + 1\}).$$

Clearly, $\mathscr{I}' \models A$ (just ignore the function), but $\mathscr{I}' \not\models A'$ since it is not true that $n > n + 1$ for all integers (in fact, for any integer). Therefore, $A' \not\equiv A$.

However, there *is* a model for A', for example:

$$\mathscr{I}'' = (\mathscr{Z}, \{>\}, \{F(x) = x - 1\}).$$

∎

The introduction of function symbols narrows the choice of models. The relations that interpret predicate symbols are *many-many*, that is, each x may be related to several y, while functions are *many-one*, that is, each x is related (mapped) to a single y, although different x's may be mapped into a single y. For example, if:

$$R = \{(1, 1), (1, 2), (1, 3), (2, 1), (2, 2), (2, 3)\},$$

then when trying to satisfy A, the whole relation R can be used, but for the clausal form A', only a functional subset of R such as $\{(1, 2), (2, 3)\}$ or $\{(1, 2), (2, 2)\}$ can be used to satisfy A'.

9.2.2 Skolem's Algorithm

We now give an algorithm to transform a formula A into a formula A' in clausal form and then prove that $A \approx A'$. The description of the transformation will be accompanied by a running example using the formula:

$$\forall x (p(x) \to q(x)) \to (\forall x p(x) \to \forall x q(x)).$$

Algorithm 9.15
Input: A closed formula A of first-order logic.
Output: A formula A' in clausal form such that $A \approx A'$.

- Rename bound variables so that no variable appears in two quantifiers.

$$\forall x(p(x) \to q(x)) \to (\forall y p(y) \to \forall z q(z)).$$

- Eliminate all binary Boolean operators other than $\vee$ and $\wedge$.

$$\neg \forall x(\neg p(x) \vee q(x)) \vee \neg \forall y p(y) \vee \forall z q(z).$$

- Push negation operators inward, collapsing double negation, until they apply to atomic formulas only. Use the equivalences:

$$\neg \forall x A(x) \equiv \exists x \neg A(x), \qquad \neg \exists x A(x) \equiv \forall x \neg A(x).$$

The example formula is transformed to:

$$\exists x(p(x) \wedge \neg q(x)) \vee \exists y \neg p(y) \vee \forall z q(z).$$

- Extract quantifiers from the matrix. Choose an *outermost* quantifier, that is, a quantifier in the matrix that is not within the scope of another quantifier still in the matrix. Extract the quantifier using the following equivalences, where Q is a quantifier and *op* is either $\vee$ or $\wedge$:

$$A \text{ op } Qx B(x) \equiv Qx(A \text{ op } B(x)), \qquad Qx A(x) \text{ op } B \equiv Qx(A(x) \text{ op } B).$$

Repeat until all quantifiers appear in the prefix and the matrix is quantifier-free. The equivalences are applicable because since no variable appears in two quantifiers. In the example, no quantifier appears within the scope of another, so we can extract them in any order, for example, x, y, z:

$$\exists x \exists y \forall z((p(x) \wedge \neg q(x)) \vee \neg p(y) \vee q(z)).$$

- Use the distributive laws to transform the matrix into CNF. The formula is now in PCNF.

$$\exists x \exists y \forall z((p(x) \vee \neg p(y) \vee q(z)) \wedge (\neg q(x) \vee \neg p(y) \vee q(z))).$$

- For every existential quantifier $\exists x$ in A, let $y_1, \ldots, y_n$ be the universally quantified variables *preceding* $\exists x$ and let f be a *new* n-ary function symbol. Delete $\exists x$ and replace every occurrence of x by $f(y_1, \ldots, y_n)$. If there are no universal quantifiers preceding $\exists x$, replace x by a new constant (0-ary function). These new function symbols are *Skolem functions* and the process of replacing existential quantifiers by functions is *Skolemization*. For the example formula we have:

$$\forall z((p(a) \vee \neg p(b) \vee q(z)) \wedge (\neg q(a) \vee \neg p(b) \vee q(z))),$$

where a and b are the Skolem functions (constants) corresponding to the existentially quantified variables x and y, respectively.

- The formula can be written in clausal form by dropping the (universal) quantifiers and writing the matrix as sets of clauses:

$$\{\{p(a),\ \neg p(b),\ q(z)\},\ \{\neg q(a),\ \neg p(b),\ q(z)\}\}.$$

■

Example 9.16 If we extract the quantifiers in the order z, x, y, the equivalent PCNF formula is:

$$\forall z \exists x \exists y ((p(x) \vee \neg p(y) \vee q(z)) \wedge (\neg q(x) \vee \neg p(y) \vee q(z))).$$

Since the existential quantifiers are preceded by a (single) universal quantifier, the Skolem functions are (unary) functions, not constants:

$$\forall z ((p(f(z)) \vee \neg p(g(z)) \vee q(z)) \wedge (\neg q(f(z)) \vee \neg p(g(z)) \vee q(z))),$$

which is:

$$\{\{p(f(z)),\ \neg p(g(z)),\ q(z)\},\ \{\neg q(f(z)),\ \neg p(g(z)),\ q(z)\}\}$$

in clausal form. ■

Example 9.17 Let us follow the entire transformation on another formula.

Original formula	$\exists x \forall y p(x, y) \rightarrow \forall y \exists x p(x, y)$
Rename bound variables	$\exists x \forall y p(x, y) \rightarrow \forall w \exists z p(z, w)$
Eliminate Boolean operators	$\neg \exists x \forall y p(x, y) \vee \forall w \exists z p(z, w)$
Push negation inwards	$\forall x \exists y \neg p(x, y) \vee \forall w \exists z p(z, w)$
Extract quantifiers	$\forall x \exists y \forall w \exists z (\neg p(x, y) \vee p(z, w))$
Distribute matrix	(no change)
Replace existential quantifiers	$\forall x \forall w (\neg p(x, f(x)) \vee p(g(x, w), w))$
Write in clausal form	$\{\{\neg p(x, f(x)),\ p(g(x, w), w)\}\}.$

f is unary because $\exists y$ is preceded by one universal quantifier $\forall x$, while g is binary because $\exists z$ is preceded by two universal quantifiers $\forall x$ and $\forall w$. ■

9.2.3 Proof of Skolem's Theorem

Proof of Skolem's Theorem The first five transformations of the algorithm can easily be shown to preserve equivalence. Consider now the replacement of an existential quantifier by a Skolem function. Suppose that:

$$\mathscr{I} \models \forall y_1 \cdots \forall y_n \exists x p(y_1, \ldots, y_n, x).$$

We need to show that there exists an interpretation $\mathscr{I}'$ such that:

$$\mathscr{I}' \models \forall y_1 \cdots \forall y_n\, p(y_1, \ldots, y_n, f(y_1, \ldots, y_n)).$$

$\mathscr{I}'$ is constructed by extending $\mathscr{I}$. Add a n-ary function F defined by: For all:

$$\{c_1, \ldots, c_n\} \subseteq D,$$

let $F(c_1, \ldots, c_n) = c_{n+1}$ for some $c_{n+1} \in D$ such that:

$$(c_1, \ldots, c_n, c_{n+1}) \in R_p,$$

where R_p is assigned to p in $\mathscr{I}$. Since $\mathscr{I} \models A$, there must be at least one element d of the domain such that $(c_1, \ldots, c_n, d) \in R_p$. We simply choose one of them arbitrarily and assign it to be the value of $F(c_1, \ldots, c_n)$. The Skolem function f was chosen to be a new function symbol not in A so the definition of F does not clash with any existing function in $\mathscr{I}$.

To show that:

$$\mathscr{I}' \models \forall y_1 \cdots \forall y_n\, p(y_1, \ldots, y_n, f(y_1, \ldots, y_n)),$$

let $\{c_1, \ldots, c_n\}$ be arbitrary domain elements. By construction, $F(c_1, \ldots, c_n) = c_{n+1}$ for some $c_{n+1} \in D$ and $v_{\mathscr{I}'}(p(c_1, \ldots, c_n, c_{n+1})) = T$. Since $c_1, \ldots, c_n$ were arbitrary:

$$v_{\mathscr{I}'}(\forall y_1 \cdots \forall y_n\, p(y_1, \ldots, y_n, f(y_1, \ldots, y_n))) = T.$$

This completes one direction of the proof of Skolem's Theorem. The proof of the converse (A is satisfiable if A' is satisfiable) is left as an exercise. ■

In practice, it is better to use a different transformation of a formula to clausal form. First, push all quantifiers *inward*, then replace existential quantifiers by Skolem functions and finally extract the remaining (universal) quantifiers. This ensures that the number of universal quantifiers preceding an existential quantifier is minimal and thus the arity of the Skolem functions is minimal.

Example 9.18 Consider again the formula of Example 9.17:

Original formula	$\exists x \forall y\, p(x, y) \to \forall y \exists x\, p(x, y)$
Rename bound variables	$\exists x \forall y\, p(x, y) \to \forall w \exists z\, p(z, w)$
Eliminate Boolean operators	$\neg \exists x \forall y\, p(x, y) \vee \forall w \exists z\, p(z, w)$
Push negation inwards	$\forall x \exists y \neg\, p(x, y) \vee \forall w \exists z\, p(z, w)$
Replace existential quantifiers	$\forall x \neg\, p(x, f(x)) \vee \forall w\, p(g(w), w)$
Extract universal quantifiers	$\forall x \forall w (\neg\, p(x, f(x)) \vee p(g(w), w))$
Write in clausal form	$\{\{\neg\, p(x, f(x)),\ p(g(w), w)\}\}.$

■

9.3 Herbrand Models

When function symbols are used to form terms, there is no easy way to describe the set of possible interpretations. The domain could be a numerical domain or a domain of data structures or almost anything else. The definition of even one function can choose to assign an arbitrary element of the domain to an arbitrary subset of arguments. In this section, we show that *for sets of clauses* there are canonical interpretations called *Herbrand models*, which are a relatively limited set of interpretations that have the following property: If a set of clauses has a model then it has an Herbrand model. Herbrand models will be central to the theoretical development of resolution in first-order logic (Sects. 10.1, 11.2); they also have interesting theoretical properties of their own (Sect. 9.4).

Herbrand Universes

The first thing that an interpretation needs is a domain. For this we use the set of syntactical terms that can be built from the symbols in the formula.

Definition 9.19 Let S be a set of clauses, $\mathscr{A}$ the set of constant symbols in S, and $\mathscr{F}$ the set of function symbols in S. H_S, *the Herbrand universe of S*, is defined inductively:

$$
\begin{aligned}
a_i &\in H_S & &\text{for } a_i \in \mathscr{A}, \\
f_i^0 &\in H_S & &\text{for } f_i^0 \in \mathscr{F}, \\
f_i^n(t_1, \ldots, t_n) &\in H_S & &\text{for } n > 1, \, f_i^n \in \mathscr{F}, t_j \in H_S.
\end{aligned}
$$

If there are no constant symbols or 0-ary function symbols in S, initialize the inductive definition of H_S with an arbitrary constant symbol a. ■

The Herbrand universe is just the set of ground terms that can be formed from symbols in S. Obviously, if S contains a function symbol, the Herbrand universe is infinite since $f(f(\ldots(a)\ldots)) \in H_S$.

Example 9.20 Here are some examples of Herbrand universes:

$$
\begin{aligned}
S_1 &= \{\{p(a), \neg p(b), q(z)\}, \{\neg p(b), \neg q(z)\}\} \\
H_{S_1} &= \{a, b\}
\end{aligned}
$$

$$
\begin{aligned}
S_2 &= \{\{\neg p(x, f(y))\}, \{p(w, g(w))\}\} \\
H_{S_2} &= \{a, f(a), g(a), f(f(a)), g(f(a)), f(g(a)), g(g(a)), \ldots\}
\end{aligned}
$$

$$
\begin{aligned}
S_3 &= \{\{\neg p(a, f(x, y))\}, \{p(b, f(x, y))\}\} \\
H_{S_3} &= \{a, b, f(a, a), f(a, b), f(b, a), f(b, b), f(a, f(a, a)), \ldots\}.
\end{aligned}
$$

■

Herbrand Interpretations

Now that we have a domain, an interpretation needs to specify assignments for the predicate, function and constant symbols. Clearly, we can let function and constant symbols be themselves: When interpreting $p(x, f(a))$, we interpret the term a by the domain element a and the term $f(a)$ by the domain element $f(a)$. Of course, this is somewhat confusing because we are using the same symbols for two purposes! Herbrand interpretations have complete flexibility in how they assign relations over the Herbrand universe to predicate symbols.

Definition 9.21 Let S be a formula in clausal where $\mathscr{P}_S = \{p_1, \ldots, p_k\}$ are the predicate symbols, $\mathscr{F}_S = \{f_1, \ldots, f_l\}$ the function symbols and $\mathscr{A}_S = \{a_1, \ldots, a_m\}$ the constant symbols appearing in S.

An *Herbrand interpretation* for S is:

$$\mathscr{I} = \{H_S, \{R_1, \ldots, R_k\}, \{f_1, \ldots, f_l\}, \mathscr{A}_S\},$$

where $\{R_1, \ldots, R_k\}$ are arbitrary relations of the appropriate arities over the domain H_S.

If f_i is a *function symbol* of arity j_i, then the *function* f_i is defined as follows: Let $\{t_1, \ldots, t_{j_i}\} \in H_S$; then $f_i(t_1, \ldots, t_{j_i}) = f_i(t_1, \ldots, t_{j_i})$.

An assignment in $\mathscr{I}$ is defined by:

$$
\begin{aligned}
v_{\mathscr{I}}(a) &= a, \\
v_{\mathscr{I}}(f(t_1, \ldots, t_n)) &= f(v_{\mathscr{I}}(t_1), \ldots, v_{\mathscr{I}}(t_n)).
\end{aligned}
$$

If $\mathscr{I} \models S$, then $\mathscr{I}$ is an *Herbrand model* for S.   ∎

Herbrand Bases

An alternate way of defining Herbrand models uses the following definition:

Definition 9.22 Let H_S be the Herbrand universe for S. B_S, *the Herbrand base for S*, is the set of *ground* atomic formulas that can be formed from predicate symbols in S and terms in H_S.   ∎

A relation over the Herbrand universe is simply a subset of the Herbrand base.

Example 9.23 The Herbrand base for S_3 from Example 9.20 is:

$$
\begin{aligned}
B_{S_3} = \ & \{p(a, f(a, a)), \ p(a, f(a, b)), \ p(a, f(b, a)), \ p(a, f(b, b)), \ \ldots, \\
& p(a, f(a, f(a, a))), \ \ldots, \\
& p(b, f(a, a)), \ p(b, f(a, b)), \ p(b, f(b, a)), \ p(b, f(b, b)), \ \ldots, \\
& p(b, f(a, f(a, a))), \ \ldots \}.
\end{aligned}
$$

An Herbrand interpretation for S_3 can be defined by giving the subset of the Herbrand base where the relation R_p holds, for example:

$$\{p(b, f(a, a)), \ p(b, f(a, b)), \ p(b, f(b, a)), \ p(b, f(b, b))\}.$$

∎

Herbrand Models Are Canonical

Theorem 9.24 *A set of clauses S has a model iff it has an Herbrand model.*

Proof Let:

$$\mathscr{I} = (D, \{R_1, \ldots, R_l\}, \{F_1, \ldots, F_m\}, \{d_1, \ldots, d_n\})$$

be an arbitrary model for S. Define the Herbrand interpretation $\mathscr{H}_{\mathscr{I}}$ by the following subset of the Herbrand base:

$$\{p_i(t_1, \ldots, t_n) \mid (v_{\mathscr{I}}(t_1), \ldots, v_{\mathscr{I}}(t_n)) \in R_i\},$$

where R_i is the relation assigned to p_i in $\mathscr{I}$. That is, a ground atom is in the subset of the Herbrand base if its value $v_{\mathscr{I}}(p_i(t_1, \ldots, t_n))$ is true when interpreted in the model $\mathscr{I}$.

We need to show that $\mathscr{H}_{\mathscr{I}} \models S$.

A set of clauses is a closed formula that is a conjunction of disjunctions of literals, so it suffices to show that one literal of each disjunction is in the subset, for each assignment of elements of the Herbrand universe to the variables.

Since $\mathscr{I} \models S$, $v_{\mathscr{I}}(S) = T$ so for all assignments by $\mathscr{I}$ to the variables and for *all* clauses $C_i \in S$, $v_{\mathscr{I}}(C_i) = T$. Thus for all clauses $C_i \in S$, there is *some* literal D_{ij} in the clause such that $v_{\mathscr{I}}(D_{ij}) = T$. But, by definition of the $\mathscr{H}_{\mathscr{I}}$, $v_{\mathscr{H}_{\mathscr{I}}}(D_{ij}) = T$ iff $v_{\mathscr{I}}(D_{ij}) = T$, from which follows $v_{\mathscr{H}_{\mathscr{I}}}(C_i) = T$ for all clauses $C_i \in S$, and $v_{\mathscr{H}_{\mathscr{I}}}(S) = T$. Thus $\mathscr{H}_{\mathscr{I}}$ is an Herbrand model for S.

The converse is trivial. ∎

Theorem 9.24 is *not* true if S is an arbitrary formula.

Example 9.25 Let $S = p(a) \wedge \exists x \neg p(x)$. Then

$$(\{0, 1\}, \{\{0\}\}, \{\}, \{0\})$$

is a model for S since $v(p(0)) = T$, $v(p(1)) = F$.

S has no Herbrand models since there are only two Herbrand interpretations and neither is a model:

$$(\{a\}, \{\{a\}\}, \{\}, \{a\}), \qquad (\{a\}, \{\{\}\}, \{\}, \{a\}).$$

∎

9.4 Herbrand's Theorem *

Herbrand's Theorem shows that questions of validity and provability in first-order logic can be reduced to questions about finite sets of ground atomic formulas. Although these results can now be obtained directly from the theory of semantic tableaux and Gentzen systems, we bring these results here (without proof) for their historical interest.

Consider a semantic tableau for an *unsatisfiable* formula in clausal form. The formula is implicitly a universally quantified formula:

$$A = \forall x_1 \cdots \forall x_n A'(x_1, \ldots, x_n)$$

whose matrix is a conjunction of disjunctions of literals. The only rules that can be used are the propositional rules for α- and β-formulas and the rule for γ-formulas with universal quantifiers. Since the closed tableau is finite, there will be a finite number of applications of the rule for γ-formulas.

Suppose that we construct the tableau by initially applying the rule for γ-formulas repeatedly for some sequence of ground terms, and only then apply the rule for α-formulas repeatedly in order to 'break up' each instantiation of the matrix A' into separate clauses. We obtain a node n labeled with a *finite* set of clauses. Repeated use of the rule for β-formulas on each clause (disjunction) will cause the tableau to eventually close because each leaf contains clashing literals. This sketch motivates the following theorem.

Theorem 9.26 (Herbrand's Theorem, semantic form 1) *A set of clauses S is unsatisfiable if and only if a finite set of ground instances of clauses of S is unsatisfiable.*

Example 9.27 The clausal form of the formula:

$$\neg [\forall x (p(x) \rightarrow q(x)) \rightarrow (\forall x p(x) \rightarrow \forall x q(x))]$$

(which is the negation of a valid formula) is:

$$S = \{\{\neg p(x), q(x)\}, \{p(y)\}, \{\neg q(z)\}\}.$$

The set of ground instances obtained by substituting a for each variable is:

$$S' = \{\{\neg p(a), q(a)\}, \{p(a)\}, \{\neg q(a)\}\}.$$

Clearly, S' is unsatisfiable because an application of the rule for the β-formula gives two nodes containing pairs of clashing literals: $\{\neg p(a), p(a), \neg q(a)\}$ and $\{q(a), p(a), \neg q(a)\}$. Theorem 9.26 states that the unsatisfiability of S' implies that S is unsatisfiable.   ∎

Since a formula is satisfiable if and only if its clausal form is satisfiable, the theorem can also be expressed as follows.

Theorem 9.28 (Herbrand's Theorem, semantic form 2) *A formula A is unsatisfiable if and only if a formula built from a finite set of ground instances of subformulas of A is unsatisfiable.*

Herbrand's Theorem transforms the problem of satisfiability within first-order logic into a problem of finding an appropriate set of ground terms and then checking satisfiability within propositional logic.

A syntactic form of Herbrand's theorem easily follows from the fact that a tableau can be turned upside-down to obtain a Gentzen proof of the formula.

Theorem 9.29 (Herbrand's Theorem, syntactic form) *A formula A of first-order logic is provable if and only if a formula built from a finite set of ground instances of subformulas of A is provable using only the axioms and inference rules of propositional logic.*

From Herbrand's theorem we obtain a relatively efficient *semi*-decision procedure for validity of formulas in first-order logic:

1. Negate the formula;
2. Transform into clausal form;
3. Generate a finite set of ground clauses;
4. Check if the set of ground clauses is unsatisfiable.

The first two steps are trivial and the last is not difficult because any convenient decision procedure for the propositional logic can be used by treating each distinct *ground* atomic formula as a distinct propositional letter. Unfortunately, we have no efficient way of generating a set of ground clauses that is likely to be unsatisfiable.

Example 9.30 Consider the formula $\exists x \forall y p(x, y) \rightarrow \forall y \exists x p(x, y)$.

Step 1: Negate it:

$$\neg (\exists x \forall y p(x, y) \rightarrow \forall y \exists x p(x, y)).$$

Step 2: Transform into clausal form:

$$\neg (\exists x \forall y p(x, y) \rightarrow \forall w \exists z p(z, w)))$$
$$\exists x \forall y p(x, y) \wedge \neg \forall w \exists z p(z, w)$$
$$\exists x \forall y p(x, y) \wedge \exists w \forall z \neg p(z, w)$$
$$\forall y p(a, y) \wedge \forall z \neg p(z, b)$$
$$\forall y \forall z (p(a, y) \wedge \neg p(z, b))$$
$$\{\{p(a, y)\}, \{\neg p(z, b)\}\}.$$

Step 3: Generate a finite set of ground clauses. In fact, there are only eight different ground clauses, so let us generate the entire set:

$$\{ \{p(a, a)\}, \{\neg p(a, b)\}, \quad \{p(a, b)\}, \{\neg p(b, b)\},$$
$$\{p(a, b)\}, \{\neg p(a, b)\}, \quad \{p(a, a)\}, \{\neg p(b, b)\} \}.$$

Step 4: Check if the set is unsatisfiable. Clearly, a set of clauses containing the clashing unit clauses $\{\neg\, p(a, b)\}$ and $\{p(a, b)\}$ is unsatisfiable.   ∎

The general resolution procedure described in the next chapter is a better approach because it does not need to generate a large number of ground clauses before checking for unsatisfiability. Instead, it generates clashing *non-ground* clauses and resolves them.

9.5 Summary

First-order logic with functions and terms is used to formalize mathematics. The theory of this logic (semantic tableaux, deductive systems, completeness, undecidability) is very similar to that of first-order logic without functions.

The clausal form of a formula in first-order logic is obtained by transforming the formula into an equivalent formula in prenex conjunctive normal form (PCNF) and then replacing existential quantifiers by Skolem functions. A formula in clausal form is satisfiable iff it has an Herbrand model, which is a model whose domain is the set of ground terms built from the function and constant symbols that appear in the formula. Herbrand's theorem states that questions of unsatisfiability and provability can be expressed in propositional logic applied to finite sets of ground formulas.

9.6 Further Reading

Functions and terms are used in all standard treatments of first-order logic such as Mendelson (2009) and Monk (1976). Herbrand models are discussed in texts on theorem-proving ((Fitting, 1996), (Lloyd, 1987)).

9.7 Exercises

9.1 Transform each of the following formulas to clausal form:

$$\forall x(p(x) \to \exists y q(y)),$$
$$\forall x \forall y(\exists z p(z) \wedge \exists u(q(x, u) \to \exists v q(y, v))),$$
$$\exists x(\neg \exists y p(y) \to \exists z(q(z) \to r(x))).$$

9.2 For the formulas of the previous exercise, describe the Herbrand universe and the Herbrand base.

9.3 Prove the converse direction of Skolem's Theorem (Theorem 9.13).

9.4 Prove:

$$\models \forall x A(x, f(x)) \ \to \ \forall x \exists y A(x, y),$$
$$\not\models \forall x \exists y A(x, y) \ \to \ \forall x A(x, f(x)).$$

9.5 Let $A(x_1, \ldots, x_n)$ be a formula with no quantifiers and no function symbols. Prove that $\forall x_1 \cdots \forall x_n A(x_1, \ldots, x_n)$ is satisfiable if and only if it is satisfiable in an interpretation whose domain has only one element.

References

M. Fitting. *First-Order Logic and Automated Theorem Proving (Second Edition)*. Springer, 1996.
J.W. Lloyd. *Foundations of Logic Programming (Second Edition)*. Springer, Berlin, 1987.
E. Mendelson. *Introduction to Mathematical Logic (Fifth Edition)*. Chapman & Hall/CRC, 2009.
J.D. Monk. *Mathematical Logic*. Springer, 1976.

Chapter 10
First-Order Logic: Resolution

Resolution is a sound and complete algorithm for propositional logic: a formula in clausal form is unsatisfiable if and only if the algorithm reports that it is unsatisfiable. For propositional logic, the algorithm is also a decision procedure for unsatisfiability because it is guaranteed to terminate. When generalized to first-order logic, resolution is still sound and complete, but it is not a decision procedure because the algorithm may not terminate.

The generalization of resolution to first-order logic will be done in two stages. First, we present *ground resolution* which works on ground literals as if they were propositional literals; then we present the *general resolution* procedure, which uses a highly efficient matching algorithm called *unification* to enable resolution on non-ground literals.

10.1 Ground Resolution

Rule 10.1 (Ground resolution rule) *Let C_1, C_2 be ground* clauses such that $l \in C_1$ *and* $l^c \in C_2$. C_1, C_2 *are said to be* clashing clauses *and to* clash *on the complementary literals* l, l^c. C, *the* resolvent *of* C_1 *and* C_2, *is the clause*:

$$Res(C_1, C_2) = (C_1 - \{l\}) \cup (C_2 - \{l^c\}).$$

C_1 *and* C_2 *are the* parent clauses *of* C. ■

M. Ben-Ari, *Mathematical Logic for Computer Science*,
DOI 10.1007/978-1-4471-4129-7_10, © Springer-Verlag London 2012

Example 10.2 Here is a tree representation of the ground resolution of two clauses. They clash on the literal $q(f(b))$:

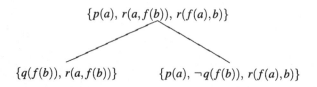

Theorem 10.3 *The resolvent C is satisfiable if and only if the parent clauses C_1 and C_2 are both satisfiable.*

Proof Let C_1 and C_2 be satisfiable clauses which clash on the literals l, l^c. By Theorem 9.24, they are satisfiable in an Herbrand interpretation $\mathcal{H}$. Let B be the subset of the Herbrand base that defines $\mathcal{H}$, that is,

$$B = \{p(c_1, \ldots, c_k) \mid v_H(p(c_1, \ldots, c_k)) = T\}$$

for ground terms c_i. Obviously, two complementary ground literals cannot both be elements of B. Suppose that $l \in B$. For C_2 to be satisfied in $\mathcal{H}$ there must be some *other* literal $l' \in C_2$ such that $l' \in B$. By construction of the resolvent C using the resolution rule, $l' \in C$, so $v_{\mathcal{H}}(C) = T$, that is, $\mathcal{H}$ is a model for C. A symmetric argument holds if $l^c \in B$.

Conversely, if C is satisfiable, it is satisfiable in an Herbrand interpretation $\mathcal{H}$ defined by a subset B of the Herbrand base. For some literal $l' \in C$, $l' \in B$. By the construction of the resolvent clause in the rule, $l' \in C_1$ or $l' \in C_2$ (or both). Suppose that $l' \in C_1$. We can extend the $\mathcal{H}$ to $\mathcal{H}'$ by defining $B' = B \cup \{l^c\}$. Again, by construction, $l \notin C$ and $l^c \notin C$, so $l \notin B$ and $l^c \notin B$ and therefore B' is well defined.

We need to show that C_1 and C_2 are both satisfied by $\mathcal{H}'$ defined by the Herbrand base B'. Clearly, since $l' \in C$, $l' \in B \subseteq B'$, so C_1 is satisfied in $\mathcal{H}'$. By definition, $l^c \in B'$, so C_2 is satisfied in $\mathcal{H}'$.

A symmetric argument holds if $l' \in C_2$.   ∎

The ground resolution procedure is defined like the resolution procedure for propositional logic. Given a set of ground clauses, the resolution step is performed repeatedly. The set of ground clauses is unsatisfiable iff some sequence of resolution steps produces the empty clause. We leave it as an exercise to show that ground resolution is a sound and complete refutation procedure for first-order logic.

Ground resolution is not a useful refutation procedure for first-order logic because the set of ground terms is infinite (assuming that there is even one function symbols). Robinson (1965) showed that how to perform resolution on clauses that are not ground by looking for substitutions that create clashing clauses. The definitions and algorithms are rather technical and are described in detail in the next two sections.

10.2 Substitution

We have been somewhat informal about the concept of substituting a term for a variable. In this section, the concept is formally defined.

Definition 10.4 A *substitution* of terms for variables is a set:

$$\{x_1 \leftarrow t_1, \ldots, x_n \leftarrow t_n\},$$

where each x_i is a distinct variable and each t_i is a term which is not identical to the corresponding variable x_i. The *empty substitution* is the empty set. ∎

Lower-case Greek letters $\{\lambda, \mu, \sigma, \theta\}$ will be used to denote substitutions. The empty substitution is denoted ε.

Definition 10.5 An *expression* is a term, a literal, a clause or a set of clauses. Let E be an expression and let $\theta = \{x_1 \leftarrow t_1, \ldots, x_n \leftarrow t_n\}$ be a substitution. An *instance* $E\theta$ of E is obtained by *simultaneously* replacing each occurrence of x_i in E by t_i. ∎

Example 10.6 Here is an expression (clause) $E = \{p(x), q(f(y))\}$ and a substitution $\theta = \{x \leftarrow y, \ y \leftarrow f(a)\}$, the instance obtained by performing the substitution is:

$$E\theta = \{p(y), q(f(f(a)))\}.$$

The word *simultaneously* in Definition 10.5 means that one does *not* substitute y for x in E to obtain:

$$\{p(y), q(f(y))\},$$

and *then* substitute $f(a)$ for y to obtain:

$$\{p(f(a)), q(f(f(a)))\}.$$

∎

The result of a substitution need not be a ground expression; at the extreme, a substitution can simply rename variables: $\{x \leftarrow y, z \leftarrow w\}$. Therefore, it makes sense to apply a substitution to an instance, because the instance may still have variables. The following definition shows how substitutions can be composed.

Definition 10.7 Let:

$$\theta = \{x_1 \leftarrow t_1, \ldots, x_n \leftarrow t_n\},$$
$$\sigma = \{y_1 \leftarrow s_1, \ldots, y_k \leftarrow s_k\}$$

be two substitutions and let $X = \{x_1, \ldots, x_n\}$ and $Y = \{y_1, \ldots, y_k\}$ be the sets of variables substituted for in θ and σ, respectively. $\theta\sigma$, the *composition of θ and σ*, is the substitution:

$$\theta\sigma = \{x_i \leftarrow t_i\sigma \mid x_i \in X, \ x_i \neq t_i\sigma\} \ \cup \ \{y_j \leftarrow s_j \mid y_j \in Y, \ y_j \notin X\}.$$

In words: apply the substitution σ to the terms t_i of θ (provided that the resulting substitutions do not collapse to $x_i \leftarrow x_i$) and then append the substitutions from σ whose variables do not already appear in θ.   ■

Example 10.8 Let:

$$
\begin{aligned}
E &= p(u, v, x, y, z), \\
\theta &= \{x \leftarrow f(y),\ y \leftarrow f(a),\ z \leftarrow u\}, \\
\sigma &= \{y \leftarrow g(a),\ u \leftarrow z,\ v \leftarrow f(f(a))\}.
\end{aligned}
$$

Then:

$$
\theta\sigma = \{x \leftarrow f(g(a)),\ y \leftarrow f(a),\ u \leftarrow z,\ v \leftarrow f(f(a))\}.
$$

The vacuous substitution $z \leftarrow z = (z \leftarrow u)\sigma$ has been deleted. The substitution $y \leftarrow g(a) \in \sigma$ has also been deleted since y already appears in θ. Once the substitution $y \leftarrow f(a)$ is performed, no occurrences of y remain in the expression. The instance obtained from the composition is:

$$
E(\theta\sigma) = p(z, f(f(a)), f(g(a)), f(a), z).
$$

Alternatively, we could have performed the substitution in two stages:

$$
\begin{aligned}
E\theta &= p(u, v, f(y), f(a), u), \\
(E\theta)\sigma &= p(z, f(f(a)), f(g(a)), f(a), z).
\end{aligned}
$$

We see that $E(\theta\sigma) = (E\theta)\sigma$.   ■

The result of performing two substitutions one after the other is the same as the result of computing the composition followed by a single substitution.

Lemma 10.9 *For any expression E and substitutions θ, σ, $E(\theta\sigma) = (E\theta)\sigma$.*

Proof Let E be a variable z. If z is not substituted for in θ or σ, the result is trivial. If $z = x_i$ for some $\{x_i \leftarrow t_i\}$ in θ, then $(z\theta)\sigma = t_i\sigma = z(\theta\sigma)$ by the definition of composition. If $z = y_j$ for some $\{y_j \leftarrow s_j\}$ in σ and $z \neq x_i$ for all i, then $(z\theta)\sigma = z\sigma = s_j = z(\theta\sigma)$.

The result follows by induction on the structure of E.   ■

We leave it as an exercise to show that composition is associative.

Lemma 10.10 *For any substitutions θ, σ, λ, $\theta(\sigma\lambda) = (\theta\sigma)\lambda$.*

10.3 Unification

The two literals $p(f(x), g(y))$ and $\neg p(f(f(a)), g(z))$ do not clash. However, under the substitution:

$$\theta_1 = \{x \leftarrow f(a), \ y \leftarrow f(g(a)), \ z \leftarrow f(g(a))\},$$

they become clashing (ground) literals:

$$p(f(f(a)), g(f(g(a)))), \qquad \neg p(f(f(a)), g(f(g(a)))).$$

The following simpler substitution:

$$\theta_2 = \{x \leftarrow f(a), \ y \leftarrow a, \ z \leftarrow a\}$$

also makes these literals clash:

$$p(f(f(a)), g(a)), \qquad \neg p(f(f(a)), g(a)).$$

Consider now the substitution:

$$\mu = \{x \leftarrow f(a), \ z \leftarrow y\}.$$

The literals that result are:

$$p(f(f(a)), g(y)), \qquad \neg p(f(f(a)), g(y)).$$

Any further substitution of a ground term for y will produce clashing ground literals.

The general resolution algorithm allows resolution on clashing literals that contain variables. By finding the simplest substitution that makes two literals clash, the resolvent is the most general result of a resolution step and is more likely to clash with another clause after a suitable substitution.

Definition 10.11 Let $U = \{A_1, \ldots, A_n\}$ be a set of atoms. A *unifier* θ is a substitution such that:

$$A_1\theta = \cdots = A_n\theta.$$

A *most general unifier (mgu)* for U is a unifier μ such that any unifier θ of U can be expressed as:

$$\theta = \mu\lambda$$

for some substitution λ. ∎

Example 10.12 The substitutions θ_1, θ_2, μ, above, are unifiers of the set of two atoms $\{p(f(x), g(y)), p(f(f(a)), g(z))\}$. The substitution μ is an mgu. The first two substitutions can be expressed as:

$$\theta_1 = \mu\{y \leftarrow f(g(a))\}, \qquad \theta_2 = \mu\{y \leftarrow a\}.$$

∎

Not all atoms are unifiable. It is clearly impossible to unify atoms whose predicate symbols are different such as $p(x)$ and $q(x)$, as well as atoms with terms whose outer function symbols are different such as $p(f(x))$ and $p(g(y))$. A more tricky case is shown by the atoms $p(x)$ and $p(f(x))$. Since x *occurs* within the larger term $f(x)$, any substitution—which must substitute simultaneously in both atoms—cannot unify them. It turns out that as long as these conditions do not hold the atoms will be unifiable.

We now describe and prove the correctness of an algorithm for unification by Martelli and Montanari (1982). Robinson's original algorithm is presented briefly in Sect. 10.3.4.

10.3.1 The Unification Algorithm

Trivially, two atoms are unifiable only if they have the same predicate letter of the same arity. Thus the unifiability of atoms is more conveniently described in terms of the unifiability of the arguments, that is, the *unifiability of a set of terms*. The set of terms to be unified will be written as a set of term equations.

Example 10.13 The unifiability of $\{p(f(x), g(y)),\ p(f(f(a)), g(z))\}$ is expressed by the set of term equations:

$$f(x) \;=\; f(f(a)),$$
$$g(y) \;=\; g(z).$$

∎

Definition 10.14 A set of term equations is in *solved form* iff:

- All equations are of the form $x_i = t_i$ where x_i is a variable.
- Each variable x_i that appears on the left-hand side of an equation does not appear elsewhere in the set.

A set of equations in solved form defines a substitution:

$$\{x_1 \leftarrow t_1,\ \ldots,\ x_n \leftarrow t_n\}.$$

∎

The following algorithm transforms a set of term equations into a set of equations in solved form, or reports if it is impossible to do so. In Sect. 10.3.3, we show that the substitution defined by the set in solved form is a most general unifier of the original set of term equations, and hence of the set of atoms from which the terms were taken.

Algorithm 10.15 (Unification algorithm)
Input: A set of term equations.
Output: A set of term equations in solved form or report *not unifiable*.

Perform the following transformations on the set of equations as long as any one of them is applicable:

1. Transform $t = x$, where t is not a variable, to $x = t$.
2. Erase the equation $x = x$.
3. Let $t' = t''$ be an equation where t', t'' are not variables.

   - If the outermost function symbols of t' and t'' are not identical, terminate the algorithm and report *not unifiable*.
   - Otherwise, replace the equation $f(t'_1, \ldots, t'_k) = f(t''_1, \ldots, t''_k)$ by the k equations $t'_1 = t''_1, \ldots, t'_k = t''_k$.

4. Let $x = t$ be an equation such that x has another occurrence in the set.

   - If x occurs in t and x differs from t, terminate the algorithm and report *not unifiable*.
   - Otherwise, transform the set by replacing all occurrences of x in other equations by t. ∎

Example 10.16 Consider the following set of two equations:

$$g(y) = x,$$
$$f(x, h(x), y) = f(g(z), w, z).$$

Apply rule 1 to the first equation and rule 3 to the second equation:

$$x = g(y),$$
$$x = g(z),$$
$$h(x) = w,$$
$$y = z.$$

Apply rule 4 to the second equation by replacing occurrences of x in other equations by $g(z)$:

$$g(z) = g(y),$$
$$x = g(z),$$
$$h(g(z)) = w,$$
$$y = z.$$

Apply rule 3 to the first equation:

$$z = y,$$
$$x = g(z),$$
$$h(g(z)) = w,$$
$$y = z.$$

Apply rule 4 to the last equation by replacing y by z in the first equation; next, erase the result $z = z$ using rule 2:

$$
\begin{aligned}
x &= g(z), \\
h(g(z)) &= w, \\
y &= z.
\end{aligned}
$$

Finally, transform the second equation by rule 1:

$$
\begin{aligned}
x &= g(z), \\
w &= h(g(z)), \\
y &= z.
\end{aligned}
$$

This successfully terminates the algorithm. We claim that:

$$
\mu = \{x \leftarrow g(z),\ w \leftarrow h(g(z)),\ y \leftarrow z\}
$$

is a most general unifier of the original set of equations. We leave it to the reader to check that the substitution does in fact unify the original set of term equations and further to check that the unifier:

$$
\theta = \{x \leftarrow g(f(a)),\ w \leftarrow h(g(f(a))),\ y \leftarrow f(a),\ z \leftarrow f(a)\}
$$

can be expressed as $\theta = \mu\{z \leftarrow f(a)\}$.   ■

10.3.2 The Occurs-Check

Algorithms for unification can be extremely inefficient because of the need to check the condition in rule 4, called the *occurs-check*.

Example 10.17 To unify the set of equations:

$$
\begin{aligned}
x_1 &= f(x_0, x_0), \\
x_2 &= f(x_1, x_1), \\
x_3 &= f(x_2, x_2), \\
&\cdots
\end{aligned}
$$

we successively create the equations:

$$
\begin{aligned}
x_2 &= f(f(x_0, x_0), f(x_0, x_0)), \\
x_3 &= f(f(f(x_0, x_0), f(x_0, x_0)), f(f(x_0, x_0), f(x_0, x_0))), \\
&\cdots
\end{aligned}
$$

The equation for x_i contains 2^i variables.   ■

In the application of unification to logic programming (Chap. 11), the occurs-check is simply ignored and the risk of an illegal substitution is taken.

10.3.3   The Correctness of the Unification Algorithm *

Theorem 10.18

- *Algorithm* 10.15 *terminates with the set of equations in solved form or it reports* not unifiable.
- *If the algorithm reports* not unifiable, *there is no unifier for the set of term equations.*
- *If the algorithm terminates successfully, the resulting set of equations is in solved form and defines the mgu:*

$$\mu = \{x_1 \leftarrow t_1, \ldots, x_n \leftarrow t_n\}.$$

Proof Obviously, rules 1–3 can be used only finitely many times without using rule 4. Let m be the number of distinct variables in the set of equations. Rule 4 can be used at most m times since it removes all occurrences, except one, of a variable and can never be used twice on the same variable. Thus the algorithm terminates.

The algorithm terminates with failure in rule 3 if the function symbols are distinct, and in rule 4 if a variable occurs within a term in the same equation. In both cases there can be no unifier.

It is easy to see that if it terminates successfully, the set of equations is in solved form. It remains to show that μ is a most general unifier.

Define a transformation as an *equivalence transformation* if it preserves sets of unifiers of the equations. Obviously, rules 1 and 2 are equivalence transformations. Consider now an application of rule 3 for $t' = f(t'_1, \ldots, t'_k)$ and $t'' = f(t''_1, \ldots, t''_k)$. If $t'\sigma = t''\sigma$, by the inductive definition of a term this can only be true if $t'_i\sigma = t''_i\sigma$ for all i. Conversely, if some unifier σ makes $t'_i = t''_i$ for all i, then σ is a unifier for $t' = t''$. Thus rule 3 is an equivalence transformation.

Suppose now that $t_1 = t_2$ was transformed into $u_1 = u_2$ by rule 4 on $x = t$. After applying the rule, $x = t$ remains in the set. So any unifier σ for the set must make $x\sigma = t\sigma$. Then, for $i = 1, 2$:

$$u_i\sigma = (t_i\{x \leftarrow t\})\sigma = t_i(\{x \leftarrow t\}\sigma) = t_i\sigma$$

by the associativity of substitution and by the definition of composition of substitution using the fact that $x\sigma = t\sigma$. So if σ is a unifier of $t_1 = t_2$, then $u_1\sigma = t_1\sigma = t_2\sigma = u_2\sigma$ and σ is a unifier of $u_1 = u_2$; it follows that rule 4 is an equivalence transformation.

Finally, the substitution defined by the set is an mgu. We have just proved that the original set of equations and the solved set of equations have the *same* set of unifiers. But the solved set itself defines a substitution (replacements of terms for variables) which is a unifier. Since the transformations were equivalence transformations, no equation can be removed from the set without destroying the property that it is a unifier. Thus any unifier for the set can only substitute more complicated terms for the same variables or substitute for other variables. That is, if μ is:

$$\mu = \{x_1 \leftarrow t_1, \ldots, x_n \leftarrow t_n\},$$

any other unifier σ can be written:

$$\sigma = \{x_1 \leftarrow t_1', \ldots, x_n \leftarrow t_n'\} \cup \{y_1 \leftarrow s_1, \ldots, y_m \leftarrow s_m\},$$

which is $\sigma = \mu\lambda$ for some substitution λ by definition of composition. Therefore, μ is an mgu.   ∎

The algorithm is nondeterministic because we may choose to apply a rule to any equation to which it is applicable. A deterministic algorithm can be obtained by specifying the order in which to apply the rules. One such deterministic algorithm is obtained by considering the set of equations as a queue. A rule is applied to the first element of the queue and then that equation goes to the end of the queue. If new equations are created by rule 3, they are added to the beginning of the queue.

Example 10.19 Here is Example 10.16 expressed as a queue of equations:

$$
\begin{array}{llll}
\langle\, g(y) = x, & f(x, h(x), y) = f(g(z), w, z) & & \rangle \\
\langle\, f(g(y), h(g(y))), y) = f(g(z), w, z), & x = g(y) & & \rangle \\
\langle\, g(y) = g(z), & h(g(y)) = w, & y = z, & x = g(y) \quad \rangle \\
\langle\, y = z, & h(g(y)) = w, & y = z, & x = g(y) \quad \rangle \\
\langle\, h(g(z)) = w, z = z, & x = g(z), & y = z \quad\quad\quad \rangle \\
\langle\, z = z, & x = g(z), & y = z, & w = h(g(z)) \,\rangle \\
\langle\, x = g(z), & y = z, & w = h(g(z)) & \rangle
\end{array}
$$

∎

10.3.4 Robinson's Unification Algorithm *

Robinson's algorithm appears in most other works on resolution so we present it here without proof (see Lloyd (1987, Sect. 1.4) for a proof).

Definition 10.20 Let A and A' be two atoms with the same predicate symbols. Considering them as sequences of symbols, let k be the leftmost position at which the sequences are different. The pair of terms $\{t, t'\}$ beginning at position k in A and A' is the *disagreement set* of the two atoms.   ∎

Algorithm 10.21 (Robinson's unification algorithm)
Input: Two atoms A and A' with the same predicate symbol.
Output: A most general unifier for A and A' or report *not unifiable*.

Initialize the algorithm by letting $A_0 = A$ and $A'_0 = A'$. Perform the following step repeatedly:

- Let $\{t, t'\}$ be the disagreement set of A_i, A'_i. If one term is a variable x_{i+1} and the other is a term t_{i+1} such that x_{i+1} does not occur in t_{i+1}, let $\sigma_{i+1} = \{x_{i+1} \leftarrow t_{i+1}\}$ and $A_{i+1} = A_i \sigma_{i+1}$, $A'_{i+1} = A'_i \sigma_{i+1}$.

If it is impossible to perform the step (because both elements of the disagreement set are compound terms or because the occurs-check fails), the atoms are not unifiable. If after some step $A_n = A'_n$, then A, A' are unifiable and the mgu is $\mu = \sigma_i \cdots \sigma_n$.   ∎

Example 10.22 Consider the pair of atoms:

$$A = p(g(y), \ f(x, h(x), y)), \qquad A' = p(x, \ f(g(z), w, z)).$$

The initial disagreement set is $\{x, g(y)\}$. One term is a variable which does not occur in the other so $\sigma_1 = \{x \leftarrow g(y)\}$, and:

$$\begin{aligned} A\sigma_1 &= p(g(y), \ f(g(y), h(g(y)), y)), \\ A'\sigma_1 &= p(g(y), \ f(g(z), w, z)). \end{aligned}$$

The next disagreement set is $\{y, z\}$ so $\sigma_2 = \{y \leftarrow z\}$, and:

$$\begin{aligned} A\sigma_1\sigma_2 &= p(g(z), \ f(g(z), h(g(z)), z)), \\ A'\sigma_1\sigma_2 &= p(g(z), \ f(g(z), w, z)). \end{aligned}$$

The third disagreement set is $\{w, h(g(z))\}$ so $\sigma_3 = \{w \leftarrow h(g(z))\}$, and:

$$\begin{aligned} A\sigma_1\sigma_2\sigma_3 &= p(g(z), \ f(g(z), h(g(z)), z)), \\ A'\sigma_1\sigma_2\sigma_3 &= p(g(z), \ f(g(z), h(g(z)), z)). \end{aligned}$$

Since $A\sigma_1\sigma_2\sigma_3 = A'\sigma_1\sigma_2\sigma_3$, the atoms are unifiable and the mgu is:

$$\mu = \sigma_1\sigma_2\sigma_3 = \{x \leftarrow g(z), \ y \leftarrow z, \ w \leftarrow h(g(z))\}.$$

∎

10.4 General Resolution

The resolution rule can be applied directly to non-ground clauses by performing unification as an integral part of the rule.

Definition 10.23 Let $L = \{l_1, \ldots, l_n\}$ be a set of literals. Then $L^c = \{l_1^c, \ldots, l_n^c\}$. ∎

Rule 10.24 (General resolution rule) *Let C_1, C_2 be clauses with no variables in common. Let $L_1 = \{l_1^1, \ldots, l_{n_1}^1\} \subseteq C_1$ and $L_2 = \{l_1^2, \ldots, l_{n_2}^2\} \subseteq C_2$ be subsets of literals such that L_1 and L_2^c can be unified by an mgu σ. C_1 and C_2 are said to be clashing clauses and to clash on the sets of literals L_1 and L_2. C, the resolvent of C_1 and C_2, is the clause:*

$$Res(C_1, C_2) = (C_1\sigma - L_1\sigma) \cup (C_2\sigma - L_2\sigma).$$

∎

Example 10.25 Given the two clauses:

$$\{p(f(x), g(y)), q(x, y)\}, \qquad \{\neg p(f(f(a)), g(z)), q(f(a), z)\},$$

an mgu for $L_1 = \{p(f(x), g(y))\}$ and $L_2^c = \{p(f(f(a)), g(z))\}$ is:

$$\{x \leftarrow f(a), y \leftarrow z\}.$$

The clauses resolve to give:

$$\{q(f(a), z), q(f(a), z)\} \quad = \quad \{q(f(a), z)\}.$$

∎

Clauses are *sets* of literals, so when taking the union of the clauses in the resolution rule, identical literals will be collapsed; this is called *factoring*.

The general resolution rule requires that the clauses have no variables in common. This is done by *standardizing apart*: renaming all the variables in one of the clauses before it is used in the resolution rule. All variables in a clause are implicitly universally quantified so renaming does not change satisfiability.

Example 10.26 To resolve the two clauses $p(f(x))$ and $\neg p(x)$, first rename the variable x of the second clause to x': $\neg p(x')$. An mgu is $\{x' \leftarrow f(x)\}$, and $p(f(x))$ and $\neg p(f(x))$ resolve to $\square$.

The clauses represent the formulas $\forall x p(f(x))$ and $\forall x \neg p(x)$, and it is obvious that their conjunction $\forall x p(f(x)) \wedge \forall x \neg p(x)$ is unsatisfiable.   ∎

Example 10.27 Let $C_1 = \{p(x), p(y)\}$ and $C_2 = \{\neg p(x), \neg p(y)\}$. Standardize apart so that $C_2' = \{\neg p(x'), \neg p(y')\}$. Let $L_1 = \{p(x), p(y)\}$ and let $L_2^c = \{p(x'), p(y')\}$; these sets have an mgu:

$$\sigma = \{y \leftarrow x, x' \leftarrow x, y' \leftarrow x\}.$$

The resolution rule gives:

$$\begin{aligned}
Res(C_1, C_2) &= (C_1\sigma - L_1\sigma) \cup (C_2'\sigma - L_2\sigma) \\
&= (\{p(x)\} - \{p(x)\}) \cup (\{\neg p(x)\} - \{\neg p(x)\}) \\
&= \square.
\end{aligned}$$

∎

In this example, the empty clause cannot be obtained without factoring, but we will talk about clashing literals rather than clashing sets of literals when no confusion will result.

Algorithm 10.28 (General Resolution Procedure)
Input: A set of clauses S.
Output: If the algorithm terminates, report that the set of clauses is *satisfiable* or *unsatisfiable*.
    Let $S_0 = S$. Assume that S_i has been constructed. *Choose* clashing clauses $C_1, C_2 \in S_i$ and let $C = Res(C_1, C_2)$. If $C = \square$, terminate and report that S is *unsatisfiable*. Otherwise, construct $S_{i+1} = S_i \cup \{C\}$. If $S_{i+1} = S_i$ for all possible pairs of clashing clauses, terminate and report S is *satisfiable*.   ∎

    While an unsatisfiable set of clauses will eventually produce $\square$ under a suitable systematic execution of the procedure, the existence of infinite models means that the resolution procedure on a satisfiable set of clauses may never terminate, so general resolution is not a decision procedure.

Example 10.29 Lines 1–7 contain a set of clauses. The resolution refutation in lines 8–15 shows that the set of clauses is unsatisfiable. Each line contains the resolvent, the mgu and the numbers of the parent clauses.

1.	$\{\neg p(x),\ q(x),\ r(x, f(x))\}$		
2.	$\{\neg p(x),\ q(x),\ r'(f(x))\}$		
3.	$\{p'(a)\}$		
4.	$\{p(a)\}$		
5.	$\{\neg r(a, y),\ p'(y)\}$		
6.	$\{\neg p'(x),\ \neg q(x)\}$		
7.	$\{\neg p'(x),\ \neg r'(x)\}$		
8.	$\{\neg q(a)\}$	$x \leftarrow a$	3, 6
9.	$\{q(a),\ r'(f(a))\}$	$x \leftarrow a$	2, 4
10.	$\{r'(f(a))\}$		8, 9
11.	$\{q(a),\ r(a, f(a))\}$	$x \leftarrow a$	1, 4
12.	$\{r(a, f(a))\}$		8, 11
13.	$\{p'(f(a))\}$	$y \leftarrow f(a)$	5, 12
14.	$\{\neg r'(f(a))\}$	$x \leftarrow f(a)$	7, 13
15.	$\{\square\}$		10, 14

                                                                            ∎

Example 10.30 Here is another example of a resolution refutation showing variable renaming and mgu's which do not produce ground clauses. The first four clauses form the set of clauses to be refuted.

1. $\{\neg p(x, y),\ p(y, x)\}$
2. $\{\neg p(x, y),\ \neg p(y, z),\ p(x, z)\}$
3. $\{p(x, f(x))\}$
4. $\{\neg p(x, x)\}$
3′. $\{p(x', f(x'))\}$ Rename 3
5. $\{p(f(x), x)\}$ $\sigma_1 = \{y \leftarrow f(x), x' \leftarrow x\}$ 1, 3′
3″. $\{p(x'', f(x''))\}$ Rename 3
6. $\{\neg p(f(x), z),\ p(x, z)\}$ $\sigma_2 = \{y \leftarrow f(x), x'' \leftarrow x\}$ 2, 3″
5‴. $\{p(f(x'''), x''')\}$ Rename 5
7. $\{p(x, x)\}$ $\sigma_3 = \{z \leftarrow x, x''' \leftarrow x\}$ 6, 5‴
4⁗. $\{\neg p(x'''', x'''')\}$ Rename 4
8. $\{\square\}$ $\sigma_4 = \{x'''' \leftarrow x\}$ 7, 4⁗

If we concatenate the substitutions, we get:

$$\sigma = \sigma_1\sigma_2\sigma_3\sigma_4 = \{y \leftarrow f(x), z \leftarrow x, x' \leftarrow x, x'' \leftarrow x, x''' \leftarrow x, x'''' \leftarrow x\}.$$

Restricted to the variables of the original clauses, $\sigma = \{y \leftarrow f(x), z \leftarrow x\}$.   ∎

10.5 Soundness and Completeness of General Resolution *

10.5.1 Proof of Soundness

We now show the soundness and completeness of resolution. The reader should review the proofs in Sect. 4.4 for propositional logic as we will just give the modifications that must be made to those proofs.

Theorem 10.31 (Soundness of resolution) *Let S be a set of clauses. If the empty clause $\square$ is derived when the resolution procedure is applied to S, then S is unsatisfiable.*

Proof We need to show that if the parent clauses are (simultaneously) satisfiable, so is the resolvent; since $\square$ is unsatisfiable, this implies that S must also be unsatisfiable. If parent clauses are satisfiable, there is an Herbrand interpretation $\mathscr{H}$ such that $v_{\mathscr{H}}(C_i) = T$ for $i = 1, 2$. The elements of the Herbrand base that satisfy C_1 and C_2 have the same form as ground atoms, so there must be a substitutions λ_i such that $C_i' = C_i\lambda_i$ are ground clauses and $v_{\mathscr{H}}(C_i') = T$.

Let C be the resolvent of C_1 and C_2. Then there is an mgu μ for C_1 and C_2 that was used to resolve the clauses. By definition of an mgu, there must substitutions θ_i such that $\lambda_i = \sigma\theta_i$. Then $C_i' = C_i\lambda_i = C_i(\sigma\theta_i) = (C_i\sigma)\theta_i$, which shows that $C_i\sigma$ is satisfiable in the same interpretation.

Let $l_1 \in C_1$ and $l_2^c \in C_2$ be the clashing literals used to derive C. Exactly one of $l_1\sigma$, $l_2^c\sigma$ is satisfiable in $\mathcal{H}$. Without loss of generality, suppose that $v_{\mathcal{H}}(l_1\sigma) = T$. Since $C_2\sigma$ is satisfiable, there must be a literal $l' \in C_2$ such that $l' \neq l_2^c$ and $v_{\mathcal{H}}(l'\sigma) = T$. But by the construction of the resolvent, $l' \in C$ so $v_{\mathcal{H}}(C) = T$.   ■

10.5.2 Proof of Completeness

Using Herbrand's theorem and semantic trees, we can prove that there is a *ground resolution refutation* of an unsatisfiable set of clauses. However, this does not generalize into a proof for general resolution because the concept of semantic trees does not generalize since the variables give rise to a potentially infinite number of elements of the Herbrand base. The difficulty is overcome by taking a ground resolution refutation and lifting it into a more abstract general refutation.

The problem is that several literals in C_1 or C_2 might collapse into one literal under the substitutions that produce the ground instances C_1' and C_2' to be resolved.

Example 10.32 Consider the clauses:

$$C_1 = \{p(x),\ p(f(y)),\ p(f(z)),\ q(x)\},$$
$$C_2 = \{\neg p(f(u)),\ \neg p(w),\ r(u)\}$$

and the substitution:

$$\{x \leftarrow f(a),\ y \leftarrow a,\ z \leftarrow a,\ u \leftarrow a,\ w \leftarrow f(a)\}.$$

The substitution results in the ground clauses:

$$C_1' = \{p(f(a)),\ q(f(a))\}, \qquad C_2' = \{\neg p(f(a)),\ r(a)\},$$

which resolve to: $C' = \{q(f(a)),\ r(a)\}$. The lifting lemma claims that there is a clause $C = \{q(f(u)),\ r(u)\}$ which is the resolvent of C_1 and C_2, such that C' is a ground instance of C. This can be seen by using the unification algorithm to obtain an mgu:

$$\{x \leftarrow f(u),\ y \leftarrow u,\ z \leftarrow u,\ w \leftarrow f(u)\}$$

of C_1 and C_2, which then resolve giving C.   ■

Theorem 10.33 (Lifting Lemma) *Let C_1', C_2' be ground instances of C_1, C_2, respectively. Let C' be a ground resolvent of C_1' and C_2'. Then there is a resolvent C of C_1 and C_2 such that C' is a ground instance of C.*

The relationships among the clauses are displayed in the following diagram.

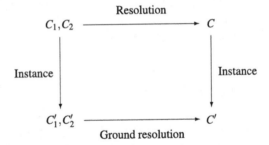

Proof The steps of the proof for Example 10.32 are shown in Fig. 10.1.

First, standardize apart so that the names of the variables in C_1 are different from those in C_2.

Let $l \in C'_1$, $l^c \in C'_2$ be the clashing literals in the ground resolution. Since C'_1 is an instance of C_1 and $l \in C'_1$, there must be a set of literals $L_1 \subseteq C_1$ such that l is an instance of each literal in L_1. Similarly, there must a set $L_2 \subseteq C_2$ such that l^c is an instance of each literal in L_2. Let λ_1 and λ_2 mgu's for L_1 and L_2, respectively, and let $\lambda = \lambda_1 \cup \lambda_2$. λ is a well-formed substitution since L_1 and L_2 have no variables in common.

By construction, $L_1\lambda$ and $L_2\lambda$ are sets which contain a single literal each. These literals have clashing ground instances, so they have a mgu σ. Since $L_i \subseteq C_i$, we have $L_i\lambda \subseteq C_i\lambda$. Therefore, $C_1\lambda$ and $C_2\lambda$ are clauses that can be made to clash under the mgu σ. It follows that they can be resolved to obtain clause C:

$$C = ((C_1\lambda)\sigma - (L_1\lambda)\sigma) \cup ((C_2\lambda)\sigma - (L_2\lambda)\sigma).$$

By the associativity of substitution (Theorem 10.10):

$$C = (C_1(\lambda\sigma) - L_1(\lambda\sigma)) \cup (C_2(\lambda\sigma) - (L_2(\lambda\sigma)).$$

C is a resolvent of C_1 and C_2 provided that $\lambda\sigma$ is an mgu of L_1 and L_2^c. But λ is already reduced to equations of the form $x \leftarrow t$ for distinct variables x and σ is constructed to be an mgu, so $\lambda\sigma$ is a reduced set of equations, all of which are necessary to unify L_1 and L_2^c. Hence $\lambda\sigma$ is an mgu.

Since C'_1 and C'_2 are ground instances of C_1 and C_2:

$$C'_1 = C_1\theta_1 = C_1\lambda\sigma\theta'_1 \qquad C'_2 = C_2\theta_2 = C_2\lambda\sigma\theta'_2$$

for some substitutions $\theta_1, \theta_2, \theta'_1, \theta'_2$. Let $\theta' = \theta'_1 \cup \theta'_2$. Then $C' = C\theta'$ and C' is a ground instance of C. ∎

Theorem 10.34 (Completeness of resolution) *If a set of clauses is unsatisfiable, the empty clause □ can be derived by the resolution procedure.*

Proof The proof is by induction on the semantic tree for the set of clauses S. The definition of semantic tree is modified as follows:

$$
\begin{aligned}
C_1 &= \{p(x),\ p(f(y)),\ p(f(z)),\ q(x)\} \\
C_2 &= \{\neg\, p(f(u)),\ \neg\, p(w),\ r(u)\}
\end{aligned}
$$

$$
\begin{aligned}
\theta_1 &= \{x \leftarrow f(a),\ y \leftarrow a,\ z \leftarrow a\} \\
\theta_2 &= \{u \leftarrow a,\ w \leftarrow f(a)\}
\end{aligned}
$$

$$
\begin{aligned}
C_1' &= C_1\theta_1 = \{p(f(a)),\ q(f(a))\} \\
C_2' &= C_2\theta_2 = \{\neg\, p(f(a)),\ r(a)\} \\
C' &= Res(C_1, C_2) = \{q(f(a)),\ r(a)\}
\end{aligned}
$$

$$
\begin{aligned}
L_1 &= \{p(x),\ p(f(y)),\ p(f(z))\} \\
\lambda_1 &= \{x \leftarrow f(y),\ z \leftarrow y\} \\
L_1\lambda_1 &= \{p(f(y))\}
\end{aligned}
$$

$$
\begin{aligned}
L_2 &= \{\neg\, p(f(u)),\ \neg\, p(w)\} \\
\lambda_2 &= \{w \leftarrow f(u)\} \\
L_2\lambda_2 &= \{\neg\, p(f(u))\}
\end{aligned}
$$

$$
\begin{aligned}
\lambda &= \lambda_1 \cup \lambda_2 = \{x \leftarrow f(y),\ z \leftarrow y,\ w \leftarrow f(u)\} \\
L_1\lambda &= \{p(f(y))\} \\
C_1\lambda &= \{p(f(y)),\ q(f(y))\} \\
L_2\lambda &= \{\neg\, p(f(u))\} \\
C_2\lambda &= \{\neg\, p(f(u)),\ r(u)\}
\end{aligned}
$$

$$
\begin{aligned}
\sigma &= \{u \leftarrow y\} \\
C &= Res(C_1\lambda, C_2\lambda) = \{q(f(y)),\ r(y)\},\ \text{using } \sigma
\end{aligned}
$$

$$
\begin{aligned}
\lambda\sigma &= \{x \leftarrow f(y),\ z \leftarrow y,\ w \leftarrow f(y),\ u \leftarrow y\} \\
C_1\lambda\sigma &= \{p(f(y)),\ q(f(y))\} \\
C_2\lambda\sigma &= \{\neg\, p(f(y)),\ r(y)\} \\
C &= Res(C_1, C_2) = \{q(f(y)),\ r(y)\},\ \text{using } \lambda\sigma
\end{aligned}
$$

$$
\begin{aligned}
\theta_1' &= \{y \leftarrow a\} \\
C_1' &= C_1\theta_1 = \{p(f(a)),\ q(f(a))\} = C_1\lambda\sigma\theta_1 \\
\theta_2' &= \{y \leftarrow a\} \\
C_2' &= C_2\theta_2 = \{\neg\, p(f(a)),\ r(a)\} = C_2\lambda\sigma\theta_2
\end{aligned}
$$

$$
\begin{aligned}
\theta' &= \{y \leftarrow a\} \\
C' &= Res(C_1',\ C_2') = \{q(f(a)),\ r(a)\}
\end{aligned}
$$

Fig. 10.1 Example for the lifting lemma

A node is a failure node if the (partial) interpretation defined by a branch falsifies some *ground instance* of a clause in S.

The critical step in the proof is showing that an inference node n can be associated with the resolvent of the clauses on the two failure nodes n_1, n_2 below it. Suppose that C_1, C_2 are associated with the failure nodes. Then there must be ground instances C'_1, C'_2 which are falsified at the nodes. By construction of the semantic tree, C'_1 and C'_2 are clashing clauses. Hence they can be resolved to give a clause C' which is falsified by the interpretation at n. By the Lifting Lemma, there is a clause C such that C is the resolvent of C'_1 and C'_2, and C' is a ground instance of C. Hence C is falsified at n and n (or an ancestor of n) is a failure node.   ∎

10.6 Summary

General resolution has proved to be a successful method for automated theorem proving in first-order logic. The key to its success is the unification algorithm. There is a large literature on strategies for choosing which clauses to resolve, but that is beyond the scope of this book. In Chap. 11 we present *logic programming*, in which programs are written as formulas in a restricted clausal form. In logic programming, unification is used to compose and decompose data structures, and computation is carried out by an appropriately restricted form of resolution that is very efficient.

10.7 Further Reading

Loveland (1978) is a classic book on resolution; a more modern one is Fitting (1996). Our presentation of the unification algorithm is taken from Martelli and Montanari (1982). Lloyd (1987) presents resolution in the context of logic programming that is the subject of the next chapter.

10.8 Exercises

10.1 Prove that ground resolution is sound and complete.

10.2 Let:
$$\theta = \{x \leftarrow f(g(y)),\ y \leftarrow u,\ z \leftarrow f(y)\},$$
$$\sigma = \{u \leftarrow y,\ y \leftarrow f(a),\ x \leftarrow g(u)\},$$
$$E = p(x, f(y), g(u), z).$$

Show that $E(\theta\sigma) = (E\theta)\sigma$.

10.3 Prove that the composition of substitutions is associative (Lemma 10.10).

10.4 Unify the following pairs of atomic formulas, if possible.

$$p(a, x, f(g(y))), \qquad p(y, f(z), f(z)),$$
$$p(x, g(f(a)), f(x)), \qquad p(f(a), y, y),$$
$$p(x, g(f(a)), f(x)), \qquad p(f(y), z, y),$$
$$p(a, x, f(g(y))), \qquad p(z, h(z, u), f(u)).$$

10.5 A substitution $\theta = \{x_1 \leftarrow t_1, \ldots, x_n \leftarrow t_n\}$ is *idempotent* iff $\theta = \theta\theta$. Let V be the set of variables occurring in the terms $\{t_1, \ldots, t_n\}$. Prove that θ is idempotent iff $V \cap \{x_1, \ldots, x_n\} = \emptyset$. Show that the mgu's produced by the unification algorithm is idempotent.

10.6 Try to unify the set of term equations:

$$x = f(y), \qquad y = g(x).$$

What happens?

10.7 Show that the composition of substitutions is not commutative: $\theta_1\theta_2 \neq \theta_2\theta_2$ for some θ_1, θ_2.

10.8 Unify the atoms in Example 10.13 using both term equations and Robinson's algorithm.

10.9 Let S be a finite set of expressions and θ a unifier of S. Prove that θ is an idempotent mgu iff for every unifier σ of S, $\sigma = \theta\sigma$.

10.10 Prove the validity of (some of) the equivalences in by resolution refutation of their negations.

References

M. Fitting. *First-Order Logic and Automated Theorem Proving (Second Edition)*. Springer, 1996.

J.W. Lloyd. *Foundations of Logic Programming (Second Edition)*. Springer, Berlin, 1987.

D.W. Loveland. *Automated Theorem Proving: A Logical Basis*. North-Holland, Amsterdam, 1978.

A. Martelli and U. Montanari. An efficient unification algorithm. *ACM Transactions on Programming Languages and Systems*, 4:258–282, 1982.

J.A. Robinson. A machine-oriented logic based on the resolution principle. *Journal of the ACM*, 12:23–41, 1965.

Chapter 11
First-Order Logic: Logic Programming

Resolution was originally developed as a method for automatic theorem proving. Later, it was discovered that a restricted form of resolution can be used for programming a computation. This approach is called *logic programming*. A program is expressed as a set of clauses and a query is expressed as an additional clause that can clash with one or more of the program clauses. The query is assumed to be the negation of result of the program. If a refutation succeeds, the query is not a logical consequence of the program, so its negation must be a logical consequence. Unifications done during the refutation provide answers to the query in addition to the simple fact that the negation of the query is true.

In this chapter we give an overview of logic programming. First, we work through an example for motivation. In the following section, we define SLD-resolution, which is the formal system most often used in logical programming. Section 11.4 is an introduction to Prolog, a widely used language for logic programming. The supplementary materials that can be downloaded contain Prolog implementations of most of the algorithms in this book.

11.1 From Formulas in Logic to Logic Programming

Consider a deductive system with axioms of two forms. One form is a universally-closed predicate:

$$\forall x(x + 0 = x).$$

The other form is a universally-closed implication where the premise is a conjunction:

$$\forall x \forall y \forall z(x \leq y \land y \leq z \rightarrow x \leq z).$$

In clausal form, the first form is a single positive literal:

$$x + 0 = x,$$

M. Ben-Ari, *Mathematical Logic for Computer Science*,
DOI 10.1007/978-1-4471-4129-7_11, © Springer-Verlag London 2012

whereas the second form is a clause all of whose literals are negative except for the last one which is positive:

$$\neg (x \le y) \lor \neg (y \le z) \lor (x \le z).$$

These types of clauses are called *program clauses*.

Suppose now that we have a set of program clauses and we want to prove that some formula:

$$G_1 \land \cdots \land G_n$$

is a logical consequence of the set. This can be done by taking the negation of the formula:

$$\neg (G_1 \land \cdots \land G_n) \equiv \neg G_1 \lor \cdots \lor \neg G_n$$

and refuting it by resolution with the program clauses.

The formula $\neg G_1 \lor \cdots \lor \neg G_n$, called a *goal clause*, consists entirely of negative literals, so it can only clash on the single positive literal of a program clause. Let:

$$B_1 \lor \neg B_2 \lor \cdots \lor \neg B_m$$

be a program clause such that G_1 and B_1 can be unified by mgu σ. The resolvent is:

$$(\neg G_2 \lor \cdots \lor \neg G_n \ \lor \ \neg B_2 \lor \cdots \lor \neg B_m)\sigma,$$

which is again a goal clause with no positive literals. We can continue resolving goal clauses with the program clauses until a unit (negative) goal clause remains that clashes with a unit (positive) program clause, resulting in the empty clause and terminating the refutation.

The sequence of resolution steps will generate a sequence of substitutions used to unify the literals and these substitutions become the answer to the query. Let us see how this is done in an example.

Refuting a Goal Clause

Consider a fragment of the theory of strings with a single binary function symbol for concatenation denoted by the infix operator · and three predicates:

- *substr*(x, y)—x is a substring of y,
- *prefix*(x, y)—x is a prefix of y,
- *suffix*(x, y)—x is a suffix of y.

The axioms of the theory are:

1. $\forall x\; substr(x, x)$,
2. $\forall x \forall y\; suffix(x, y \cdot x)$,
3. $\forall x \forall y\; prefix(x, x \cdot y)$,
4. $\forall x \forall y \forall z\; (substr(x, y) \wedge suffix(y, z) \rightarrow substr(x, z))$,
5. $\forall x \forall y \forall z\; (substr(x, y) \wedge prefix(y, z) \rightarrow substr(x, z))$.

They can be written in clausal form as:

1. $substr(x, x)$,
2. $suffix(x, y \cdot x)$,
3. $prefix(x, x \cdot y)$,
4. $\neg substr(x, y) \vee \neg suffix(y, z) \vee substr(x, z)$,
5. $\neg substr(x, y) \vee \neg prefix(y, z) \vee substr(x, z)$.

We can prove the formula:

$$substr(a \cdot b \cdot c,\; a \cdot a \cdot b \cdot c \cdot c)$$

by refuting its negation:

$$\neg substr(a \cdot b \cdot c,\; a \cdot a \cdot b \cdot c \cdot c).$$

Here is a refutation, where the parent clauses of each resolvent are given in the right-hand column, together with the substitutions needed to unify the clashing clauses:

6. $\neg substr(a \cdot b \cdot c,\; a \cdot a \cdot b \cdot c \cdot c)$
7. $\neg substr(a \cdot b \cdot c,\; y1) \vee \neg suffix(y1,\; a \cdot a \cdot b \cdot c \cdot c)$

$$6, 4, \{x \leftarrow a \cdot b \cdot c,\; y \leftarrow y1,\; z \leftarrow a \cdot a \cdot b \cdot c \cdot c\}$$

8. $\neg substr(a \cdot b \cdot c,\; a \cdot b \cdot c \cdot c)$

$$7, 2, \{x \leftarrow a \cdot b \cdot c \cdot c,\; y \leftarrow a,\; y1 \leftarrow a \cdot b \cdot c \cdot c\}$$

9. $\neg substr(a \cdot b \cdot c,\; y2) \vee \neg prefix(y2,\; a \cdot b \cdot c \cdot c)$

$$8, 5, \{x \leftarrow a \cdot b \cdot c,\; y \leftarrow y2,\; z \leftarrow a \cdot b \cdot c \cdot c\}$$

10. $\neg substr(a \cdot b \cdot c,\; a \cdot b \cdot c)$

$$9, 3, \{x \leftarrow a \cdot b \cdot c,\; y \leftarrow c,\; y2 \leftarrow a \cdot b \cdot c\}$$

11. $\square$

$$10, 1, \{x \leftarrow a \cdot b \cdot c\}$$

Answer Substitutions

This refutation is not very exciting; all it does is check if $substr(a \cdot b \cdot c,\; a \cdot a \cdot b \cdot c \cdot c)$ is true or not. Suppose, however, that instead of determining whether a *ground* goal clause is a logical consequence of the axioms, we try to determine if the existentially quantified formula $\exists w\; substr(w,\; a \cdot a \cdot b \cdot c \cdot c)$ is a logical consequence of the axioms. In terms of resolution we try to refute the negation of the formula:

$$\neg (\exists w\; substr(w,\; a \cdot a \cdot b \cdot c \cdot c)) \equiv \forall w \neg\; substr(w,\; a \cdot a \cdot b \cdot c \cdot c).$$

A universally quantified literal is a clause so a resolution refutation of this clause together with the clauses from the axioms can be attempted:

6. $\neg\, substr(w,\ a \cdot a \cdot b \cdot c \cdot c)$

7. $\neg\, substr(w,\ y1) \vee \neg\, suffix(y1,\ a \cdot a \cdot b \cdot c \cdot c)$

$$6, 4, \{x \leftarrow w,\ y \leftarrow y1,\ z \leftarrow a \cdot a \cdot b \cdot c \cdot c\}$$

8. $\neg\, substr(w,\ a \cdot b \cdot c \cdot c)$

$$7, 2, \{x \leftarrow a \cdot b \cdot c \cdot c,\ y \leftarrow a,\ y1 \leftarrow a \cdot b \cdot c \cdot c\}$$

9. $\neg\, substr(w,\ y2) \vee \neg\, prefix(y2,\ a \cdot b \cdot c \cdot c)$

$$8, 5, \{x \leftarrow w,\ y \leftarrow y2,\ z \leftarrow a \cdot b \cdot c \cdot c\}$$

10. $\neg\, substr(w,\ a \cdot b \cdot c)$

$$9, 3, \{x \leftarrow a \cdot b \cdot c,\ y \leftarrow c,\ y2 \leftarrow a \cdot b \cdot c\}$$

11. □

$$10, 1, \{x \leftarrow w,\ w \leftarrow a \cdot b \cdot c\}$$

The unification in the final step of the resolution causes w to receive the substitution $\{w \leftarrow a \cdot b \cdot c\}$. Not only have we proved that $\exists w\ substr(w,\ a \cdot a \cdot b \cdot c \cdot c)$ is a logical consequence of the axioms, but we have also *computed* a value $a \cdot b \cdot c$ for w such that $substr(w,\ a \cdot a \cdot b \cdot c \cdot c)$ is true.

Refutations as Computations

Given a set of *program clauses* and a *query* expressed as a goal clause with no positive literals, the result of a successful refutation is an *answer* obtained from the substitutions carried out during unifications. In ordinary programming languages, control of the computation is *explicitly* constructed by the programmer as part of the program. This can be instantly recognized by the central place occupied by the *control structures*:

```
if ( ... ) { ... } else { ... }
while ( ... ) { ... }
for ( ... ) { ... }
```

In logic programming, the programmer writes declarative formulas (program and goal clauses) that describe the relationship between the input and output. The resolution inference engine supplies a uniform *implicit* control structure, thus relieving the programmer of the task of explicitly specifying it. Logic programming abstracts away from the control structure in the same way that a programming language abstracts away from the explicit memory and register allocation that must be done when writing assembler.

The computation of a logic program is highly nondeterministic:

- Given a goal clause:

$$\neg \, substr(w, \; y1) \vee \neg \, suffix(y1, \; a \cdot a \cdot b \cdot c \cdot c),$$

  it is possible that several literals clash with a positive literal of a program clause. The *computation rule* of a logic programming language must specify how a literal in the goal clause is chosen.
- Once a literal has been chosen, it is possible that (after unification) it clashes with the positive literal of several program clauses. The literal $\neg \, substr(w, \; y1)$ in the goal clause above can be made to clash with both clauses 4 and 5 after unification. The *search rule* of a logic programming language must specify how a program clause is chosen.

11.2   Horn Clauses and SLD-Resolution

In this section we present the theoretical basis of logic programming. We start by defining Horn clauses, the restricted form of clauses used in logic programming. Refutations of Horn clauses are done by a restriction of the resolution procedure called SLD-resolution, which is sound and complete for Horn clauses.

11.2.1   Horn Clauses

Definition 11.1 A *Horn clause* is a clause of the form:

$$A \leftarrow B_1, \; \ldots, \; B_n \equiv A \vee \neg B_1, \; \ldots, \; \neg B_n$$

with at most one positive literal. The positive literal A is the *head* and the negative literals B_i are the *body*. The following terminology is used with Horn clauses:

- A *fact* is a positive unit Horn clause $A \leftarrow$.
- A *goal clause* is a Horn clause with no positive literals $\leftarrow B_1, \; \ldots, \; B_n$.
- A *program clause* is a Horn clause with one positive literal and one or more negative literals. ∎

Logic programming prefers the use of $\leftarrow$, the reverse implication operator, to the familiar forward implication operator $\rightarrow$. The reverse operator in $A \leftarrow B_1, \; \ldots, \; B_n$ has the natural reading:

  To prove A, prove $B_1, \ldots, B_n$.

We can interpret this computationally as a procedure executing a sequence of statements or calling other procedures: To compute A, compute $B_1, \ldots, B_n$.

Definition 11.2

- A set of non-goal Horn clauses whose heads have the same predicate letter is a *procedure*.
- A set of procedures is a *(logic) program*.
- A procedure composed of ground facts only is a *database*.   ∎

Example 11.3 The following program has two procedures p and q; p is also a database.

$$
\begin{array}{ll}
1. & q(x, y) \leftarrow p(x, y) \\
2. & q(x, y) \leftarrow p(x, z), q(z, y)
\end{array}
$$

$$
\begin{array}{llll}
3. & p(b, a) & \quad 7. & p(f, b) \\
4. & p(c, a) & \quad 8. & p(h, g) \\
5. & p(d, b) & \quad 9. & p(i, h) \\
6. & p(e, b) & \quad 10. & p(j, h)
\end{array}
$$

∎

11.2.2 Correct Answer Substitutions for Horn Clauses

Definition 11.4 Let P be a program and G a goal clause. A substitution θ for the variables in G is a *correct answer substitution* if $P \models \forall (\neg G\theta)$, where the universal quantification is taken over all the free variables in $\neg G\theta$.   ∎

Example 11.5 Let P be a set of axioms for arithmetic.

- Let G be the goal clause $\neg (6 + y = 13)$ and θ the substitution $\{y \leftarrow 7\}$:

$$
\begin{aligned}
\forall (\neg G\theta) & \equiv \forall (\neg\neg (6 + y = 13)\{y \leftarrow 7\}) \\
& \equiv \forall (6 + 7 = 13) \\
& \equiv (6 + 7 = 13).
\end{aligned}
$$

Since $P \models (6 + 7 = 13)$, θ is a correct answer substitution for G.
- Let G be the goal clause $\neg (x = y + 13)$ and $\theta = \{y \leftarrow x - 13\}$:

$$
\begin{aligned}
\forall (\neg G\theta) & \equiv \forall (\neg\neg (x = y + 13)\{y \leftarrow x - 13\}) \\
& \equiv \forall x (x = x - 13 + 13).
\end{aligned}
$$

Since $P \models \forall x (x = x - 13 + 13)$, θ is a correct answer substitution for G.
- Let G be the goal clause $\neg (x = y + 13)$ and $\theta = \varepsilon$, the empty substitution:

$$
\begin{aligned}
\forall (\neg G\theta) & \equiv \forall (\neg\neg (x = y + 13)\varepsilon) \\
& \equiv \forall x \forall y (x = y + 13).
\end{aligned}
$$

Since $P \not\models \forall x \forall y (x = y + 13)$, θ is not a correct answer substitution.   ∎

Given a program P, goal clause $G = \neg G_1 \vee \cdots \vee \neg G_n$, and a correct answer substitution θ, by definition $P \models \forall(\neg G)\theta$, so:

$$P \models \forall(G_1 \wedge \cdots \wedge G_n)\theta.$$

Therefore, for any substitution σ that makes the conjunction into a ground formula, $(G_1 \wedge \cdots \wedge G_n)\theta\sigma$ is true in any model of P. This explains the terminology because the substitution $\theta\sigma$ gives an answer to the query expressed in the goal clause.

11.2.3 SLD-Resolution

Before defining the resolution procedure for logic programs, let use work through an example.

Example 11.6 Let $\leftarrow q(y, b)$, $q(b, z)$ be a goal clause for the program in Example 11.3. At each step we must choose a literal within the clause and a clause whose head clashes with the literal. (For simplicity, the only substitutions shown are those to the original variables of the goal clause.)

1. Choose $q(y, b)$ and resolve with clause 1 giving $\leftarrow p(y, b)$, $q(b, z)$.
2. Choose $p(y, b)$ and resolve with clause 5 giving $\leftarrow q(b, z)$.
   This requires the substitution $\{y \leftarrow d\}$.
3. There is only one literal and we resolve it with clause 1 giving $\leftarrow p(b, z)$.
4. There is only one literal and we resolve it with clause 3 giving $\square$.
   This requires the substitution $\{z \leftarrow a\}$.

Therefore, we have a refutation of $\leftarrow q(y, b)$, $q(b, z)$ under the substitution $\theta = \{y \leftarrow d, z \leftarrow a\}$. By the soundness of resolution:

$$P \models \forall \neg (\neg q(y, b) \vee \neg q(b, z))\theta),$$

so that θ is a correct answer substitution and $q(d, b) \wedge q(b, a)$ is true in any model of P.   ∎

Definition 11.7 (SLD-resolution) Let P be a set of program clauses, R a computation rule and G a goal clause. A *derivation by SLD-resolution* is a sequence of resolution steps between goal clauses and the program clauses. The first goal clause G_0 is G. G_{i+1} is derived from G_i *selecting* a literal $A_i^j \in G_i$, *choosing* a clause $C_i \in P$ such that the head of C_i unifies with A_i^j by mgu θ_i and resolving:

$$
\begin{aligned}
G_i &= \leftarrow A_i^1, \ldots, A_i^{j-1}, A_i^j, A_i^{j+1}, \ldots, A_i^{n_i} \\
C_i &= B_i^0 \leftarrow B_i^1, \ldots, B_i^{k_i} \\
A_i^j \theta_i &= B_i^0 \theta_i \\
G_{i+1} &= \leftarrow (A_i^1, \ldots, A_i^{j-1}, B_i^1, \ldots, B_i^{k_i}, A_i^{j+1}, \ldots, A_i^{n_i})\theta_i.
\end{aligned}
$$

An *SLD-refutation* is an SLD-derivation of $\square$.

The rule for selecting a literal A_i^j from a goal clause G_i is the *computation rule*. The rule for choosing a clause $C_i \in P$ is the *search rule*.   ∎

Soundness of SLD-Resolution

Theorem 11.8 (Soundness of SLD-resolution) *Let P be a set of program clauses, R a computation rule and G a goal clause. Suppose that there is an SLD-refutation of G. Let $\theta = \theta_1 \cdots \theta_n$ be the sequence of unifiers used in the refutation and let σ be the restriction of θ to the variables of G. Then σ is a correct answer substitution for G.*

Proof By definition of σ, $G\theta = G\sigma$, so $P \cup \{G\sigma\} = P \cup \{G\theta\}$ which is unsatisfiable by the soundness of resolution. But $P \cup \{G\sigma\}$ is unsatisfiable implies that $P \models \neg G\sigma$. Since this is true for any substitution into the free variables of $G\sigma$, $P \models \forall(\neg G\sigma)$.   ∎

Completeness of SLD-Resolution

SLD-refutation is complete for sets of Horn clauses but not in general.

Example 11.9 Consider the unsatisfiable set of clauses S:

$$
\begin{array}{ll}
1. & p \vee q \\
2. & \neg p \vee q \\
3. & p \vee \neg q \\
4. & \neg p \vee \neg q
\end{array}
$$

S is not a set of Horn clauses since $p \vee q$ has two positive literals. S has an unrestricted resolution refutation, of course, since it is unsatisfiable and resolution is complete:

$$
\begin{array}{lll}
4. & q & 1,2 \\
5. & \neg q & 3,4 \\
6. & \square & 4,5
\end{array}
$$

However, this is not an SLD-refutation because the final step resolves two goal clauses, not a goal clause with one of the program clauses in S.   ∎

Theorem 11.10 (Completeness of SLD-resolution) *Let P be a set of program clauses, R a computation rule, G a goal clause, and σ be a correct answer substitution. There is an SLD-refutation of G from P such that σ is the restriction of the sequence of unifiers $\theta = \theta_1 \cdots \theta_n$ to the variables in G.*

Proof We will give an outline of the proof which can be found in Lloyd (1987, Sect. 2.8).

The proof is by induction on the depth of the terms in the goal clause. Consider the program P:

$$p(a)$$
$$p(f(x)) \leftarrow p(x).$$

Obviously there is a one-step refutation of the goal clause $\leftarrow p(a)$ and just as obviously $p(a)$ is a logical consequence of P.

Given a goal clause $G_i = \leftarrow p(f(f(\cdots(a)\cdots)))$, we can resolve it with the second program clause to obtain $G_{i-1} = \leftarrow p(f(\cdots(a)\cdots))$, reducing the depth of the term. By induction, G_{i-1} can be refuted and $p(f(\cdots(a)\cdots))$ is a logical consequence of P. From G_{i-1} and the second clause, it follows that $p(f(f(\cdots(a)\cdots)))$ is a logical consequence of P.

This bottom-up inductive construction—starting from facts in the program and resolving with program clauses—defines an Herbrand interpretation. Given a ground goal clause whose atoms are in the Herbrand base of the interpretation, it can be proved by induction that it has a refutation and that its negation is a logical consequence of P. To prove that a non-ground clause has a refutation, technical lemmas are needed which keep track of the unifiers. The final step is a proof that there exists a refutation regardless of the choice of computation rule.   ∎

11.3 Search Rules in SLD-Resolution

Theorem 11.10 states that *some* SLD-refutation of a program exists regardless of the computation rule that is used. The same is not true of the choice of the search rule. In this section we explore the effect that the search rule can have on a refutation.

11.3.1 Possible Outcomes when Attempting a Refutation

The discussion will be based upon the program in Example 11.3, repeated here for convenience:

1. $q(x, y) \leftarrow p(x, y)$
2. $q(x, y) \leftarrow p(x, z), q(z, y)$

3. $p(b, a)$ 7. $p(f, b)$
4. $p(c, a)$ 8. $p(h, g)$
5. $p(d, b)$ 9. $p(i, h)$
6. $p(e, b)$ 10. $p(j, h)$

In Example 11.6, we showed that there is a refutation for the goal $\leftarrow q(y, b), q(b, z)$ with correct answer substitution $\theta = \{y \leftarrow d, z \leftarrow a\}$. Consider now the following refutation, where we have omitted the steps of standardizing apart the variables of the program clauses and the substitutions to these new variables:

11. $\leftarrow q(y,b),\, q(b,z)$

12. $\leftarrow p(y,b),\, q(b,z)$ 1, 11

13. $\leftarrow q(b,z)$ 6, 12, $\{y \leftarrow e\}$

14. $\leftarrow p(b,z)$ 1, 13

15. $\square$ 3, 14, $\{z \leftarrow a\}$

The goal clause has been refuted with the substitution $\{y \leftarrow e, z \leftarrow a\}$, showing that there may be more than one correct answer substitution for a given goal clause and the answer obtained depends on the search rule.

Suppose now that the computation rule is to always choose the *last* literal in a goal clause, in this case $q(b,z)$, and suppose that the search rule always chooses to resolve literals with the predicate symbol q first with clause 2 and only then with clause 1. The SLD-derivation becomes:

11. $\leftarrow q(y,b), q(b,z)$

12. $\leftarrow q(y,b), p(b,z'), q(z',z)$ 2, 11

13. $\leftarrow q(y,b), p(b,z'), p(z',z''), q(z'',z)$ 2, 12

14. $\leftarrow q(y,b), p(b,z'), p(z',z''), p(z'',z'''), q(z''',z)$ 2, 13

  $\cdots$

Even though a correct answer substitution exists for the goal clause, this specific attempt at constructing a refutation does not terminate.

Returning to the computation rule that always chooses the *first* literal in the goal clause, we have the following attempt at a refutation:

11. $\leftarrow q(y,b),\, q(b,z)$

12. $\leftarrow p(y,z'), q(z',b), q(b,z)$ 2, 11

13. $\leftarrow q(b,b), q(b,z)$ 6, 12, $\{y \leftarrow e, z' \leftarrow b\}$

14. $\leftarrow p(b,b), q(b,z)$ 1, 13

15. ???

Even though a correct answer substitution exists, the refutation has failed, because *no* program clause unifies with $p(b,b)$.

SLD-resolution is very sensitive to the computation and search rules that are used. Even if there are one or more correct answer substitutions, the resolution procedure may fail to terminate or terminate without finding an answer.

11.3.2 SLD-Trees

The set of SLD-derivations for a logic program can be displayed as a tree.

Definition 11.11 Let P be a set of program clauses, R a computation rule and G a goal clause. An *SLD-tree* is generated as follows: The root is labeled with the goal

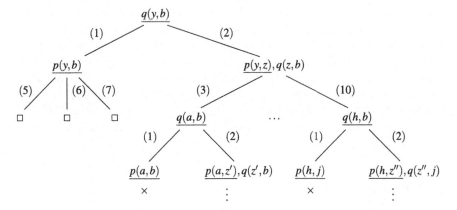

Fig. 11.1 SLD-tree for selection of leftmost literal

clause G. Given a node n labeled with a goal clause G_n, create a child n_i for each new goal clause G_{n_i} that can be obtained by resolving *the literal chosen by R* with the head of a clause in P.   ∎

Example 11.12 An SLD-tree for the program clauses in Example 11.3 and the goal clause $\leftarrow q(y, b)$ is shown in Fig. 11.1. The computation rule is always to choose the leftmost literal of the goal clause. This is indicated by underlining the chosen literal. The number on an edge refers to the number of the program clause resolved with the goal clause.   ∎

Definition 11.13 In an SLD-tree, a branch leading to a refutation is a *success branch*. A branch leading to a goal clause whose selected literal does not unify with any clause in the program is a *failure branch*. A branch corresponding to a non-terminating derivation is an *infinite branch*.   ∎

There are many different SLD-trees, one for each computation rule; nevertheless, we have the following theorem which shows that all trees are similar. The proof can be found in Lloyd (1987, Sect. 2.10).

Theorem 11.14 *Let P be a program and G be a goal clause. Then every SLD-tree for P and G has infinitely many success branches or they all have the same finite number of success branches.*

Definition 11.15 A *search rule* is a procedure for searching an SLD-tree for a refutation. An *SLD-refutation procedure* is the SLD-resolution algorithm together with the specification of a computation rule and a search rule.   ∎

Theorem 11.10 states that SLD-resolution is complete regardless of the choice of the computation rule, but it only says that some refutation exists. The search rule will determine if the refutation is found or not and how efficient the search will be. A *breadth-first search* of an SLD-tree, where the nodes at each depth are checked before searching deeper in the tree, is guaranteed to find a success branch if one exists, while a *depth-first search* can choose to head down a non-terminating branch if one exists. In practice, depth-first search is preferred because it needs much less memory: a stack of the path being searched, where each element in the stack records which branch taken at each node and the substitutions done at that node. In a breadth-first search, this information must be stored for all the leaves of the search.

11.4 Prolog

Prolog was the first logic programming language. There are high-quality implementations that make Prolog a practical tool for software development.

The computation rule in Prolog is to choose the *leftmost* literal in the goal clause. The search rule is to choose clauses from *top to bottom* in the list of the clauses of a procedure. The notation of Prolog is different from the mathematical notation that we have been using: (a) variables begin with upper-case letters, (b) predicates begin with lower-case letters (as do functions and constants), and (c) the symbol : – is used for ←.

Let us rewrite program of Example 11.3 using the notation of Prolog. We have also replaced the arbitrary symbols by symbols that indicate the intended meaning of the program:

```
ancestor(X,Y) :- parent(X,Y).
ancestor(X,Y) :- parent(X,Z), ancestor(Z,Y).

parent(bob,allen).           parent(fred,dave).
parent(catherine,allen).     parent(harry,george).
parent(dave,bob).            parent(ida,george).
parent(ellen,bob).           parent(joe,harry).
```

The database contains facts that we are assuming to be true, such as *catherine is a parent of allen*. The procedure for ancestor gives a declarative meaning to this concept in terms of the parent relation:

- X is an ancestor of Y if X is a parent of Y.
- X is an ancestor of Y if there are Z's such that X is a parent of Z and Z is an ancestor of Y.

Using the Prolog computation and search rules, the goal clause:

```
:- ancestor(Y,bob), ancestor(bob,Z).
```

will succeed and return the correct answer substitution Y=dave, Z=allen, meaning that dave is an ancestor of bob who in turn is an ancestor of allen. Here is the refutation:

```
:- ancestor(Y,bob), ancestor(bob,Z).
:- parent(Y, bob), ancestor(bob, Z). { Y <- dave }
:- ancestor(bob, Z).
:- parent(bob, Z).                    { Z <-allen }
:-
```

11.4.1 Depth-First Search

The search in the proof tree is depth-first, which can lead to non-termination of the computation even if a terminating computation exists. A Prolog programmer must carefully order clauses within a procedure and literals within clauses to avoid non-termination.

Since failure may occur at any step, the Prolog implementation must store a list of *backtrack points*. These backtrack points represent previous nodes in the SLD-tree where additional branches exist.

Example 11.16 Consider the program consisting of four facts:

```
p(a).
p(b).
p(c).
q(c).
```

and the goal clause:

```
:- p(X), q(X).
```

Here is the SLD-tree for this program:

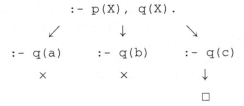

The depth-first search attempts to resolve the first literal p(X) from the goal clause with p(a). While this succeeds, the goal clause q(a) which results cannot be resolved. The search must backtrack and try the next clause in the procedure for p, namely, p(b). Here too, the computation fails and must backtrack again to find a successful refutation.   ■

An important concept in Prolog programming is forcing failure. This is implemented by the predicate fail for which no program clauses are defined. Consider the goal clause:

```
:- ancestor(Y,bob), ancestor(bob,Z), fail.
```

Once the answer Y=dave, Z=allen is obtained, backtracking will force the refutation to continue and produce a second answer Y=ellen, Z=allen. Prolog lacks iterative structures such as for- and while-loops, so recursion and forced failure are fundamental programming techniques in the language.

11.4.2 Prolog and the Theory of Logic Programming

The designers of Prolog added a number of constructs to the language to enable it become a practical programming language, even though these constructs are not consistent with the theory of logic programming that we presented in the previous sections.

Non-logical Predicates

Non-logical predicates are predicates whose main or only purpose is the side-effects they generate. Obvious examples are the I/O predicates read and write that have no declarative meaning as logical formulas. As literals in a goal clause, they always succeed (except that read may fail at end of file), but they have side-effects causing data to be read into a variable or displayed on a screen.

Arithmetic

Prolog departs from theoretical logic programming in its treatment of numeric data types. As we show in Sect. 12.4, it is possible to formalize arithmetic in first-order logic, but there are two problems with the formalism. The first is that it would be unfamiliar, to say the least, to execute a query on the number of employees in a department and to receive as an answer the term $f(f(f(f(f(a)))))$ instead of 5. The second problem is the inefficiency of resolution as a method for numeric computation.

Prolog supports standard arithmetic computation. The syntax is that of a predicate with an infix operator Result is Expression. The following clause retrieves the list price and the discount from a database and computes the value of Price after applying the discount:

```
selling_price(Item, Price) :-
  list_price(Item, List),
  discount_percent(Item, Discount),
  Price is List - List * Discount / 100.
```

Arithmetic predicates differ from ordinary predicates, because they are one-way, unlike unification. If 10 is X+Y were a logical predicate, X and Y could be unified with say, 0 and 10, and upon backtracking with 1 and 9, and so on. However, this is illegal. In `Result is Expression`, `Expression` must evaluate to a numeric value, which is then unified with `Result` (usually an uninstantiated variable).

Arithmetic predicates are *not* assignment statements. The following program is not correct:

```
selling_price(Item,Price) :-
  list_price(Item,List),
  discount_percent(Item,Discount),
  Price is List - List * Discount / 100,
  tax_percent(Item,Tax),
  Price is Price * (1 + Tax / 100).
```

Once `Price` has been unified with the result of the computation `List - List * Discount / 100`, any attempt to unify again will fail, just as a variable x in a logical formula cannot be modified once a ground substitution such as $\{x \leftarrow a\}$ has been applied. An additional variable must be used to hold the intermediate value:

```
selling_price(Item,Price) :-
  list_price(Item,List),
  discount_percent(Item,Discount),
  Price1 is List - List * Discount / 100,
  tax_percent(Item,Tax),
  Price is Price1 * (1 + Tax / 100).
```

Cuts

The most controversial modification of logic programming introduced into Prolog is the *cut*. Consider the following program for computing the factorial of a number N:

```
factorial(0, 1).
factorial(N, F) :-
  N1 is N - 1,
  factorial(N1, F1),
  F is N * F1.
```

This is a translation into Prolog of the recursive formula:

$$f(0) = 1, \qquad f(n) = n \cdot f(n-1).$$

Now assume that `factorial` is called in another procedure, perhaps for checking a property of numbers that are factorials:

```
check(N) :- factorial(N, F), property(F).
```

If `check` is called with N=0, it will call `factorial(0, F)` which will compute F=1 and call `property(1)`. Suppose that this call fails. Then the SLD-resolution procedure will backtrack, *undo* the substitution F=1, and try the second clause in the procedure for factorial. The recursive call `factorial(-1,F1)` will initiate a non-terminating computation. A call to `factorial` with the argument 0 has only one possible answer; if we backtrack through it, the goal clause should fail.

This can be avoided by introducing a *cut*, denoted by an exclamation point, into the first clause:

```
factorial(0, 1) :- !.
```

The cut prevents backtracking in this procedure. Once a cut is executed it cuts away a portion of the SLD-tree and prevents unwanted backtracking. In the following diagram, the rightmost branch is cut away, so that if property(1) fails, there is no longer a backtrack point in its parent node:

```
                    :- check(0).

                         ↓

        :- factorial(0, F), property(F).

             ↙                        ↘

    :- property(1)        :- factorial(-1, F1),
                             property(F1).

            ×                          ...
```

In the case of the factorial procedure there is a better solution, namely, adding a predicate to the body of the procedure that explicitly prevents the unwanted behavior:

```
factorial(0, 1).
factorial(N, F) :-
  N > 0,
  N1 is N - 1,
  factorial(N1, F1),
  F is N * F1.
```

11.5 Summary

A Horn clause is a clause that has at most one positive literal. A fact is a unit Horn clause with one positive literal; a program clause is a Horn clause with one positive literal and one or more negative literals; a goal clause is a Horn clause with no positive literals. A logic program consists of a set of program clauses and facts. Given a logic program and a goal clause, SLD-resolution (which is sound and complete) can be used to search for a refutation. If a refutation exists, then the negation of the goal clause is a logical consequence of the program clauses and facts, and the substitutions made during the refutation form the answer of the program.

Prolog is a logic programming language written as Horn clauses. Computation in Prolog is by SLD-resolution with a specific computation rule—choose the leftmost literal of a goal—and a specific search rule—choose the program clauses in textual order.

11.6 Further Reading

Lloyd (1987) presents the theory of SLD resolution in full detail. For more on Prolog programming, see textbooks Sterling and Shapiro (1994), Bratko (2011) and Clocksin and Mellish (2003).

11.7 Exercises

11.1 Let P be the program $p(a) \leftarrow$ and G be the goal clause $\leftarrow p(x)$. Is the empty substitution a correct answer substitution? Explain.

11.2 Draw an SLD-tree similar to that of Fig. 11.1 except that the computation rule is to select the rightmost literal in a clause.

11.3 Given the logic program

$$p(a, b)$$
$$p(c, b)$$
$$p(x, y) \leftarrow p(x, y), p(y, z)$$
$$p(x, y) \leftarrow p(y, x),$$

and the goal clause $\leftarrow p(a, c)$, show that if any clause is omitted from the program then there is no refutation. From this prove that if a depth-first search rule is used with any fixed order of the clauses, there is no refutation no matter what computation rule is used.

11.4 Given the logic program

$$p \leftarrow q(x, x)$$
$$q(x, f(x)),$$

and the goal clause $\leftarrow p$, prove that there is a refutation if and only if the occurs-check is omitted. Show that omitting the occurs-check invalidates the soundness of SLD-resolution.

11.5 Given the logic program

$$p \leftarrow q(x, x)$$
$$q(x, f(x)) \leftarrow q(x, x),$$

and the goal clause $\leftarrow p$, what happens if a refutation is attempted without using the occurs-check?

11.6 Write a logic program for the *Slowsort* algorithm by directly implementing the following specification of sorting: $sort(L1, L2)$ is true if $L2$ is a permutation of $L1$ and $L2$ is ordered.

11.7 (Assumes a knowledge of lists.) In Prolog, [] denotes the empty list and [Head | Tail] denotes the list whose head is Head and whose tail is Tail. Consider the Prolog program for appending one list to another:

```
append([], List, List).
append ([Head | Tail], List, [Head | NewTail]) :-
  append (Tail, List, NewTail).
```

It is common to add a cut to the first clause of the program:

```
append([], List, List) :- !.
```

Compare the execution of the programs with and without the cut for the goal clauses:

```
:- append([a,b,c], [d,e,f], List).
:- append([a,b,c], List1, List2).
:- append(List1, List2, [a,b,c]).
```

11.8 A set of clauses S is *renamable-Horn* iff there is a set of propositional letters U such that $R_U(S)$ is a set of Horn clauses. (Recall Definition 6.12 and Lemma 6.13). Prove the following theorem:

Theorem 11.9 (Lewis) *Let $S = \{C_1, \ldots, C_m\}$ be a set of clauses where $C_i = l_1^i \vee \cdots \vee l_{n_i}^i$, and let*

$$S^* = \bigcup_{i=1}^{m} \bigcup_{1 \le j < k \le n_i} (l_j^i \vee l_k^i).$$

Then S is renamable-Horn if and only if S^* is satisfiable.

References

I. Bratko. *Prolog Programming for Artificial Intelligence (Fourth Edition).* Addison-Wesley, Boston, 2011.

W.F. Clocksin and C.S. Mellish. *Programming in Prolog: Using the ISO Standard.* Springer, Berlin, 2003.

J.W. Lloyd. *Foundations of Logic Programming (Second Edition).* Springer, Berlin, 1987.

L. Sterling and E. Shapiro. *The Art of Prolog: Advanced Programming Techniques (Second Edition).* MIT Press, Cambridge, MA, 1994.

Chapter 12
First-Order Logic: Undecidability and Model Theory *

The chapter surveys several important theoretical results in first-order logic. In Sect. 12.1 we prove that validity in first-order logic is undecidable, a result first proved by Alonzo Church. Validity is decidable for several classes of formulas defined by syntactic restrictions on their form (Sect. 12.2). Next, we introduce model theory (Sect. 12.3): the fact that a semantic tableau has a countable number of nodes leads to some interesting results. Finally, Sect. 12.4 contains an overview of Gödel's surprising incompleteness result.

12.1 Undecidability of First-Order Logic

We show the undecidability of validity in first-order logic by reduction from a problem whose undecidability is already known, the *halting problem*: to decide whether a Turing machine will halt if started on a blank tape (Minsky (1967, Sect. 8.3.3), Manna (1974, Sect. 1-5.2)). The proof that there is no decision procedure for validity describes an algorithm that takes an arbitrary Turing machine T and generates a formula S_T in first-order logic, such that S_T is valid if and only if T halts on an blank tape. If there were a decision procedure for validity, this construction would give us an decision procedure for the halting problem.

12.1.1 Two-Register Machines

Instead of working directly with Turing machines, we work with a simpler form of automata: two-register machines. The halting problem for two-register machines is undecidable because there is a reduction from Turing machines to two-register machines.

M. Ben-Ari, *Mathematical Logic for Computer Science*,
DOI 10.1007/978-1-4471-4129-7_12, © Springer-Verlag London 2012

Definition 12.1 A *two-register machine* M consists of two registers x and y which can store natural numbers, and a program $P = \{L_0, \ldots, L_n\}$, where L_n is the instruction halt and for $0 \leq i < n$, L_i is one of the instructions:

- x = x + 1;
- y = y + 1;
- if (x == 0) goto L_j; else x = x - 1;
- if (y == 0) goto L_j; else y = y - 1;

An execution sequence of M is a sequence of states $s_k = (L_i, x, y)$, where L_i is the current instruction and x, y are the contents of the registers x and y. s_{k+1} is obtained from s_k by executing L_i. The initial state is $s_0 = (L_0, m, 0)$ for some m. If for some k, $s_k = (L_n, x, y)$, the computation of M halts and M has computed $y = f(m)$. ∎

Theorem 12.2 *Let T be a Turing machine that computes a function f. Then there is a two-register machine M_T that computes the function f.*

Proof Minsky (1967, Sect. 14.1), Hopcroft et al. (2006, Sect. 7.8). ∎

The proof shows how the contents of the tape of a Turing machine can be encoded in an (extremely large) natural number and how the modifications to the tape can be carried out when copying the contents of one register into another. Clearly, two-register machines are even more impractical than Turing machines, but it is the theoretical result that is important.

12.1.2 Church's Theorem

Theorem 12.3 (Church) *Validity in first-order logic is undecidable.*

Proof Let M be an arbitrary two-register machine. We will construct a formula S_M such that S_M is valid iff M terminates when started in the state $(L_0, 0, 0)$. The formula is:

$$S_M = \left(p_0(a, a) \wedge \bigwedge_{i=0}^{n-1} S_i \right) \rightarrow \exists z_1 \exists z_2 p_n(z_1, z_2),$$

where S_i is defined by cases of the instruction L_i:

L_i	S_i
x = x + 1;	$\forall x \forall y (p_i(x, y) \rightarrow p_{i+1}(s(x), y))$
y = y + 1;	$\forall x \forall y (p_i(x, y) \rightarrow p_{i+1}(x, s(y)))$
if (x == 0) goto Lj;	$\forall x (p_i(a, x) \rightarrow p_j(a, x)) \wedge$
else x = x - 1;	$\forall x \forall y (p_i(s(x), y) \rightarrow p_{i+1}(x, y))$
if (y == 0) then goto Lj;	$\forall x (p_i(x, a) \rightarrow p_j(x, a)) \wedge$
else y = y - 1;	$\forall x \forall y (p_i(x, s(y)) \rightarrow p_{i+1}(x, y))$

The predicates are $p_0, \ldots, p_n$, one for each statement in M. The intended meaning of $p_i(x, y)$ is that the computation of M is at the label L_i and the values x, y are in the two registers. The constant a is intended to mean 0 and the function s is intended to mean the successor function $s(m) = m + 1$.

s is used both for the function symbol in the formula S_M and for states in the execution of M. The meaning will be clear from the context.

We have to prove that M halts if and only if S_M is valid.

If M Halts then S_M Is Valid

Let $s_0, \ldots, s_m$ be a computation of M that halts after m steps; we need to show that S_M is valid, that is, that it is true under any interpretation for the formula. However, we need not consider every possible interpretation. If $\mathscr{I}$ is an interpretation for S_M such that $v_{\mathscr{I}}(S_i) = F$ for some $0 \le i \le n - 1$ or such that $v_{\mathscr{I}}(p_0(a, a)) = F$, then trivially $v_{\mathscr{I}}(S_M) = T$ since the antecedent of S_M is false. Therefore, we need only consider interpretations that satisfy the antecedent of S_M. For such interpretations, we need to show that $v_{\mathscr{I}}(\exists z_1 \exists z_2 p_n(z_1, z_2)) = T$. By induction on k, we show that $v_{\mathscr{I}}(\exists z_1 \exists z_2 p_k(z_1, z_2)) = T$.

If $k = 0$, the result is trivial since $p_0(a, a) \to \exists z_1 \exists z_2 p_0(z_1, z_2)$ is valid.

Let us assume the inductive hypothesis for $k - 1$ (provided that $k > 0$) and prove $v_{\mathscr{I}}(\exists z_1 \exists z_2 p_k(z_1, z_2)) = T$. We will work through the details when L_k is x=x+1 and leave the other cases to the reader.

By assumption the antecedent is true, in particular, its subformula S_{k-1}:

$$v_{\mathscr{I}}(\forall x \forall y (p_{k-1}(x, y) \to p_k(s(x), y))) = T,$$

and by the inductive hypothesis:

$$v_{\mathscr{I}}(\exists z_1 \exists z_2 p_{k-1}(z_1, z_2)) = T,$$

from which:

$$v_{\mathscr{I}}(\exists z_1 \exists z_2 p_k(s(z_1), z_2)) = T$$

follows by reasoning in first-order logic.

Let c_1 and c_2 be the domain elements assigned to z_1 and z_2, respectively, such that $(succ(c_1), c_2) \in P_k$, where P_k is the interpretation of p_k and $succ$ is the interpretation of s. Since $c_3 = succ(c_1)$ for some domain element c_3, the existentially quantified formula in the consequent is true:

$$v_{\mathscr{I}}(\exists z_1 \exists z_2 p_k(z_1, z_2)) = T.$$

If S_M Is Valid then M Halts

Suppose that S_M is valid and consider the interpretation:

$$\mathscr{I} = (\mathscr{N}, \{P_0, \ldots, P_n\}, \{succ\}, \{0\}),$$

where *succ* is the successor function on $\mathscr{N}$, and $(x, y) \in P_i$ iff (L_i, x, y) is reached by the register machine when started in $(L_0, 0, 0)$.

We show by induction on the length of the computation that the antecedent of S_M is true in $\mathscr{I}$. The initial state is $(L_0, 0, 0)$, so $(a, a) \in P_0$ and $v_{\mathscr{I}}(p_0(a, a)) = T$. The inductive hypothesis is that in state $s_{k-1} = (L_i, x_i, y_i)$, $(x_i, y_i) \in P_i$. The inductive step is again by cases on the type of the instruction L_i. For x=x+1, $s_k = (L_{i+1}, succ(x_i), y_i)$ and $(succ(x_i), y_i) \in P_{i+1}$ by the definition of P_{i+1}.

Since S_M is valid, $v_{\mathscr{I}}(\exists z_1 \exists z_2 p_n(z_1, z_2)) = T$ and $v_{\mathscr{I}}(p_n(m_1, m_2)) = T$ for some $m_1, m_2 \in \mathscr{N}$. By definition, $(m_1, m_2) \in P_n$ means that M halts and computes $m_2 = f(0)$.   ∎

Church's Theorem holds even if the structure of the formulas is restricted:

- The formulas contain only binary predicate symbols, one constant and one unary function symbol. This follows from the structure of S_M in the proof.
- The formulas are logic programs: a set of program clauses, a set of facts and a goal clause (Chap. 11). This follows immediately since S_M is of this form.
- The formulas are pure (Mendelson, 2009, 3.6).

Definition 12.4 A formula of first-order logic is *pure* if it contains no function symbols (including constants which are 0-ary function symbols).   ∎

12.2 Decidable Cases of First-Order Logic

Theorem 12.5 *There are decision procedures for the validity of pure PCNF formulas whose prefixes are of one of the forms (where $m, n \geq 0$):*

$$\forall x_1 \cdots \forall x_n \exists y_1 \cdots \exists y_m,$$
$$\forall x_1 \cdots \forall x_n \exists y \forall z_1 \cdots \forall z_m,$$
$$\forall x_1 \cdots \forall x_n \exists y_1 \exists y_2 \forall z_1 \cdots \forall z_m.$$

These classes are conveniently abbreviated $\forall^* \exists^*$, $\forall^* \exists \forall^*$, $\forall^* \exists \exists \forall^*$.

The decision procedures can be found in Dreben and Goldfarb (1979). This is the best that can be done because the addition of existential quantifiers makes validity undecidable. See Lewis (1979) for proofs of the following result.

Theorem 12.6 *There are no decision procedures for the validity of pure PCNF formulas whose prefixes are of one of the forms*:

$$\exists z \, \forall x_1 \cdots \forall x_n \, \exists y_1 \cdots \exists y_m,$$
$$\forall x_1 \cdots \forall x_n \, \exists y_1 \exists y_2 \exists y_3 \, \forall z_1 \cdots \forall z_m.$$

For the first prefix, the result holds even if $n = m = 1$:

$$\exists z \, \forall x_1 \, \exists y_1,$$

and for the second prefix, the result holds even if $n = 0, m = 1$:

$$\exists y_1 \exists y_2 \exists y_3 \, \forall z_1.$$

Even if the matrix is restricted to contain only binary predicate symbols, there is still no decision procedure.

There are other restrictions besides those on the prefix that enable decision procedures to be given (see Dreben and Goldfarb (1979)):

Theorem 12.7 *There is a decision procedure for PCNF formulas whose matrix is of one of the forms*:

1. *All conjunctions are single literals.*
2. *All conjunctions are either single atomic formulas or consists entirely of negative literals.*
3. *All atomic formulas are monadic, that is, all predicate letters are unary.*

12.3 Finite and Infinite Models

Definition 12.8 A set of formulas U has the *finite model property* iff: U is satisfiable iff it is satisfiable in an interpretation whose domain is a finite set.

Theorem 12.9 *Let U be a set of pure formulas of the form*:

$$\exists x_1 \cdots \exists x_k \forall y_1 \cdots \forall y_l A(x_1, \ldots, x_k, y_1, \ldots, y_l),$$

where A is quantifier-free. Then U has the finite model property.

Proof In a tableau for U, once the δ-rules have been applied to the existential quantifiers, no more existential quantifiers remain. Thus the set of constants will be finite and the tableau will terminate once all substitutions using these constants have been made for the universal quantifiers. ∎

Theorem 12.10 (Löwenheim) *If a formula is satisfiable then it is satisfiable in a countable domain.*

Proof The domain D defined in the proof of completeness is countable. ∎

Löwenheim's Theorem can be generalized to countable sets of formulas $U = \{A_0, A_1, A_2, \ldots\}$. Start the tableaux with formula A_0 at the root. Whenever constructing a node at depth d, add the formula A_d into its label in addition to whatever formulas are specified by the tableau rule. If the tableau does not close, eventually, every A_i will appear on the branch, and the labels will form a Hintikka set. Hintikka's Lemma and completeness can be proved as before.

Theorem 12.11 (Löwenheim–Skolem) *If a countable set of formulas is satisfiable then it is satisfiable in a countable domain.*

Uncountable sets such as the real numbers can be described by countably many axioms (formulas). Thus formulas that describe real numbers also have a countable model in addition to the standard uncountable model! Such models are called *non-standard* models.

As in propositional logic (Theorem 3.48), compactness holds.

Theorem 12.12 (Compactness) *Let U be a countable set of formulas. If all finite subsets of U are satisfiable then so is U.*

12.4 Complete and Incomplete Theories

Definition 12.13 Let $\mathcal{T}(U)$ be a theory. $\mathcal{T}(U)$ is *complete* if and only if for every closed formula A, $U \vdash A$ or $U \vdash \neg A$. $\mathcal{T}(U)$ is *incomplete* iff it is not complete, that is, iff for some closed formula A, $U \not\vdash A$ and $U \not\vdash \neg A$.   ∎

It is important not to confuse a complete *theory* with the completeness of a *deductive system*. The latter relates the syntactic concept of proof to the semantic concept of validity: a closed formula can be proved if and only if it is valid. Completeness of a theory looks at what formulas are logical consequences of a set of formulas.

In one of the most surprising results of mathematical logic, Kurt Gödel proved that *number theory* is incomplete. Number theory, first developed by Guiseppe Peano, is a first-order logic with one constant symbol 0, one binary predicate symbol $=$, one unary function symbol s representing the successor function and two binary function symbols $+$, $*$. A set of axioms for number theory $\mathcal{N}\mathcal{T}$ consists of eight axioms and one axiom scheme for induction (Mendelson, 2009, 3.1).

Theorem 12.14 (Gödel's Incompleteness Theorem) *If $\mathcal{T}(\mathcal{N}\mathcal{T})$ is consistent then $\mathcal{T}(\mathcal{N}\mathcal{T})$ is incomplete.*

If $\mathcal{T}(\mathcal{N}\mathcal{T})$ were inconsistent, that is, if a theorem and its negation were both provable, then by Theorem 3.43, *every* formula would be a theorem so the theory would have be of no interest whatsoever.

The detailed proof of Gödel's theorem is tedious but not too difficult. An informal justification can be found in Smullyan (1978). Here we give a sketch of the formal proof (Mendelson, 2009, 3.4–3.5). The idea is to define a mapping, called a *Gödel numbering*, from logical objects such as formulas and proofs to natural numbers, and then to prove the following theorem.

Theorem 12.15 *There exists a formula $A(x, y)$ in $\mathcal{N}\mathcal{T}$ with the following property: For any numbers i, j, $A(i, j)$ is true if and only if i is the Gödel number associated with some formula $B(x)$ with one free variable x, and j is the Gödel number associated with the proof of $B(i)$. Furthermore, if $A(i, j)$ is true then a proof can be constructed for these specific integers $\vdash A(i, j)$.*

Consider now the formula $C(x) = \forall y \neg A(x, y)$ which has one free variable x, and let m be the Gödel number of this formula $C(x)$. Then $C(m) = \forall y \neg A(m, y)$ means that for no y is y the Gödel number of a proof of $C(m)$!

Theorem 12.16 (Gödel) *If $\mathcal{N}\mathcal{T}$ is consistent then $\not\vdash C(m)$ and $\not\vdash \neg C(m)$.*

Proof We show that assuming either $\vdash C(m)$ or $\vdash \neg C(m)$ contradicts the consistency of $\mathcal{N}\mathcal{T}$.

- Suppose that $\vdash C(m) = \forall y \neg A(m, y)$ and compute n, the Gödel number of this proof. Then $A(m, n)$ is true and by Theorem 12.15, $\vdash A(m, n)$. Now apply Axiom 4 of first-order logic to $C(m)$ to obtain $\vdash \neg A(m, n)$. But $\vdash A(m, n)$ and $\vdash \neg A(m, n)$ contradict the consistency of $\mathcal{N}\mathcal{T}$.
- Suppose that $\vdash \neg C(m) = \neg \forall y \neg A(m, y) = \exists y A(m, y)$. Then for some n, $A(m, n)$ is true, where n is the Gödel number of a proof of $C(m)$, that is, $\vdash C(m)$. But we assumed $\vdash \neg C(m)$ so $\mathcal{N}\mathcal{T}$ is inconsistent. ∎

12.5 Summary

The decidability of validity for first-order logic has been investigated in detail and it is possible to precisely demarcate restricted classes of formulas which are decidable from less restricted classes that are not decidable. The Löwenheim-Skolem Theorem is surprising since it means that it is impossible to characterize uncountable structures in first-order logic. Even more surprising is Gödel's incompleteness result, since it demonstrates that there are true formulas of mathematical theories that cannot be proved in the theories themselves.

12.6 Further Reading

The two sides of the decidability question are comprehensively presented by Dreben and Goldfarb (1979) and Lewis (1979). The details of Gödel numbering can be found in (Mendelson, 2009, Chap. 3) and (Monk, 1976, Chap. 3). For an introduction to model theory see (Monk, 1976, Part 4).

12.7 Exercises

12.1 Prove that a formula is satisfiable iff it is satisfiable in an infinite model.

12.2 Prove the Löwenheim-Skolem Theorem (12.11) using the construction of semantic tableaux for infinite sets of formulas.

12.3 A closed pure formula A is *n-condensable* iff every unsatisfiable conjunction of instances of the matrix of A contains an unsatisfiable subconjunction made up of n or fewer instances.

- Let A be a PCNF formula whose matrix is a conjunction of literals. Prove that A is 2-condensable.
- Let A be a PCNF formula whose matrix is a conjunction of positive literals and disjunctions of negative literals. Prove that A is $n + 1$-condensable, where n is the maximum number of literals in a disjunction.

12.4 * Prove Church's Theorem by reducing Post's Correspondence Problem to validity in first-order logic.

References

B. Dreben and W.D. Goldfarb. *The Decision Problem: Solvable Classes of Quantificational Formulas*. Addison-Wesley, Reading, MA, 1979.

J.E. Hopcroft, R. Motwani, and J.D. Ullman. *Introduction to Automata Theory, Languages and Computation (Third Edition)*. Addison-Wesley, 2006.

H.R. Lewis. *Unsolvable Classes of Quantificational Formulas*. Addison-Wesley, Reading, MA, 1979.

Z. Manna. *Mathematical Theory of Computation*. McGraw-Hill, New York, NY, 1974. Reprinted by Dover, 2003.

E. Mendelson. *Introduction to Mathematical Logic (Fifth Edition)*. Chapman & Hall/CRC, 2009.

M.L. Minsky. *Computation: Finite and Infinite Machines*. Prentice-Hall, Englewood Cliffs, NJ, 1967.

J.D. Monk. *Mathematical Logic*. Springer, 1976.

R.M. Smullyan. *What Is the Name of This Book?—The Riddle of Dracula and Other Logical Puzzles*. Prentice-Hall, 1978.

Chapter 13
Temporal Logic: Formulas, Models, Tableaux

Temporal logic is a formal system for reasoning about time. Temporal logic has found extensive application in computer science, because the behavior of both hardware and software is a function of time. This section will follow the same approach that we used for other logics: we define the syntax of formulas and their interpretations and then describe the construction of semantic tableaux for deciding satisfiability.

Unlike propositional and first-order logics whose variants have little theoretical or practical significance, there are many temporal logics that are quite different from each other. A survey of this flexibility is presented in Sect. 13.3, but you can skim it and go directly to Sect. 13.4 that presents the logic we focus on: *linear temporal logic*.

13.1 Introduction

Example 13.1 Here are some examples of specifications that use temporal concepts (italicized):

- *After* the reset-line of a flip-flop is asserted, the zero-line is asserted. The output lines *maintain* their values *until* the set-line is asserted; *then* they are complemented.
- If a request is made to print a file, *eventually* the file will be printed.
- The operating system will *never* deadlock.

The temporal aspects of these specification can be expressed in first-order logic using quantified variables for points in time:

$$\forall t_1 (reset(t_1) \rightarrow \exists t_2 (t_2 \geq t_1 \wedge zero(t_2))),$$

$$\forall t_1 \exists n (output(t_1) = n \wedge$$
$$\exists t_2 (t_2 \geq t_1 \wedge set(t_2) \wedge output(t_2 + 1) = 1 - n \wedge$$
$$\forall t_3 (t_1 \leq t_3 < t_2 \rightarrow output(t_3) = n))),$$

$$\forall t_1 (RequestPrint(t_1) \rightarrow \exists t_2 (t_2 \geq t_1 \wedge PrintedAt(t_2))),$$

$$\forall t \neg deadlocked(t).$$

∎

The use of explicit variables for points of time is awkward, especially since the specifications do not actually refer to concrete values of time. 'Eventually' simply means at any later time; the specification does not require that the file be printed within one minute or ten minutes. Temporal logic introduces new operators that enable abstract temporal relations like 'eventually' to be expressed directly within the logic.

Temporal logics are related to formal systems called *modal logics*. Modal logics express the distinction between what is *necessarily* true and what is *possibly* true. For example, the statement '7 *is a prime number*' is necessarily true because—given the definitions of the concepts in the statement—the statement is true always and everywhere. In contrast, the statement *the head of state of this country is a king* is possibly true, because its truth changes from place to place and from time to time. Temporal logic and modal logic are related because 'always' is similar to 'necessarily' and 'eventually' to 'possibly'.

Although temporal and modal logics first appeared in Greek philosophy, their vague concepts proved difficult to formalize and an acceptable formal semantics for modal logic was first given by Saul Kripke in 1959. In 1977, Amir Pnueli showed that temporal logic can specify properties of concurrent programs and that Kripke's semantics could be adapted to develop a formal theory of the temporal logic of programs. In this chapter and the next one we present the theory of linear temporal logic. Chapter 16 shows how the logic can be used for the specification of correctness properties of concurrent programs and for the verification of these properties. In that chapter, we will describe another temporal logic called computational tree logic that is also widely used in computer science.

13.2 Syntax and Semantics

13.2.1 Syntax

The initial presentation of the syntax and semantics of temporal logic will follow that used for general modal logics. We do this so that the presentation will be useful for readers who have a broader interest in modal logic and so that temporal logic can be seen within this wider context. Later, we specialize the presentation to a specific temporal logic that is used for the specification and verification of programs.

Definition 13.2 The syntax of *propositional temporal logic (PTL)* is defined like the syntax of propositional logic (Definition 2.1), except for the addition of two additional unary operators:

- $\Box$, read *always*,
- $\Diamond$, read *eventually*.   ∎

The discussion of syntax in Sect. 2.1 is extended appropriately: formulas of PTL are trees so they are unambiguous and various conventions are used to write the formulas as linear text. In particular, the two unary temporal logic operators have the same precedence as negation.

Example 13.3 The following are syntactically correct formulas in PTL:

$$p \wedge q, \quad \Box p, \quad \Diamond(p \wedge q) \to \Diamond p, \quad \Box\Box p \leftrightarrow \Box p, \quad \Diamond\Box p \leftrightarrow \Box\Diamond p, \quad \neg\Diamond p \wedge \Box\neg q.$$

The formula $\neg\Diamond p \wedge \Box\neg q$ is not ambiguous because the temporal operators and negation have higher precedence than the conjunction operator. The formula can be written $(\neg\Diamond p) \wedge (\Box\neg q)$ to distinguish it from $\neg(\Diamond p \wedge \Box\neg q)$.   ∎

13.2.2 Semantics

Informally, $\Box$ is a universal operator meaning 'for *any* time t in the future', while $\Diamond$ is an existential operator meaning 'for *some* time t in the future'. Two of the formulas from Example 13.1 can be written as follows in PTL:

$$\Box(reset \to \Diamond zero), \qquad \Box\neg deadlocked.$$

Interpretations of PTL formulas are based upon state transition diagrams. The intuitive meaning is that each state represents a *world* and a formula can have different truth values in different worlds. The transitions represent changes from one world to another.

Fig. 13.1 State transition
diagram

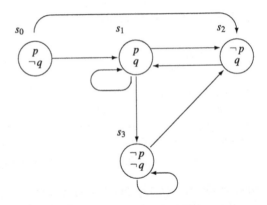

Definition 13.4 A *state transition diagram* is a directed graph. The nodes are *states*
and the edges are *transitions*. Each state is labeled with a set of propositional literals
such that clashing literals do not appear in any state.   ■

Example 13.5 Figure 13.1 shows a state transition diagram where states are circles
labeled with literals and transitions are arrows.   ■

In modal logic, *necessarily* means in *all* (reachable) worlds, whereas *possibly*
means in *some* (reachable) world. If a formula is possibly true, it can be true in
some worlds and false in another.

Example 13.6 Consider the formula $A = $ *the head of state of this country is a king.*
The formula is possibly true but not necessarily true. If the possible worlds are the
different countries, then at the present time A is true in Spain, false in Denmark
(because the head of state is a queen) and false in France (which does not have a
royal house). Even in a single country, the truth of A can change over time if a king
is succeeded by a queen or if a monarchy becomes a republic.   ■

Temporal logic is similar to modal logic except that the states are considered
to specify what is true at a particular point of time and the transitions define the
passage of time.

Example 13.7 Consider the formula $A = $ *it is raining in London today.* On the day
that this is being written, A is false. Let us consider each day as a state and the
transitions to be the passage of time from one day to the next. Even in London $\Box A$
(meaning *every day, it rains in London*) is not true, but $\Diamond A$ (meaning *eventually,
London will have a rainy day*) is certainly true.   ■

We are now ready to define the semantics of PTL. An interpretation is a state transition diagram and the truth value of a formula is computed using the assignments to
atomic propositions in each state and their usual meaning of the propositional operators. A formula that contains a temporal operator is interpreted using the transitions
between the states.

Fig. 13.2 Alternate
representation of the state
transition diagram in
Fig. 13.1

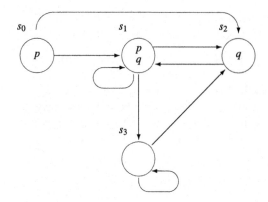

Definition 13.8 An *interpretation* $\mathscr{I}$ for a formula A in PTL is a pair $(\mathscr{S}, \rho)$, where $\mathscr{S} = \{s_1, \ldots, s_n\}$ is a set of states each of which is an assignment of truth values to the atomic propositions in A, $s_i : \mathscr{P} \to \{T, F\}$, and ρ is a binary relation on the states, $\rho \subseteq S \times S$.   ∎

  When displaying an interpretation graphically, the states are usually labeled only with the atomic propositions that are assigned T (Fig. 13.2). If an atom is not shown in the label of a state, it is assumed to be assigned F. Since it is clear how to transform one representation to the other, we will use whichever one is convenient.

  A binary relation can be considered to be a mapping from a state to a set of states $\rho : S \to 2^S$, so the relational notation $(s_1, s_2) \in \rho$ will usually be written functionally as $s_2 \in \rho(s_1)$.

Example 13.9 In Fig. 13.2:

$$s_0(p) = T, \quad s_0(q) = F,$$
$$s_1(p) = T, \quad s_1(q) = T,$$
$$s_2(p) = F, \quad s_2(q) = T,$$
$$s_3(p) = F, \quad s_3(q) = F.$$

$$\rho(s_0) = \{s_1, s_2\},$$
$$\rho(s_1) = \{s_1, s_2, s_3\},$$
$$\rho(s_2) = \{s_1\},$$
$$\rho(s_3) = \{s_2, s_3\}.$$

                                                                                    ∎

Definition 13.10 Let A be a formula in PTL. $v_{\mathscr{I},s}(A)$, the *truth value of A in s*, is defined by structural induction as follows:

- If A is $p \in \mathscr{P}$, then $v_{\mathscr{I},s}(A) = s(p)$.
- If A is $\neg A'$ then $v_{\mathscr{I},s}(A) = T$ iff $v_{\mathscr{I},s}(A') = F$.
- If A is $A' \vee A''$ then $v_{\mathscr{I},s}(A) = T$ iff $v_{\mathscr{I},s}(A') = T$ or $v_{\mathscr{I},s}(A'') = T$, and similarly for the other Boolean operators.
- If A is $\Box A'$ then $v_{\mathscr{I},s}(A) = T$ iff $v_{\mathscr{I},s'}(A') = T$ for *all* states $s' \in \rho(s)$.
- If A is $\Diamond A'$ then $v_{\mathscr{I},s}(A) = T$ iff $v_{\mathscr{I},s'}(A') = T$ for *some* state $s' \in \rho(s)$.

The notation $s \models_{\mathscr{I}} A$ is used for $v_{\mathscr{I},s}(A) = T$. When $\mathscr{I}$ is clear from the context, it can be omitted $s \models A$ iff $v_s(A) = T$. ∎

Example 13.11 Let us compute the truth value of the formula $\Box p \vee \Box q$ for each state s in Fig. 13.2.

- $\rho(s_0) = \{s_1, s_2\}$. Since $s_1 \models q$ and $s_2 \models q$, it follows that $s_0 \models \Box q$. By the semantics of $\vee$, $s_0 \models \Box p \vee \Box q$.
- $s_3 \in \rho(s_1)$, but $s_3 \not\models p$ and $s_3 \not\models q$, so $s_1 \not\models \Box p$ and $s_1 \not\models \Box q$. Therefore, $s_1 \not\models \Box p \vee \Box q$.
- $\rho(s_2) = \{s_1\}$. Since $s_1 \models p$, we have $s_2 \models \Box p$ and $s_2 \models \Box p \vee \Box q$.
- $s_3 \in \rho(s_3)$. $s_3 \not\models \Box p \vee \Box q$ by the same argument used for s_1. ∎

13.2.3 Satisfiability and Validity

The definition of semantic properties in PTL is more complex than it is in propositional or first-order logic, because an interpretation consists of both states and truth values.

Definition 13.12 Let A be a formula in PTL.

- A is *satisfiable* iff there is an interpretation $\mathscr{I} = (\mathscr{S}, \rho)$ for A and a state $s \in \mathscr{S}$ such that $s \models_{\mathscr{I}} A$.
- A is *valid* iff for all interpretations $\mathscr{I} = (\mathscr{S}, \rho)$ for A and for all states $s \in \mathscr{S}$, $s \models_{\mathscr{I}} A$. Notation: $\models A$. ∎

Example 13.13 The analysis we did for the formula $A = \Box p \vee \Box q$ in Example 13.11 shows that A is satisfiable because $s_0 \models_{\mathscr{I}} A$ or because $s_2 \models_{\mathscr{I}} A$. The formulas A is not valid because $s_1 \not\models_{\mathscr{I}} A$ or because $s_3 \not\models_{\mathscr{I}} A$. ∎

We leave it as an exercise to show that any valid formula of propositional logic is a valid formula of PTL, as is any *substitution instance* of a valid propositional formula obtained by substituting PTL formulas uniformly for propositional letters. For example, $\Box p \rightarrow (\Box q \rightarrow \Box p)$ is valid since it is a substitution instance of the valid propositional formula $A \rightarrow (B \rightarrow A)$.

There are other formulas of PTL that are valid because of properties of temporal logic and not as instances of propositional validities. We will prove the validity of two formulas directly from the semantic definition. The first establishes a duality between $\Box$ and $\Diamond$, and the second is the distribution of $\Box$ over $\rightarrow$, similar to the distribution of $\forall$ over $\rightarrow$.

Theorem 13.14 (Duality) $\models \Box p \leftrightarrow \neg \Diamond \neg p$.

Proof Let $\mathscr{I} = (\mathscr{S}, \rho)$ be an arbitrary interpretation for the formula and let s be an arbitrary state in $\mathscr{S}$. Assume that $s \models \Box p$, and suppose that $s \models \Diamond \neg p$. Then there exists a state $s' \in \rho(s)$ such that $s' \models \neg p$. Since $s \models \Box p$, for all states $t \in \rho(s)$, $t \models p$, in particular, $s' \models p$, contradicting $s' \models \neg p$. Therefore, $s \models \neg \Diamond \neg p$. Since $\mathscr{I}$ and s were arbitrary we have proved that $\models \Box p \rightarrow \neg \Diamond \neg p$. We leave the converse as an exercise. ∎

Theorem 13.15 $\models \Box(p \rightarrow q) \rightarrow (\Box p \rightarrow \Box q)$.

Proof Suppose, to the contrary, that there is an interpretation $\mathscr{I} = (S, \rho)$ and a state $s \in S$, such that $s \models \Box(p \rightarrow q)$ and $s \models \Box p$, but $s \models \neg \Box q$. By Theorem 13.14, $s \models \neg \Box q$ is equivalent to $s \models \Diamond \neg q$, so there exists a state $s' \in \rho(s)$ such that $s' \models \neg q$. By the first two assumptions, $s' \models p \rightarrow q$ and $s' \models p$, which imply $s' \models q$, a contradiction. ∎

13.3 Models of Time

In modal and temporal logics, different logics can be obtained by placing restrictions on the transition relation. In this section, we discuss the various restrictions, leading up to the ones that are appropriate for the temporal logics used in computer science. For each restriction on the transition relation, we give a formula that characterizes interpretations with that restriction. Proofs of the characterizations are given in a separate subsection.

Reflexivity

Definition 13.16 An interpretation $\mathscr{I} = (\mathscr{S}, \rho)$ is *reflexive* iff ρ is a reflexive relation: for all $s \in \mathscr{S}$, $(s, s) \in \rho$, or $s \in \rho(s)$ in functional notation. ∎

Consider the formula $\Diamond running$, whose intuitive meaning is *eventually the program is in the state 'running'*. Obviously, if a program is running *now*, then there is an reachable state (namely, *now*) in which the program is running. Thus it is reasonable to require that interpretations for properties of programs be reflexive.

Theorem 13.17 *An interpretation with a reflexive relation is characterized by the formula $\Box A \rightarrow A$ (or, by duality, by the formula $A \rightarrow \Diamond A$).*

Transitivity

Definition 13.18 An interpretation $\mathscr{I} = (\mathscr{S}, \rho)$ is *transitive* iff ρ is a transitive relation: for all $s_1, s_2, s_3 \in \mathscr{S}$, $s_2 \in \rho(s_1) \wedge s_3 \in \rho(s_2) \rightarrow s_3 \in \rho(s_1)$.   ∎

It is natural to require that interpretations be transitive. Consider a situation where we have proved that $s_1 \models \Diamond running$ because $s_2 \models running$ for $s_2 \in \rho(s_1)$, and, furthermore, we have proved $s_2 \models \Diamond running$ because $s_3 \models running$ for $s_3 \in \rho(s_2)$. It would be very strange if $s_3 \notin \rho(s_1)$ and could not be used to prove $s_1 \models \Diamond running$.

Theorem 13.19 *An interpretation with a transitive relation is characterized by the formula $\Box A \rightarrow \Box\Box A$ (or by the formula $\Diamond\Diamond A \rightarrow \Diamond A$).*

Example 13.20 In Fig. 13.2, ρ is not transitive since $s_1 \in \rho(s_2)$ and $s_3 \in \rho(s_1)$ but $s_3 \notin \rho(s_2)$. This leads to the anomalous situation where $s_2 \models \Box p$ but $s_2 \not\models \Box\Box p$.   ∎

Corollary 13.21 *In an interpretation that both is reflexive and transitive, $\models \Box A \leftrightarrow \Box\Box A$ and $\models \Diamond A \leftrightarrow \Diamond\Diamond A$.*

Linearity

Definition 13.22 An interpretation $\mathscr{I} = (\mathscr{S}, \rho)$ is *linear* if ρ is a function, that is, for all $s \in \mathscr{S}$, there is at most one $s' \in \mathscr{S}$ such that $s' \in \rho(s)$.

It might appear that a linear temporal logic would be limited to expressing properties of sequential programs and could not express properties of concurrent programs, where each state can have several possible successors depending on the interleaving of the statements of the processes. However, linear temporal logic is successful precisely in the context of concurrent programs because there is an implicit universal quantification in the definitions.

Suppose we want to prove that a program satisfies a correctness property expressed as a temporal logic formula like $A = \Box\Diamond running$: in any state, the execution will eventually reach a state in which the computation is *running*. The program will be correct if this formula is true in *every* possible execution of the program obtained by interleaving the instructions of its processes. Each interleaving can be considered as a single linear interpretation, so if we prove $\models_\mathscr{I} A$ for an arbitrary linear interpretation $\mathscr{I}$, then the correctness property holds for the program.

Discreteness

Although the passage of time is often considered to be continuous and expressible by real numbers, the execution of a program is considered to be a sequence of *discrete* steps, where each step consists of the execution of a single instruction of the CPU. Thus it makes sense to express the concept of the *next* instant in time. To express discrete steps in temporal logic, an additional operator is added.

Definition 13.23 The unary operator $\bigcirc$ is called *next*.   ∎

The definition of the truth value of a formula is extended as expected:

Definition 13.24 If A is $\bigcirc A'$ then $v_{\mathscr{I},s}(A) = T$ iff $v_{\mathscr{I},s'}(A') = T$ for some $s' \in \rho(s)$.   ∎

The next operator is self-dual in a linear interpretation.

Theorem 13.25 *A linear interpretation whose relation ρ is a function is characterized by the formula* $\bigcirc A \leftrightarrow \neg \bigcirc \neg A$.

The operator $\bigcirc$ plays a crucial role in the theory of temporal logic and in algorithms for deciding properties like satisfiability, but it is rarely used to express properties of programs. In a concurrent program, not much can be said about what happens *next* since we don't know which operation will be executed in the next step. Furthermore, we want a correctness statement to hold regardless of how the interleaving selects a *next* operation. Therefore, properties are almost invariably expressed in terms of *always* and *eventually*, not in terms of *next*.

13.3.1 Proofs of the Correspondences *

The following definition enables us to talk about the structure (the states and transitions) of an entire class of interpretations while abstracting away from the assignment to atomic propositions in each state. A frame is obtained from an interpretation by ignoring the assignments in the states; conversely, a interpretation is obtained from a frame by associating an assignment with each state.

Definition 13.26 A *frame* $\mathscr{F}$ is a pair $(\mathscr{W}, \rho)$, where $\mathscr{W}$ is a set of states and ρ is a binary relation on states. An interpretation $\mathscr{I} = (\mathscr{S}, \rho)$ is *based on a frame* $\mathscr{F} = (\mathscr{W}, \rho)$ iff there is a one-to-one mapping from $\mathscr{S}$ onto $\mathscr{W}$.

A PTL formula A *characterizes* a class of frames iff for every $\mathscr{F}_i$ in the class, the set of interpretations $\mathscr{I}$ based on $\mathscr{F}_i$ is the same as the set of interpretations in which A is true.   ∎

Theorems 13.17, 13.19 and 13.25 are more precisely stated as follows: the formulas $\Box A \to A$, $\Box A \to \Box\Box A$ and $\bigcirc A \leftrightarrow \neg \bigcirc \neg A$ characterize the sets of reflexive, transitive, and linear *frames*, respectively.

Proof of Theorem 13.17 Let $\mathscr{F}_i$ be a reflexive frame, let $\mathscr{I}$ be an arbitrary interpretation based on $\mathscr{F}_i$, and suppose that $\not\models_{\mathscr{I}} \Box A \to A$. Then there is a state $s \in \mathscr{S}$ such that $s \models_{\mathscr{I}} \Box A$ and $s \not\models_{\mathscr{I}} A$. By the definition of $\Box$, for *any* state $s' \in \rho(s)$, $s' \models_{\mathscr{I}} A$. By reflexivity, $s \in \rho(s)$, so $s \models_{\mathscr{I}} A$, a contradiction.

Conversely, suppose that $\mathscr{F}_i$ is not reflexive, and let $s \in \mathscr{S}$ be a state such that $s \notin \rho(s)$. If $\rho(s)$ is empty, $\Box p$ is vacuously true in s; by assigning F to $v_s(p)$, $s \not\models_{\mathscr{I}} \Box p \to p$. If $\rho(s)$ is non-empty, let $\mathscr{I}$ be an interpretation based on $\mathscr{F}_i$ such that $v_s(p) = F$ and $v_{s'}(p) = T$ for all $s' \in \rho(s)$. These assignments are well-defined since $s \notin \rho(s)$. Then $s \not\models_{\mathscr{I}} \Box p \to p$.   ∎

Proof of Theorem 13.19 Let $\mathscr{F}_i$ be a transitive frame, let $\mathscr{I}$ be an arbitrary interpretation based on $\mathscr{F}_i$, and suppose that $\not\models_{\mathscr{I}} \Box A \to \Box\Box A$. Then there is an $s \in \mathscr{S}$ such that $s \models_{\mathscr{I}} \Box A$ and $s \not\models_{\mathscr{I}} \Box\Box A$. From the latter formula, there must be an $s' \in \rho(s)$ such that $s' \not\models_{\mathscr{I}} \Box A$, and, then, there must be an $s'' \in \rho(s')$ be such that $s'' \not\models_{\mathscr{I}} A$. But $s \models_{\mathscr{I}} \Box A$, and by transitivity, $s'' \in \rho(s)$, so $s'' \models_{\mathscr{I}} A$, a contradiction.

Conversely, suppose that $\mathscr{F}_i$ is not transitive, and let $s, s', s'' \in \mathscr{S}$ be states such that $s' \in \rho(s)$, $s'' \in \rho(s')$, but $s'' \notin \rho(s)$. Let $\mathscr{I}$ be an interpretation based on $\mathscr{F}_i$ which assigns T to p in all states in $\rho(s)$ and F to p in s'', which is well-defined since $s'' \notin \rho(s)$. Then $s \models_{\mathscr{I}} \Box p$, but $s \not\models_{\mathscr{I}} \Box\Box p$. If there are only two states, s' need not be distinct from s. A one state frame is necessarily transitive, possibly vacuously if the relation is empty.   ∎

We leave the proof of Theorem 13.25 as an exercise.

13.4 Linear Temporal Logic

In the context of programs, the natural interpretations of temporal logic formulas are discrete, reflexive, transitive and linear. There is another restriction that simplifies the presentation: the transition function must be total so that each state has exactly one next state. An interpretation for a computation that terminates in state s is assumed to have a transition from s to s.

Definition 13.27 *Linear temporal logic (LTL)* is propositional temporal logic whose interpretations are limited to transitions which are discrete, reflexive, transitive, linear and total.   ∎

These interpretations can be represented as infinite paths:

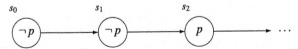

Since there is only one transition out of each state, it need not be explicitly represented, so interpretations in LTL are defined to be paths of states:

Definition 13.28 An interpretation for an LTL formula A is a *path* of *states*:

$$\sigma = s_0, s_1, s_2, \ldots,$$

where each s_i is an assignment of truth values to the atomic propositions in A, $s_i : \mathscr{P} \to \{T, F\}$. Given σ, σ_i is the path that is the ith suffix of σ:

$$\sigma_i = s_i, s_{i+1}, s_{i+2}, \ldots.$$

$v_\sigma(A)$, the *truth value of A in σ*, is defined by structural induction:

- If A is $p \in \mathcal{P}$, then $v_\sigma(A) = s_0(p)$.
- If A is $\neg A'$ then $v_\sigma(A) = T$ iff $v_\sigma(A') = F$.
- If A is $A' \vee A''$ then $v_\sigma(A) = T$ iff $v_\sigma(A') = T$ or $v_\sigma(A'') = T$, and similarly for the other Boolean operators.
- If A is $\bigcirc A'$ then $v_\sigma(A) = T$ iff $v_{\sigma_1}(A') = T$.
- If A is $\Box A'$ then $v_\sigma(A) = T$ iff $v_{\sigma_i}(A') = T$ for *all* $i \geq 0$.
- If A is $\Diamond A'$ then $v_\sigma(A) = T$ iff $v_{\sigma_i}(A') = T$ for *some* $i \geq 0$.

If $v_\sigma(A) = T$, we write $\sigma \models A$.   ∎

Definition 13.29 Let A be a formula in LTL. A is *satisfiable* iff there is an interpretation σ for A such that $\sigma \models A$. A is *valid* iff for all interpretations σ for A, $\sigma \models A$. Notation: $\models A$.   ∎

Definition 13.30 A formula of the form $\bigcirc A$ or $\neg \bigcirc A$ is a *next* formula. A formula of the form $\Diamond A$ or $\neg \Box A$ is a *future* formula.   ∎

13.4.1 Equivalent Formulas in LTL

This section presents LTL formulas that are equivalent because of their temporal properties. Since any substitution instance of a formula in propositional logic is also an LTL formula, the equivalences in Sect. 2.3.3 also hold.

The equivalences are expressed in terms of an atom p but the intention is that they hold for arbitrary LTL formulas A.

The following formulas are direct consequences of our restriction of interpretations in LTL. The first three hold because interpretations are total, while the fourth holds because of linearity.

Theorem 13.31

$$\models \Box p \rightarrow \bigcirc p, \qquad \models \bigcirc p \rightarrow \Diamond p, \qquad \models \Box p \rightarrow \Diamond p, \qquad \models \bigcirc p \leftrightarrow \neg \bigcirc \neg p.$$

Inductive

The following theorem is extremely important because it provides an method for proving properties of LTL formulas inductively.

Theorem 13.32

$$\models \Box p \leftrightarrow p \wedge \bigcirc \Box p, \qquad \models \Diamond p \leftrightarrow p \vee \bigcirc \Diamond p.$$

These formulas can be easily understood by reading them in words: For a formula to be always true, p must be true today and, in addition, p must be always true tomorrow. For a formula to be true eventually, either p is true today or it must be true in some future of tomorrow.

We prove the first formula; the second follows by duality.

Proof Let σ be an arbitrary interpretation and assume that $\sigma \models \Box p$. By definition, $\sigma_i \models p$ for all $i \geq 0$; in particular, $\sigma_0 \models p$. But σ_0 is the same as σ, so $\sigma \models p$. If $\sigma \not\models \bigcirc \Box p$, then $\sigma_1 \not\models \Box p$, so for some $i \geq 1$, $\sigma_i \not\models p$, contradicting $\sigma \models \Box p$.

Conversely, assume that $\sigma \models p \wedge \bigcirc \Box p$. We prove by induction that $\sigma_i \models p \wedge \bigcirc \Box p$ for all $i \geq 0$. Since $\models A \wedge B \rightarrow A$ is a valid formula of propositional logic, we can conclude that $\sigma_i \models p$ for all $i \geq 0$, that is, $\sigma \models \Box p$.

The base case is immediate from the assumption since $\sigma_0 = \sigma$. Assume the inductive hypothesis that $\sigma_i \models p \wedge \bigcirc \Box p$. By definition of the semantics of $\bigcirc$, $\sigma_{i+1} \models \Box p$, that is, for all $j \geq i + 1$, $\sigma_j \models p$, in particular $\sigma_{i+1} \models p$. Furthermore, for $j \geq i + 2$, $\sigma_j \models p$, so $\sigma_{i+2} \models \Box p$ and $\sigma_{i+1} \models \bigcirc \Box p$.   ∎

Induction in LTL is based upon the following valid formula:

$$\models \Box(p \rightarrow \bigcirc p) \rightarrow (p \rightarrow \Box p).$$

The base case is to show that p holds in a state. The inductive assumption is p and the inductive step is to show that $p \rightarrow \bigcirc p$. When these two steps have been performed, we can conclude that $\Box p$.

Instead of proving the following equivalences semantically as in Theorem 13.32, we will prove them deductively in Chap. 14. By the soundness of the deductive system, they are valid.

Distributivity

The operators $\Box$ and $\bigcirc$ distribute over conjunction:

$$\models \Box(p \wedge q) \leftrightarrow (\Box p \wedge \Box q),$$
$$\models \bigcirc(p \wedge q) \leftrightarrow (\bigcirc p \wedge \bigcirc q).$$

The next operator also distributes over disjunction because it is self-dual, but $\Box$ only distributes over disjunction in one direction:

$$\models (\Box p \vee \Box q) \rightarrow \Box(p \vee q),$$
$$\models \bigcirc(p \vee q) \leftrightarrow (\bigcirc p \vee \bigcirc q).$$

By duality, there are similar formulas for $\Diamond$:

$$\models \Diamond(p \vee q) \leftrightarrow (\Diamond p \vee \Diamond q),$$
$$\models \Diamond(p \wedge q) \rightarrow (\Diamond p \wedge \Diamond q).$$

Similarly, $\Box$ and $\bigcirc$ distribute over implication in one direction, while $\bigcirc$ distributes in both directions:

$$\models \Box(p \rightarrow q) \rightarrow (\Box p \rightarrow \Box q),$$
$$\models (\Diamond p \rightarrow \Diamond q) \rightarrow \Diamond(p \rightarrow q),$$
$$\models \bigcirc(p \rightarrow q) \leftrightarrow (\bigcirc p \rightarrow \bigcirc q).$$

Example 13.33 Here is a counterexample to $\models (\Diamond p \wedge \Diamond q) \rightarrow \Diamond (p \wedge q)$:

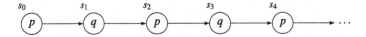

The atomic proposition p is true in even-numbered states, while q is true in odd-numbered states, but there is no state in which both are true.   ■

Commutativity

The operator $\bigcirc$ commutes with $\Box$ and $\Diamond$, but $\Box$ and $\Diamond$ commute only in one direction:

$$\models \Box \bigcirc p \leftrightarrow \bigcirc \Box p,$$
$$\models \Diamond \bigcirc p \leftrightarrow \bigcirc \Diamond p,$$
$$\models \Diamond \Box p \rightarrow \Box \Diamond p.$$

Be careful to distinguish between $\Box \Diamond p$ and $\Diamond \Box p$. The formula $\Box \Diamond p$ means *infinitely often*: p is not required to hold continuously, but at any state it will hold at some future state.

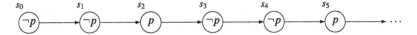

The formula $\Diamond \Box p$ means *for all but a finite number of states*: in a path $\sigma = s_0, s_1, s_2, \ldots,$ there is a natural number n such that p is true in all states in $\sigma_n = s_n, s_{n+1}, s_{n+2}, \ldots$.

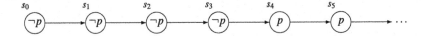

Theorem 13.34 $\models (\Diamond \Box p \wedge \Box \Diamond q) \rightarrow \Box \Diamond (p \wedge q).$

Once p becomes always true, it will be true in the (infinite number of) states where q is true. We leave the proof as an exercise.

The diagram in Example 13.33 is also a counterexample to the formula: $\models (\Box \Diamond p \wedge \Box \Diamond q) \rightarrow \Box \Diamond (p \wedge q).$

Collapsing

In a formula without the $\bigcirc$ operator, no more than two temporal operators need appear in a sequence. A sequence of identical operators $\Box$ or $\Diamond$ is equivalent to a single occurrence and a sequence of three non-identical operators collapses to a pair of operators:

$$\models \Box\Box p \leftrightarrow \Box p,$$
$$\models \Diamond\Diamond p \leftrightarrow \Diamond p,$$
$$\models \Box\Diamond\Box p \leftrightarrow \Diamond\Box p,$$
$$\models \Diamond\Box\Diamond p \leftrightarrow \Box\Diamond p.$$

13.5 Semantic Tableaux

The method of semantic tableaux is a decision procedure for satisfiability in LTL. The construction of a semantic tableau for a formula of LTL is more complex than that it is for a formula of propositional logic for two reasons:

First, to show that a formula in propositional logic is satisfiable, one need only find a *single* assignment to the atomic propositions that makes the formula evaluate to true. In LTL, however, there are many different assignments, one for each state. Therefore, we need to distinguish between ordinary nodes in the tableau used to decompose formulas such as $p \wedge q$ and $p \vee q$ from nodes that represent different states. For example, if $\bigcirc p$ is to be true in state s, then p must be assigned T in the state s' that follows s, but p could be assigned either T or F in s itself.

The second complication comes from future formulas like $\Diamond p$. For future formulas, it is not sufficient that they are consistent with the other subformulas; $\Diamond p$ requires that there actually exist a subsequent state where p is assigned T. This is similar to the case of $\exists x p(x)$ in first-order logic: we must demonstrate that a value a exists such that $p(a)$ is true. In first-order logic, this was simple, because we just chose new constant symbols from a countable set. In LTL, to establish the existence or non-existence of a state that *fulfills* a future formula requires an analysis of the graph of states constructed when the tableau is built.

13.5.1 The Tableau Rules for LTL

The tableau rules for LTL consist of the rules for propositional logic shown in Fig. 2.8, together with the following new rules, where next formulas are called X-formulas:

α	α_1	α_2	β	β_1	β_2	X	X_1
$\Box A$	A	$\bigcirc\Box A$	$\Diamond A$	A	$\bigcirc\Diamond A$	$\bigcirc A$	A
$\neg\Diamond A$	$\neg A$	$\neg\bigcirc\Diamond A$	$\neg\Box A$	$\neg A$	$\neg\bigcirc\Box A$	$\neg\bigcirc A$	$\neg A$

The Rules for α- and β-Formulas

The rules for the α- and β-formulas are based on Theorem 13.32:

- If $\Box A$ is true in a state s, then A is true in s *and* A must continue to be true in *all* subsequent states starting at the *next* state s'.
- If $\Diamond A$ is true in a state s, then either A is true in s *or* A will eventually become true in *some* subsequent state starting at the *next* state s'.

The Rule for X-Formulas

Consider now the tableau obtained for the formula $A = (p \vee q) \wedge \bigcirc(\neg p \wedge \neg q)$ after applying the rules for α- and β-formulas:

$$(p \vee q) \wedge \bigcirc(\neg p \wedge \neg q)$$
$$\downarrow$$
$$p \vee q, \ \bigcirc(\neg p \wedge \neg q)$$
$$\swarrow \qquad \searrow$$

$$\boxed{p, \ \bigcirc(\neg p \wedge \neg q)} \qquad \boxed{q, \ \bigcirc(\neg p \wedge \neg q)}$$

In a model σ for A, either $v_\sigma(p) = s_0(p) = T$ or $v_\sigma(q) = s_0(q) = T$, and this is expressed by the two leaf nodes that contain the atomic propositions. Since no more rules for α- and β-formulas are applicable, we have complete information on the assignment to atomic propositions in the initial state s_0. These nodes, therefore, define *states*, indicated by the frame around the node.

These nodes contain additional information: in order to satisfy the formula A, the formula $\bigcirc(\neg p \wedge \neg q)$ must evaluate to T in σ_0. Therefore, the formula $\neg p \wedge \neg q$ must evaluate to T in σ_1. The application of the rule for X-formulas begins the construction of the new state s_1:

$$(p \vee q) \wedge \bigcirc(\neg p \wedge \neg q)$$
$$\downarrow$$
$$p \vee q, \ \bigcirc(\neg p \wedge \neg q)$$
$$\swarrow \qquad \searrow$$

$$\boxed{p, \ \bigcirc(\neg p \wedge \neg q)} \qquad \boxed{q, \ \bigcirc(\neg p \wedge \neg q)}$$
$$\downarrow \qquad\qquad\qquad \downarrow$$
$$\neg p \wedge \neg q \qquad\qquad \neg p \wedge \neg q$$
$$\downarrow \qquad\qquad\qquad \downarrow$$
$$\boxed{\neg p, \ \neg q} \qquad\qquad \boxed{\neg p, \ \neg q}$$

The literals in s_0 are *not* copied to the labels of the nodes created by the application of the rule for the X-formula because whatever requirements exist on the assignment in s_0 are not relevant to what happens in s_1.

On both branches, the new node is labeled by the formula $\neg p \wedge \neg q$ and an application of the rule for the propositional α-formula gives $\{\neg p, \neg q\}$ as the label of the next node. Since this node no longer contains α- or β-formulas, it defines a new state s_1.

The construction of the tableau is now complete and we have two open branches. Therefore, we can conclude that any model for A must be consistent with one of the following graphs:

This structure is not an interpretation. First, it is not total since there is no transition from s_1, but this is easily fixed by adding a self-loop to the final state:

More importantly, we have not specified the value of the second literal in either of the possible states s_0. However, the structures are *Hintikka structures*, which can be extended to interpretations by specifying the values of all atoms in each state.

Future Formulas

Consider the formula $A = \neg(\Box(p \wedge q) \to \Box p)$ which is the negation of a valid formula. Here is a semantic tableau, where (by duality) we have implicitly changed $\neg\Box$ to $\Diamond\neg$ for clarity:

$$\neg(\Box(p \wedge q) \to \Box p)$$
$$\downarrow$$
$$\Box(p \wedge q),\ \Diamond\neg p$$
$$\downarrow$$
$$p \wedge q,\ \bigcirc\Box(p \wedge q),\ \Diamond\neg p$$
$$\downarrow$$
$$p,\ q,\ \bigcirc\Box(p \wedge q),\ \Diamond\neg p$$
$$\swarrow \qquad \searrow$$

$$p,\ q,\ \bigcirc\Box(p \wedge q),\ \neg p \qquad \boxed{p,\ q,\ \bigcirc\Box(p \wedge q),\ \bigcirc\Diamond\neg p}$$
$$\times$$

The left-hand branch closes, while the right-hand leaf defines a state s_0 in which p and q must be true. When rule for the X-formula is applied to this node, a new node is created that is labeled by $\{\Box(p \wedge q),\ \Diamond\neg p\}$. But this is the same set of formulas that labels the second node in the tableau. It is clear that the continuation of the construction will create an infinite structure:

Something is wrong since A is unsatisfiable and its tableau should close!

This structure is a Hintikka structure (no node contains clashing literals and for every α-, β- and X-formula, the Hintikka conditions hold). However, the structure cannot be extended to model for A, since the future subformula $\Diamond \neg p$ is not *fulfilled*; that is, the structure promises to eventually produce a state in which $\neg p$ is true but defers forever the creation of such a state.

Finite Presentation of an Interpretation

There are only a finite number of *distinct* states in an interpretation for an LTL formula A since every state is labeled with a subset of the atomic propositions appearing in A and there are a finite number of such subsets. Therefore, although an interpretation is an infinite path, it can be *finitely presented* by reusing existing states instead of creating new ones. The infinite structure above can be finitely presented as follows:

13.5.2 Construction of Semantic Tableaux

The construction of semantic tableaux for LTL formulas and the proof of an algorithm for the decidability of satisfiability is contained in the following four subsections. First, we describe the construction of the tableau; then, we show how a Hintikka structure is defined by an open tableau; third, we extract a linear structure which can be extended to an interpretation; and finally, we show how to decide if future formulas are fulfilled.

The meaning of the following definition will become clear in the following subsection, but it is given here so that we can use it in the algorithm for constructing a tableau.

Definition 13.35 A *state node* in a tableau is a node l such that its label $U(l)$ contains only literals and next formulas, and there are no complementary pairs of literals in $U(l)$. ∎

Algorithm 13.36 (Construction of a semantic tableau)
Input: An LTL formula A.
Output: A semantic tableau $\mathcal{T}$ for A.

Each node of $\mathscr{T}$ is labeled with a set of formulas. Initially, $\mathscr{T}$ consists of a single node, the root, labeled with the singleton set $\{A\}$. The tableau is built inductively as follows. Choose an unmarked leaf l labeled with a set of formulas $U(l)$ and perform one of the following steps:

- If there is a complementary pair of literals $\{p, \neg p\} \subseteq U(l)$, mark the leaf *closed* $\times$. If $U(l)$ is a set of literals but no pair is complementary, mark the leaf *open* $\odot$.
- If $U(l)$ is not a set of literals, choose $A \in U(l)$ which is an α-formula. Create a new node l' as a child of l and label l' with:

$$U(l') = (U(l) - \{A\}) \cup \{\alpha_1, \alpha_2\}.$$

(In the case that A is $\neg\neg A_1$, there is no α_2.)

- If $U(l)$ is not a set of literals, choose $A \in U(l)$ which a β-formula. Create two new nodes l' and l'' as children of l. Label l' with:

$$U(l') = (U(l) - \{A\}) \cup \{\beta_1\},$$

and label l'' with:

$$U(l'') = (U(l) - \{A\}) \cup \{\beta_2\}.$$

- If l is a state node (Definition 13.35) with at least one next formula, let:

$$\{\bigcirc A_1, \ldots, \bigcirc A_m, \neg \bigcirc A_{m+1}, \ldots, \neg \bigcirc A_n\}$$

be the set of next formulas in $U(l)$. Create a new node l' as a child of l and label l' with:

$$U(l') = \{A_1, \ldots, A_m, \neg A_{m+1}, \ldots, \neg A_n\}.$$

If $U(l') = U(l'')$ for a *state node* l'' that already exists in the tableau, do not create l'; instead connect l to l''.

The construction terminates when every leaf is marked $\times$ or $\odot$.   ∎

We leave it as an exercise to show that the construction always terminates.

Definition 13.37 A tableau whose construction has terminated is a *completed tableau*. A completed tableau is *closed* if all leaves are marked closed and there are no cycles. Otherwise, it is *open*.   ∎

Example 13.38 Here is a completed open semantic tableau with *no* leaves:

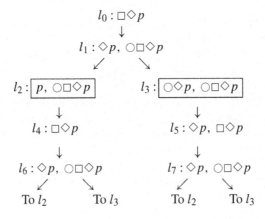

13.5.3 From a Semantic Tableau to a Hintikka Structure

The next step is to construct a structure from an open tableau, to define the conditions for a structure to be a Hintikka structure and to prove that the structure resulting from the tableau satisfies those conditions. The definition of a structure is similar to the definition of an interpretation for *PTL* formulas (Definition 13.8); the difference is that the labels of a state are sets of formulas, not just sets of atomic propositions that are assigned true. To help understand the construction, you might want to refresh your memory by re-reading Sect. 2.7.2 on the definition and use of Hintikka structures in propositional logic.

Definition 13.39 A *structure* $\mathcal{H}$ for a formula A in LTL is a pair $(\mathcal{S}, \rho)$, where $\mathcal{S} = \{s_1, \ldots, s_n\}$ is a set of states each of which is labeled by a subset of formulas built from the atomic propositions in A and ρ is a binary relation on states, $\rho \subseteq S \times S$. ∎

As before, functional notation may be used $s_2 \in \rho(s_1)$.

The states of the structure will be the state nodes of the tableau. However, the labels of the states must include more than the literals that label the nodes in the tableau. To obtain a Hintikka structure, the state in the structure must also include the formulas whose decomposition eventually led to each literal.

Example 13.40 In Example 13.38, state node l_2 will define a state in the structure that is labeled with p, since p must be assigned true in any interpretation containing that state. In addition, the state in the structure must also include $\Diamond p$ from l_1 (because p in l_2 resulted from the decomposition of $\Diamond p$), as well as $\Box \Diamond p$ from l_0 (because $\Diamond p$ in l_1 resulted from the decomposition of $\Box \Diamond p$). ∎

The transitions in the structure are defined by paths between state nodes.

Definition 13.41 A *state path* is a path $(l_0, l_1, \ldots, l_{k-1}, l_k)$ through connected nodes in the tableau, such that l_0 is a state node or the root of the tableau, l_k is a state node, and none of $\{l_1, \ldots, l_{k-1}\}$ are state nodes. It is possible that $l_0 = l_k$ so that the set $\{l_1, \ldots, l_{k-1}\}$ is empty.   ∎

Given a tableau, a structure can be defined by taking the state nodes as the states and defining the transitions by the state paths. The label of a state is the union of all formulas that appear on incoming state paths (not including the first state of the path unless it is the root). The formal definition is:

Definition 13.42 Let $\mathscr{T}$ be an open tableau for an LTL formula A. The structure $\mathscr{H}$ constructed from $\mathscr{T}$ is:

- $\mathscr{S}$ is the set of state nodes.
- Let $s \in \mathscr{S}$. Then $s = l$ for some node l in the tableau. Let $\pi^i = (l_0^i, l_1^i, \ldots, l_{k_i}^i = l)$ be a state path terminating in the node l and let:

$$U^i = U(l_1^i) \cup \cdots \cup U(l_{k_i}^i)$$

or

$$U^i = U(l_0^i) \cup \cdots \cup U(l_{k_i}^i)$$

if l_0^i is the root. Label s by the set of formulas:

$$U_i = \cup_i U^i,$$

where the union is taken over all i such that π^i is a state path terminating in $l = s$.
- $s' \in \rho(s)$ iff there is a state path from s to s'.   ∎

It is possible to obtain several disconnected structures from the tableau for a formula such as $\Diamond p \vee \Diamond q$, but this is no problem as the formula can be satisfiable if and only if at least one of the structures leads to a model.

Now that we know how the structure is constructed from the tableau, it is possible to optimize Algorithm 13.36. Change:

> For a *state node* l', if $U(l') = U(l'')$ for a *state node* l'' that already exists in the tableau, do not create l'; instead connect l to l''.

so that it applies to any node l' in the tableau, not just to state nodes, *provided* that this doesn't create a cycle not containing a state node.

Fig. 13.3 Structure for $\Box(\Diamond(p \land q) \land \Diamond(\neg p \land q) \land \Diamond(p \land \neg q))$

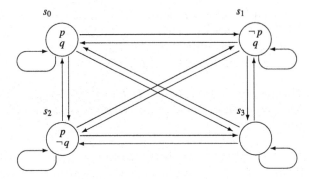

Example 13.43 Here is an optimized tableau corresponding to the one in Example 13.38:

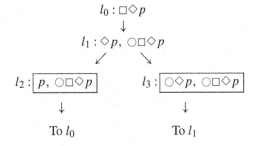

$l_0 : \Box\Diamond p$

$l_1 : \Diamond p, \bigcirc\Box\Diamond p$

$l_2 : \boxed{p, \bigcirc\Box\Diamond p}$ $l_3 : \boxed{\Diamond p, \bigcirc\Box\Diamond p}$

To l_0 To l_1

and here is the structure constructed from this semantic tableau:

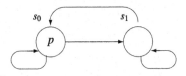

where $s_0 = l_2$ and $s_1 = l_3$. To save space, each state s_i is labeled only with the positive literals in U_i. ∎

Example 13.44 Let:

$$A = \Box(\Diamond(p \land q) \land \Diamond(\neg p \land q) \land \Diamond(p \land \neg q)).$$

The construction of the tableau for A is left as an exercise. The structure obtained from the tableau is shown in Fig. 13.3. ∎

Definition 13.45 Let $\mathscr{H} = (\mathscr{S}, \rho)$ be a structure for an LTL formula A. $\mathscr{H}$ is a *Hintikka structure for A* iff $A \in s_0$ and for all states s_i the following conditions hold for U_i, the set of formulas labeling s_i:

1. For all atomic propositions p in A, either $p \notin U_i$ or $\neg p \notin U_i$.
2. If $\alpha \in U_i$, then $\alpha_1 \in U_i$ *and* $\alpha_2 \in U_i$.
3. If $\beta \in U_i$, then $\beta_1 \in U_i$ *or* $\beta_2 \in U_i$.
4. If $X \in U_i$, then for all $s_j \in \rho(s_i)$, $X_1 \in U_j$.   ∎

Theorem 13.46 *Let A be an LTL formula and suppose that the tableau $\mathcal{T}$ for A is open. Then the structure $\mathcal{H}$ created as described in Definition 13.42 is a Hintikka structure for A.*

Proof The structure $\mathcal{H}$ is created from an open tableau, so condition (1) holds. Rules for α- and β-formulas are applied *before* rules for next formulas, so the union of the formulas on every incoming state path to a state node contains all the formulas required by conditions (2) and (3). When the rule for a next formula $\bigcirc A$ is applied, A will appear in the label of the next node (and similarly for $\neg \bigcirc A$), and hence in every state at the end of a state path that includes this node.   ∎

13.5.4 Linear Fulfilling Hintikka Structures

The construction of the tableau and the Hintikka structure is quite straightforward given the decomposition of formulas with temporal operators. Now we turn to the more difficult problem of deciding if an interpretation for an LTL formula can be extracted from a Hintikka structure. First, we need to extract a linear structure and show that it is also a Hintikka structure.

Definition 13.47 Let $\mathcal{H}$ be a Hintikka structure for an LTL formula A. $\mathcal{H}$ is a *linear* Hintikka structure iff ρ is a total function, that is, if for each s_i there is exactly one $s_j \in \rho(s_i)$.   ∎

Lemma 13.48 *Let $\mathcal{H}$ be a Hintikka structure for an LTL formula A and let $\mathcal{H}'$ be an infinite path through $\mathcal{H}$. Then $\mathcal{H}'$ is a linear Hintikka structure.*

Proof Clearly, $\mathcal{H}'$ is a linear structure. Conditions (1–3) of Definition 13.45 hold because they already held in $\mathcal{H}$. Let s be an arbitrary state and let U be the label of s. If a next formula $\bigcirc A'$ occurs in U, then by condition (4) of Definition 13.45, A' occurs in *all* states of $\rho(s)$, in particular, for the one chosen in the construction of $\mathcal{H}'$.   ∎

Next, we need to check if the linear structure fulfills all the future formulas. We define the concept of fulfilling and then show that a fulfilling Hintikka structure can be used to define a model. The algorithm for deciding if a Hintikka structure is fulfilling is somewhat complex and is left to the next subsection. To simplify the presentation, future formulas will be limited to those of the form $\Diamond A$. By duality, the same presentation is applicable to future formulas of the form $\neg \Box A$.

Recall that $\rho*$ is the transitive, reflexive closure of ρ (Definition A.21).

Definition 13.49 Let $\mathcal{H} = (\mathcal{S}, \rho)$ be a Hintikka structure. $\mathcal{H}$ is a *fulfilling* iff the following condition holds for all future formulas $\Diamond A$:

For all $s \in \mathcal{S}$, if $\Diamond A \in U_s$, then for some $s' \in \rho^*(s)$, $A \in U_{s'}$.

The state s' is said to *fulfill* $\Diamond A$. ∎

Theorem 13.50 (Hintikka's Lemma for LTL) *Let $\mathcal{H} = (\mathcal{S}, \rho)$ be a linear fulfilling Hintikka structure for an LTL formula A. Then A is satisfiable.*

Proof An LTL interpretation is a path consisting of states labeled with atomic propositions (see Definition 13.28). The path is defined simply by taking the linear Hintikka structure and restricting the labels to atomic propositions. There is thus a natural mapping between states of the interpretation and states of the Hintikka structure, so for the propositional operators and next formulas, we can use the conditions on the structure to prove that A is satisfiable using structural induction.

For future formulas, the satisfiability follows from the assumption that the Hintikka structure is fulfilling.

Consider now a formula of the form $\Box A \in U_{s_i}$. We must show that $v_{\sigma_j}(A) = T$ for all $j \geq i$. We generalize this for the inductive proof and show that $v_{\sigma_j}(A) = T$ and $v_{\sigma_j}(\bigcirc \Box A) = T$ for all $j \geq i$.

The base case is $j = i$. But $\Box A \in U_{s_i}$, so by Hintikka condition (2) $A \in U_{s_i}$ and $\bigcirc \Box A \in U_{s_i}$.

Let $k \geq i$ and assume the inductive hypothesis that $v_{\sigma_k}(A) = T$ and $\bigcirc \Box A \in U_{s_k}$. By Hintikka condition (4), $\Box A \in U_{s_{k+1}}$, so using Hintikka condition (2) again, $v_{\sigma_{k+1}}(A) = T$ and $\bigcirc \Box A \in U_{s_{k+1}}$. ∎

Here is a finite presentation of a linear fulfilling Hintikka structure constructed from the structure in Fig. 13.3:

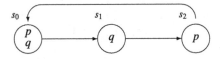

13.5.5 Deciding Fulfillment of Future Formulas *

The last link needed to obtain a decision procedure for satisfiability in LTL is an algorithm that takes an arbitrary Hintikka structure, and decides if it contains a path that is a linear fulfilling Hintikka structure. We begin with some definitions from graph theory. The concepts should be familiar, though it is worthwhile giving formal definitions.

Definition 13.51 A *graph* $G = (V, E)$ consists of a set of *vertices* $V = \{v_1, \ldots, v_n\}$ and a set of *edges* $E = \{e_1, \ldots, e_m\}$, which are pairs of vertices $e_k = \{v_i, v_j\} \subseteq V$.

Fig. 13.4 Strongly
connected components

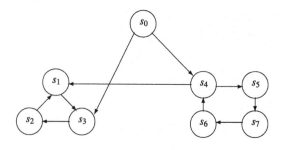

In a *directed* graph, each edge is an ordered pair, $e_k = (v_i, v_j)$. A *path* from v to v',
denoted $v \rightsquigarrow v'$, is a sequence of edges such that the second component of one edge
is the first component of the next:

$$
\begin{aligned}
e_1 &= (v = v_{i_1}, v_{i_2}), \\
e_2 &= (v_{i_2}, v_{i_3}), \\
&\cdots \\
e_{l-1} &= (v_{i_{l-2}}, v_{i_{l-1}}), \\
e_l &= (v_{i_{l-1}}, v_{i_l} = v').
\end{aligned}
$$

A subgraph $G' = (V', E')$ of a directed graph $G = (V, E)$ is a graph such that
$V' \subseteq V$ and $E' \subseteq E$, provided that $e = (v_i, v_j) \in E'$ implies $\{v_i, v_j\} \subseteq V'$. ∎

Definition 13.52 A *strongly connected component (SCC)* $G' = (V', E')$ in a di-
rected graph G is a subgraph such that $v_i \rightsquigarrow v_j$ for all $\{v_i, v_j\} \subseteq V'$. A *maximal
strongly connected component (MSCC)* is an SCC not properly contained in an-
other. A *transient SCC* is an MSCC consisting of a single vertex. A *terminal SCC* is
an MSCC with no outgoing edges. ∎

Example 13.53 The directed graph in Fig. 13.4 contains three strongly connected
components: $G_0 = \{s_0\}$, $G_1 = \{s_1, s_2, s_3\}$, $G_2 = \{s_4, s_5, s_6, s_7\}$. G_0 is transient and
G_1 is terminal. ∎

Definition 13.54 A directed graph G can be represented as a *component graph*,
which is a directed graph whose vertices are the MSCCs of G and whose edges are
edges of G pointing from a vertex of one MSCC to a vertex of another MSCC. ∎

See Even, Sect. 3.4 for an algorithm that constructs the component graph of a
directed graph and a proof of the following theorem.

Theorem 13.55 *The component graph is acyclic.*

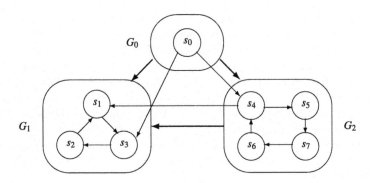

Fig. 13.5 Component graph

Example 13.56 Figure 13.5 shows the graph of Fig. 13.4 with its component graph indicated by ovals and thick arrows. ∎

Suppose that we have a Hintikka structure and a future formula in a *terminal* MSCC, such as G_1 in Fig. 13.5. Then if the formula is going to be fulfilled at all, it will be fulfilled within the terminal MSCC because there are no other reachable nodes to which the fulfillment can be deferred. If a future formula is in a *non-terminal* MSCC such as G_2, it can either be fulfilled within its own MSCC, or the fulfillment can be deferred to an reachable MSCC, in this case G_1. This suggests an algorithm for checking fulfillment: start at terminal MSCCs and work backwards.

Let $\mathscr{H} = (\mathscr{S}, \rho)$ be a Hintikka structure. $\mathscr{H}$ can be considered a graph $G = (V, E)$, where V is $\mathscr{S}$ and $(s_i, s_j) \in E$ iff $s_j \in \rho(s_i)$. We simplify the notation and write $A \in v$ for $A \in U_i$ when $v = s_i$.

Definition 13.57 Let $G = (V, E)$ be a SCC of $\mathscr{H}$. G is *self-fulfilling* iff for all $v \in V$ and for all future formulas $\Diamond A \in v$, $A \in v'$ for some $v' \in V$. ∎

Lemma 13.58 *Let $G = (V, E) \subseteq G' = (V', E')$ be SCCs of a Hintikka structure. If G is self-fulfilling, then so is G'.*

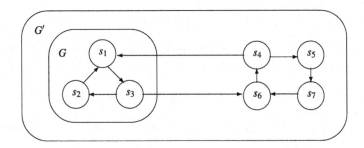

Fig. 13.6 An SCC is contained in an MSCC

Example 13.59 Let $\Diamond A$ be an arbitrary future formula that has to be fulfilled in G' in Fig. 13.6. If $\Diamond A \in s_i$ for $s_i \in G$, then by the assumption that G is self-fulfilling, $A \in s_j$ for some $s_j \in G \subset G'$ and G' is also self-fulfilling.

Suppose now that $\Diamond A \in s_7$, where $s_7 \in V' - V$. If $A \in s_7$, then s_7 itself fulfills $\Diamond A$. Otherwise, by Hintikka condition (3), $\bigcirc \Diamond A \in s_7$, so $\Diamond A \in s_6$ by Hintikka condition (4). Continuing, $A \in s_6$ or $\bigcirc \Diamond A \in s_6$; $A \in s_4$ or $\bigcirc \Diamond A \in s_4$; $A \in s_5$ or $\bigcirc \Diamond A \in s_5$. If $A \in s_j$ for one of these vertices in $V' - V$, we have the G' is self-fulfilling.

If not, then by Hintikka condition (4), $\bigcirc \Diamond A \in s_4$ implies that $\Diamond A \in s_1$, because condition (4) is a requirement on *all* immediate successors of a node. By assumption, G is self-fulfilling, so $A \in s_j$ for some $s_j \in G \subset G'$ and G' is also self-fulfilling. ∎

Proof of Lemma 13.58 Let $\Diamond A$ be an arbitrary future formula in $v' \in V' - V$. By definition of a Hintikka structure, either $A \in v'$ or $\bigcirc \Diamond A \in v'$. If $A \in v'$, then A is fulfilled in G'; otherwise, $\Diamond A \in v''$ for every $v'' \in \rho(v')$. By induction on the number of vertices in $V' - V$, either A is fulfilled in $V' - V$ or $\Diamond A \in v$ for some v in V. But G is self-fulfilling, so $\Diamond A$ is fulfilled in some state $v_A \in V \subseteq V'$. Since G' is an SCC, $v' \rightsquigarrow v_A$ and A is fulfilled in G'. ∎

Corollary 13.60 *Let G be a self-fulfilling SCC of a Hintikka structure. Then G can be extended to a self-fulfilling MSCC.*

Proof If G itself is not an MSCC, create a new graph G' by adding a vertex $v' \in V' - V$ and all edges (v', v) and (v, v'), where $v \in V$, provided that G' is an SCC. Continue this procedure until no new SCCs can be created. By Lemma 13.58, the SCC is self-fulfilling and by construction it is maximal. ∎

Lemma 13.61 *Let $G = (V, E)$ be an MSCC of $\mathcal{H}$ and let $\Diamond A \in v \in V$ be a future formula. If G is not self-fulfilling, $\Diamond A$ can only be fulfilled by some v' in an MSCC G', such that $G \rightsquigarrow G'$ in the component graph.*

Proof Since G is not self-fulfilling, $\Diamond A$ must be fulfilled by some $v' \notin V$ such that $v \rightsquigarrow v'$. But $v' \not\rightsquigarrow v$, otherwise v' could be added to the vertices of G creating a larger SCC, contradicting the assumption that G is maximal. Therefore, $v' \in G'$ for a component $G' \neq G$. ∎

This lemma directly gives the following corollary.

Corollary 13.62 *If G is a terminal MSCC and $\Diamond A \in v$ for $v \in V$, then if $\Diamond A$ cannot be fulfilled in G, it cannot be fulfilled at all.*

Algorithm 13.63 (Construction of a linear fulfilling structure)
Input: A Hintikka structure $\mathcal{H}$.
Output: A linear fulfilling Hintikka structure that is a path in $\mathcal{H}$, or a report that no such structure exists.

Construct the component graph H of $\mathcal{H}$. Since H is acyclic (Theorem 13.55), there must be a terminal MSCC G. If G is not self-fulfilling, delete G and all its incoming edges from H. Repeat until every terminal MSCC is self-fulfilling or until the component graph is empty. If every terminal MSCC is self-fulfilling, the proof of the following theorem shows how a linear fulfilling Hintikka structure can be constructed. Otherwise, if the graph is empty, the algorithm reports that no linear fulfilling Hintikka structure exists.   ∎

Theorem 13.64 *Algorithm* 13.63 *terminates with a non-empty graph iff a linear fulfilling Hintikka structure can be constructed.*

Proof Suppose that the algorithm terminates with an non-empty component graph G and let $G_1 \rightsquigarrow \cdots \rightsquigarrow G_n$ be a maximal path in G. We now define a path in $\mathscr{H}$ based upon this path in the component graph.

There must be vertices $\{v_1, \ldots, v_n\}$ in $\mathscr{H}$, such that $v_i \in G_i$, $v_{i+1} \in G_{i+1}$ and $v_i \rightsquigarrow v_{i+1}$. Furthermore, each component G_i is an SCC, so for each i there is a path $v_1^i \rightsquigarrow \cdots \rightsquigarrow v_{k_i}^i$ in $\mathscr{H}$ containing all the vertices in G_i.

Construct a path in $\mathscr{H}$ by replacing every component by a partial path and connecting them by the edges $v_i \rightsquigarrow v_{i+1}$:

- Replace a transient component by the single vertex v_1^i.
- Replace a terminal component by the closure

$$v_i \rightsquigarrow \cdots \rightsquigarrow (v_1^i \rightsquigarrow \cdots \rightsquigarrow v_{k_i}^i)^*.$$

- Replace a non-transient, non-terminal component by

$$v_i \rightsquigarrow \cdots \rightsquigarrow v_1^i \rightsquigarrow \cdots v_{k_i}^i \rightsquigarrow v_1^i \rightsquigarrow \cdots v_{k_i}^i \rightsquigarrow \cdots \rightsquigarrow v_{i+1}.$$

We leave it as an exercise to prove that this path is a fulfilling linear Hintikka structure.

Conversely, let $\mathscr{H}' = (s_1, \ldots, \ldots)$ be a fulfilling linear Hintikka structure in $\mathscr{H}$. Since $\mathscr{H}$ is finite, some suffix of $\mathscr{H}'$ must be composed of states which repeat infinitely often. These states must be contained within a self-fulfilling SCC G. By Corollary 13.60, G is contained in a self-fulfilling MSCC.   ∎

Example 13.65 There are two maximal paths in the component graph in Fig. 13.5: $G_0 \rightsquigarrow G_1$ and $G_0 \rightsquigarrow G_2 \rightsquigarrow G_1$. The paths constructed in the underlying graphs are:

$$s_0 \rightsquigarrow (s_3 \rightsquigarrow s_2 \rightsquigarrow s_1)^*$$

and

$$s_0 \rightsquigarrow s_4 \rightsquigarrow s_5 \rightsquigarrow s_7 \rightsquigarrow s_6 \rightsquigarrow s_4 \rightsquigarrow s_5 \rightsquigarrow s_7 \rightsquigarrow s_6 \rightsquigarrow s_4 \rightsquigarrow (s_1 \rightsquigarrow s_2 \rightsquigarrow s_3)^*,$$

respectively.   ∎

Theorem 13.66 *There is a decision procedure for satisfiability in LTL.*

Proof Let A be a formula in LTL. Construct a semantic tableau for A. If it closes, A is unsatisfiable. If there is an open branch, A is satisfiable. Otherwise, construct the structure from the tableau as described in Definition 13.42. By Theorem 13.46, this is a Hintikka structure. Apply Algorithm 13.63 to construct a fulfilling Hintikka structure. If the resulting graph is empty, A is unsatisfiable. Otherwise, apply the construction in Theorem 13.64 to construct a linear fulfilling Hintikka structure. By Theorem 13.50, a model can be constructed from the structure.   ∎

The following corollary is obvious since the number of possible states in a structure constructed for a particular formula is finite:

Corollary 13.67 (Finite model property) *A formula in LTL is satisfiable iff it is satisfiable in a finitely-presented model.*

13.6 Binary Temporal Operators *

Consider the following correctness specification from the introduction:

> The output lines *maintain* their values *until* the set-line is asserted.

We cannot express this in LTL as defined above because we have no binary temporal operators that can connect two propositions: *unchanged-output* and *set-asserted*. To express such properties, a binary operator $\mathscr{U}$ (read *until*) can be added to LTL. Infix notation is used:

$$unchanged\text{-}output \ \mathscr{U} \ set\text{-}asserted.$$

The semantics of the operator is defined by adding the following item to Definition 13.28:

- If A is $A_1 \mathscr{U} A_2$ then $v_\sigma(A) = T$ iff $v_{\sigma_i}(A_2) = T$ for *some* $i \geq 0$ and for *all* $0 \leq k < i$, $v_{\sigma_k}(A_1) = T$.

Example 13.68 The formula $p \mathscr{U} q$ is true in the interpretation represented by the following path:

q is true at s_2 and for all previous states $\{s_0, s_1\}$, p is true.

$p \mathscr{U} q$ is not true in the following interpretation assuming that state s_2 is repeated indefinitely:

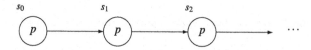

The reason is that q never becomes true.

$p \, \mathcal{U} \, q$ is also not true in the following interpretation:

because p becomes false before q becomes true.   ■

Defining the Existing Operators in Terms of $\mathcal{U}$

It is easy to see that:

$$\Diamond A \equiv \textit{true} \, \mathcal{U} \, A.$$

The definition of the semantics of $\mathcal{U}$ requires that A become true eventually just as in the semantics of $\Diamond A$. The additional requirement is that *true* evaluate to T in every previous state, but that clearly holds in every interpretation.

Since binary operators are essential for expressing correctness properties, advanced presentations of LTL take $\bigcirc$ and $\mathcal{U}$ as the primitive operators of LTL and define $\Diamond$ as an abbreviation for the above formula, and then $\square$ as an abbreviation for $\neg \Diamond \neg$.

Semantic Tableaux with $\mathcal{U}$

Constructing a semantic tableau for a formula that uses the $\mathcal{U}$ operator does not require any new concepts. The operator can be decomposed as follows:

$$A_1 \, \mathcal{U} \, A_2 \equiv A_2 \lor (A_1 \land \bigcirc(A_1 \, \mathcal{U} \, A_2)).$$

For $A_1 \, \mathcal{U} \, A_2$ to be true, either A_2 is true today, or we put off to tomorrow the requirement to satisfy $A_1 \, \mathcal{U} \, A_2$, while requiring that A_1 be true today. The decomposition shows that a $\mathcal{U}$-formula is a β-formula very similar to $\Diamond A$. The similarity goes deeper, because $A_1 \, \mathcal{U} \, A_2$ is a future formula and must be fulfilled by having A_2 appear in a state eventually.

The construction of semantic tableau is more efficient if operators have duals. The dual of $\mathcal{U}$ is the operator $\mathcal{R}$ (read *release*), defined as:

$$A_1 \mathcal{R} A_2 \equiv \neg(\neg A_1 \, \mathcal{U} \, \neg A_2).$$

We leave it as an exercise to write the definition of the semantics of $\mathcal{R}$.

The Weak Until Operator

Sometimes it is convenient to express precedence properties without actually requiring that something eventually occur. $\mathcal{W}$ (read *weak until*) is the same as the operator $\mathcal{U}$ except that it is not required that the second formula ever become true:

- If A is $A_1 \mathcal{W} A_2$ then $v_\sigma(A) = T$ iff: **if** $v_{\sigma_i}(A_2) = T$ for *some* $i \geq 0$, **then** for *all* $0 \leq k < i$, $v_{\sigma_k}(A_1) = T$.

Clearly, the following equivalence holds:

$$A_1 \mathcal{W} A_2 \equiv (A_1 \mathcal{U} A_2) \vee \Box A_1.$$

We leave it as an exercise to show:

$$
\begin{aligned}
\Box A &\equiv A \mathcal{W} \textit{ false,} \\
\neg(A_1 \mathcal{W} A_2) &\equiv (A_1 \wedge \neg A_2) \mathcal{U} (\neg A_1 \wedge \neg A_2), \\
\neg(A_1 \mathcal{U} A_2) &\equiv (A_1 \wedge \neg A_2) \mathcal{W} (\neg A_1 \wedge \neg A_2), \\
\neg(A_1 \mathcal{U} A_2) &\equiv (\neg A_2) \mathcal{W} (\neg A_1 \wedge \neg A_2).
\end{aligned}
$$

13.7 Summary

Since the state of a computation changes over time, temporal logic is an appropriate formalism for expressing correctness properties of programs. The syntax of linear temporal logic (LTL) is that of propositional logic together with the unary temporal operators $\Box$, $\Diamond$, $\bigcirc$. Interpretations are infinite sequences of states, where each state assigns truth values to atomic propositions. The meaning of the temporal operators is that some property must hold in $\Box$ *all* subsequent states, in $\Diamond$ *some* subsequent state or in the $\bigcirc$ *next* state.

Satisfiability and validity of formulas in LTL are decidable. The tableau construction for propositional logic is extended so that *next formulas* (of the form $\bigcirc A$) cause new states to be generated. A open tableau defines a Hintikka structure which can be extended to a satisfying interpretation, provided that all *future formulas* (of the form $\Diamond A$ or $\neg \Box A$) are fulfilled. By constructing the component graph of strongly connected components, the fulfillment of the future formulas can be decided.

Many important correctness properties use the binary operators $\mathcal{U}$ and $\mathcal{W}$, which require that one formula hold until a second one becomes true.

13.8 Further Reading

Temporal logic (also called *tense logic*) has a long history, but it was first applied to program verification by Pnueli (1977). The definitive reference for the specification and verification of concurrent programs using temporal logic is Manna and Pnueli (1992, 1995). The third volume was never completed, but a partial draft is available (Manna and Pnueli, 1996). Modern treatments of LTL can be found in Kröger and Merz (2008, Chap. 2), and Baier and Katoen (2008, Chap. 5). The tableau method for a different version of temporal logic first appeared in Ben-Ari et al. (1983); for a modern treatment see Kröger and Merz (2008, Chap. 2).

13.9 Exercises

13.1 Prove that in LTL every substitution instance of a valid propositional formula is valid.

13.2 Prove $\models \neg \Diamond \neg p \to \Box p$ (the converse direction of Theorem 13.14).

13.3 Prove that a linear interpretation is characterized by $\bigcirc A \leftrightarrow \neg \bigcirc \neg A$ (Theorem 13.25).

13.4 * Identify the property of a reflexive relation characterized by $A \to \Box \Diamond A$. Identify the property of a reflexive relation characterized by $\Diamond A \to \Box \Diamond A$.

13.5 Show that in an interpretation with a reflexive transitive relation, any formula (without $\bigcirc$) is equivalent to one whose only temporal operators are $\Box$, $\Diamond$, $\Diamond \Box$, $\Box \Diamond$, $\Diamond \Box \Diamond$ and $\Box \Diamond \Box$. If the relation is also characterized by the formula $\Diamond A \to \Box \Diamond A$, any formula is equivalent to one with a single temporal operator.

13.6 Prove Theorem 13.34: $\models (\Diamond \Box p \land \Box \Diamond q) \to \Box \Diamond (p \land q)$.

13.7 Construct a tableau and find a model for the negation of $\Box \Diamond p \to \Diamond \Box p$.

13.8 Prove that the construction of a semantic tableau terminates.

13.9 Prove that the construction of the path in the proof of Theorem 13.64 gives a linear fulfilling Hintikka structure.

13.10 Write the definition of the semantics of the operator $\mathscr{R}$.

13.11 Prove the equivalences on $\mathscr{W}$ at the end of Sect. 13.6.

References

C. Baier and J.-P. Katoen. *Principles of Model Checking*. MIT Press, 2008.

M. Ben-Ari, Z. Manna, and A. Pnueli. The temporal logic of branching time. *Acta Informatica*, 20:207–226, 1983.

S. Even. *Graph Algorithms*. Computer Science Press, Potomac, MD, 1979.

F. Kröger and S. Merz. *Temporal Logic and State Systems*. Springer, 2008.

Z. Manna and A. Pnueli. *The Temporal Logic of Reactive and Concurrent Systems. Vol. I: Specification*. Springer, New York, NY, 1992.

Z. Manna and A. Pnueli. *The Temporal Logic of Reactive and Concurrent Systems. Vol. II: Safety*. Springer, New York, NY, 1995.

Z. Manna and A. Pnueli. Temporal verification of reactive systems: Progress. Draft available at `http://www.cs.stanford.edu/~zm/tvors3.html`, 1996.

A. Pnueli. The temporal logic of programs. In *18th IEEE Annual Symposium on Foundations of Computer Science*, pages 46–57, 1977.

Chapter 14
Temporal Logic: A Deductive System

This chapter defines the deductive system $\mathscr{L}$ for linear temporal logic. We will prove many of the formulas presented in the previous chapter, as well as the soundness and completeness of $\mathscr{L}$.

14.1 Deductive System $\mathscr{L}$

The operators of $\mathscr{L}$ are the Boolean operators of propositional logic together with the temporal operators $\Box$ and $\bigcirc$. The operator $\Diamond$ is defined as an abbreviation for $\neg \Box \neg$.

Definition 14.1 The axioms of $\mathscr{L}$ are:

Axiom 0	**Prop**	Any substitution instance of a valid propositional formula.
Axiom 1	**Distribution of $\Box$**	$\vdash \Box(A \rightarrow B) \rightarrow (\Box A \rightarrow \Box B)$.
Axiom 2	**Distribution of $\bigcirc$**	$\vdash \bigcirc(A \rightarrow B) \rightarrow (\bigcirc A \rightarrow \bigcirc B)$.
Axiom 3	**Expansion of $\Box$**	$\vdash \Box A \rightarrow (A \wedge \bigcirc A \wedge \bigcirc \Box A)$.
Axiom 4	**Induction**	$\vdash \Box(A \rightarrow \bigcirc A) \rightarrow (A \rightarrow \Box A)$.
Axiom 5	**Linearity**	$\vdash \bigcirc A \leftrightarrow \neg \bigcirc \neg A$.

The rules of inference are *modus ponens* and *generalization*: $\dfrac{\vdash A}{\vdash \Box A}$. ∎

In order to simplify proofs of formulas in LTL, the deductive system $\mathscr{L}$ takes all substitution instances of valid formulas of propositional logic as axioms. Validity in propositional logic is decidable and by the completeness of $\mathscr{H}$ we can produce a proof of any valid formula if asked to do so. In fact, we will omit justifications of deductions in propositional logic and just write *Prop* if a step in a proof is justified by propositional reasoning.

The distributive axioms are valid in virtually all modal and temporal logics (Theorem 13.15). The expansion axiom expresses the basic properties of $\Box$ that were

M. Ben-Ari, *Mathematical Logic for Computer Science*,
DOI 10.1007/978-1-4471-4129-7_14, © Springer-Verlag London 2012

used to construct semantic tableaux, as well as $\vdash \Box A \rightarrow \bigcirc A$ (Theorem 13.25), which holds because all interpretations are infinite paths. The linearity axiom for $\bigcirc$ (Theorem 13.25) captures the restriction of LTL to linear interpretations.

The induction axiom is fundamental in $\mathscr{L}$: since interpretations in LTL are infinite paths, proofs of non-trivial formulas usually require induction. In a proof by induction, the inductive step is $A \rightarrow \bigcirc A$, that is, we assume that A is true today and prove that A is true tomorrow. If this inductive step is always true, $\Box(A \rightarrow \bigcirc A)$, then $A \rightarrow \Box A$ by the induction axiom. Finally, if A is true today (the base case), then A is always true $\Box A$.

The rules of inference are the familiar *modus ponens* and generalization using $\Box$, which is similar to generalization using $\forall$ in first-order logic.

Derived Rules

Here are some useful derived rules:

$$\frac{\vdash A \rightarrow B}{\vdash \Box A \rightarrow \Box B}, \qquad \frac{\vdash A \rightarrow B}{\vdash \bigcirc A \rightarrow \bigcirc B}, \qquad \frac{\vdash A \rightarrow \bigcirc A}{\vdash A \rightarrow \Box A}.$$

The first is obtained by applying generalization and then the distribution axiom; the second is similar except that the expansion axiom is used between the generalization and the distribution. When using these rules, we write the justification as *generalization*. The third rule will be called *induction* because it is a shortcut for generalization followed by the induction axiom.

14.2 Theorems of $\mathscr{L}$

The theorems and their proofs will be stated and proved for atomic propositions p and q although the intention is that they hold for arbitrary LTL formulas.

Distributivity

This subsection explores in more detail the distributivity of the temporal operators over proposition operators. The results will not be surprising, because $\Box$ and $\Diamond$ behave similarly to $\forall$ and $\exists$ in first-order logic. $\bigcirc$ is a special case because of linearity.

Theorem 14.2 $\vdash \bigcirc(p \wedge q) \leftrightarrow (\bigcirc p \wedge \bigcirc q)$.

Proof

1.	$\vdash (p \wedge q) \rightarrow p$	Prop
2.	$\vdash \bigcirc(p \wedge q) \rightarrow \bigcirc p$	Generalization
3.	$\vdash (p \wedge q) \rightarrow q$	Prop
4.	$\vdash \bigcirc(p \wedge q) \rightarrow \bigcirc q$	Generalization
5.	$\vdash \bigcirc(p \wedge q) \rightarrow (\bigcirc p \wedge \bigcirc q)$	2, 4, Prop
6.	$\vdash \bigcirc(p \rightarrow \neg q) \rightarrow (\bigcirc p \rightarrow \bigcirc \neg q)$	Distribution
7.	$\vdash \neg (\bigcirc p \rightarrow \bigcirc \neg q) \rightarrow \neg \bigcirc(p \rightarrow \neg q)$	6, Prop
8.	$\vdash \neg \bigcirc p \vee \bigcirc \neg q \vee \neg \bigcirc(p \rightarrow \neg q)$	7, Prop
9.	$\vdash \neg \bigcirc p \vee \neg \bigcirc q \vee \bigcirc \neg (p \rightarrow \neg q)$	8, Linearity
10.	$\vdash (\bigcirc p \wedge \bigcirc q) \rightarrow \bigcirc(p \wedge q)$	9, Prop
11.	$\vdash \bigcirc(p \wedge q) \leftrightarrow (\bigcirc p \wedge \bigcirc q)$	5, 10, Prop

∎

By linearity, $\bigcirc$ is self-dual, while $\vee$ is the dual of $\wedge$, so we immediately have $\vdash \bigcirc(p \vee q) \leftrightarrow (\bigcirc p \vee \bigcirc q)$.

Theorem 14.3 (Distribution) $\vdash \Box(p \wedge q) \leftrightarrow (\Box p \wedge \Box q)$.

The proof of the forward implication $\vdash \Box(p \wedge q) \rightarrow (\Box p \wedge \Box q)$ is similar to that of Theorem 14.2 and is left as an exercise. Before proving the converse, we need to prove the converse of the expansion axiom; the proof uses the forward implication of Theorem 14.3, which we assume that you have already proved.

Theorem 14.4 (Contraction) $\vdash p \wedge \bigcirc \Box p \rightarrow \Box p$.

Proof

1.	$\vdash \Box p \rightarrow p \wedge \bigcirc \Box p$	Expansion
2.	$\vdash \bigcirc \Box p \rightarrow \bigcirc(p \wedge \bigcirc \Box p)$	1, Generalization
3.	$\vdash p \wedge \bigcirc \Box p \rightarrow \bigcirc(p \wedge \bigcirc \Box p)$	2, Prop
4.	$\vdash p \wedge \bigcirc \Box p \rightarrow \Box(p \wedge \bigcirc \Box p)$	3, Induction
5.	$\vdash p \wedge \bigcirc \Box p \rightarrow (\Box p \wedge \Box \bigcirc \Box p)$	4, Distribution
6.	$\vdash p \wedge \bigcirc \Box p \rightarrow \Box p$	5, Prop

∎

For symmetry with the expansion axiom, $\bigcirc p$ could have been included in the premise of this theorem, but it is not needed.

Now we can prove the converse of Theorem 14.3. The structure of the proof is typical of inductive proofs in $\mathscr{L}$. An explanation of some of the more difficult steps of the formal proof is given at its end.

Proof Let $r = \Box p \wedge \Box q \wedge \neg \Box(p \wedge q)$.

1.	$\vdash r \rightarrow (p \wedge \bigcirc\Box p) \wedge (q \wedge \bigcirc\Box q) \wedge$ $\neg((p \wedge q) \wedge \bigcirc\Box(p \wedge q))$	Expansion Contraction
2.	$\vdash r \rightarrow (p \wedge \bigcirc\Box p) \wedge (q \wedge \bigcirc\Box q) \wedge$ $(\neg(p \wedge q) \vee \neg\bigcirc\Box(p \wedge q))$	1, Prop
3.	$\vdash r \rightarrow (p \wedge \bigcirc\Box p) \wedge (q \wedge \bigcirc\Box q) \wedge \neg\bigcirc\Box(p \wedge q)$	2, Prop
4.	$\vdash r \rightarrow \bigcirc\Box p \wedge \bigcirc\Box q \wedge \neg\bigcirc\Box(p \wedge q)$	3, Prop
5.	$\vdash r \rightarrow \bigcirc\Box p \wedge \bigcirc\Box q \wedge \bigcirc\neg\Box(p \wedge q)$	4, Linearity
6.	$\vdash r \rightarrow \bigcirc r$	5, Distribution
7.	$\vdash r \rightarrow \Box r$	6, Induction
8.	$\vdash r \rightarrow \Box p \wedge \Box q$	Def. of r, Prop
9.	$\vdash r \rightarrow p \wedge q$	8, Expansion
10.	$\vdash \Box r \rightarrow \Box(p \wedge q)$	9, Generalization
11.	$\vdash r \rightarrow \Box(p \wedge q)$	7, 10, Prop
12.	$\vdash r \rightarrow \neg\Box(p \wedge q)$	Def. of r, Prop
13.	$\vdash r \rightarrow \mathit{false}$	11, 12, Prop
14.	$\vdash \Box p \wedge \Box q \wedge \neg\Box(p \wedge q) \rightarrow \mathit{false}$	13, Def. of r
15.	$\vdash \Box p \wedge \Box q \rightarrow \Box(p \wedge q)$	14, Prop

∎

Steps 1–7 prove that r is invariant, meaning that r is true initially and remains true in any interpretation. The second line of Step 1 is justified by the contrapositive of contraction $\neg\Box(p \wedge q) \rightarrow \neg((p \wedge q) \wedge \bigcirc\Box(p \wedge q))$. Step 3 follows from Step 2 because $\neg(p \wedge q)$ is inconsistent with p and q that must be true by the expansion of $\Box p$ and $\Box q$.

The operator $\Box$ distributes over disjunction only in one direction. We leave the proof as an exercise, together with the task of showing that the converse is not valid.

Theorem 14.5 (Distribution) $\vdash (\Box p \vee \Box q) \rightarrow \Box(p \vee q)$.

Transitivity of $\Box$

Induction is used to prove that $\Box$ is transitive.

Theorem 14.6 (Transitivity) $\vdash \Box\Box p \leftrightarrow \Box p$

Proof

1.	$\vdash \Box\Box p \rightarrow \Box p$	Expansion
2.	$\vdash \Box p \rightarrow \bigcirc\Box p$	Expansion
3.	$\vdash \Box p \rightarrow \Box\Box p$	2, Induction
4.	$\vdash \Box\Box p \leftrightarrow \Box p$	1, 3, Prop

∎

Commutativity

Another expected result is that $\Box$ and $\bigcirc$ commute:

Theorem 14.7 (Commutativity) $\vdash \Box\bigcirc p \leftrightarrow \bigcirc\Box p$.

Proof

1.	$\vdash \Box p \rightarrow \bigcirc p$	Expansion
2.	$\vdash \Box\Box p \rightarrow \Box\bigcirc p$	1, Generalization
3.	$\vdash \Box p \rightarrow \Box\bigcirc p$	2, Transitivity
4.	$\vdash \Box p \rightarrow p$	Expansion
5.	$\vdash \Box p \rightarrow p \wedge \Box\bigcirc p$	3, 4, Prop
6.	$\vdash \bigcirc\Box p \rightarrow \bigcirc(p \wedge \Box\bigcirc p)$	5, Generalization
7.	$\vdash \bigcirc\Box p \rightarrow \bigcirc p \wedge \bigcirc\Box\bigcirc p$	6, Distribution
8.	$\vdash \bigcirc\Box p \rightarrow \Box\bigcirc p$	7, Contraction
9.	$\vdash \Box\bigcirc p \rightarrow \bigcirc p \wedge \bigcirc\Box\bigcirc p$	Expansion
10.	$\vdash p \wedge \Box\bigcirc p \rightarrow \bigcirc p \wedge \bigcirc\Box\bigcirc p$	9, Prop
11.	$\vdash p \wedge \Box\bigcirc p \rightarrow \bigcirc(p \wedge \Box\bigcirc p)$	10, Distribution
12.	$\vdash p \wedge \Box\bigcirc p \rightarrow \Box(p \wedge \Box\bigcirc p)$	11, Induction
13.	$\vdash p \wedge \Box\bigcirc p \rightarrow \Box p$	12, Distribution, Prop
14.	$\vdash \bigcirc(p \wedge \Box\bigcirc p) \rightarrow \bigcirc\Box p$	13, Generalization
15.	$\vdash \bigcirc p \wedge \bigcirc\Box\bigcirc p \rightarrow \bigcirc\Box p$	14, Distribution
16.	$\vdash \Box\bigcirc p \rightarrow \bigcirc\Box p$	9, 15, Prop
17.	$\vdash \Box\bigcirc p \leftrightarrow \bigcirc\Box p$	8, 16, Prop

∎

$\Box$ and $\Diamond$ commute in only one direction.

Theorem 14.8 $\vdash \Diamond\Box p \rightarrow \Box\Diamond p$.

We leave the proof as an exercise.

Example 14.9 Consider the interpretation where $s_i(p) = T$ for even i and $s_i(p) = F$ for odd i:

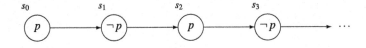

The formula $\Box\Diamond p$ is true, since for any i, $\sigma_{2i} \models p$. Obviously, $\Diamond\Box p$ is false in all states of the diagram, because for any i, $\sigma_i \models \neg p$ if i is odd and $\sigma_{i+1} \models \neg p$ if i is even. ∎

Dual Theorems for $\Diamond$

We leave it as an exercise to prove the following theorems using the duality of $\Box$ and $\Diamond$ and the linearity of $\bigcirc$.

Theorem 14.10

(a) $\vdash p \rightarrow \Diamond p$ (b) $\vdash \bigcirc p \rightarrow \Diamond p$

(c) $\vdash \Box p \rightarrow \Diamond p$ (d) $\vdash \Box(p \rightarrow q) \rightarrow (\Diamond p \rightarrow \Diamond q)$

(e) $\vdash \Diamond(p \vee q) \leftrightarrow (\Diamond p \vee \Diamond q)$ (f) $\vdash \Diamond(p \wedge q) \rightarrow (\Diamond p \wedge \Diamond q)$

(g) $\vdash \Diamond p \leftrightarrow p \vee \bigcirc \Diamond p$ (h) $\vdash \Diamond \bigcirc p \leftrightarrow \bigcirc \Diamond p$

(i) $\vdash \Diamond \Diamond p \leftrightarrow \Diamond p$

From Theorem 14.10(d), we obtain a generalization rule for $\Diamond$:

$$\frac{\vdash A \rightarrow B}{\vdash \Diamond A \rightarrow \Diamond B}.$$

Collapsing Sequences of Operators

The transitivity of $\Box$ (Theorem 14.6) and its dual for $\Diamond$ (Theorem 14.10(i)) show that any string of $\Box$'s or $\Diamond$'s can be collapsed. No expressive power is gained by using more than two operators in sequence, as shown by the following theorem.

Theorem 14.11

    (a) $\vdash \Box \Diamond \Box p \leftrightarrow \Diamond \Box p$ (b) $\vdash \Diamond \Box \Diamond p \leftrightarrow \Box \Diamond p$.

We prove (a) and then (b) follows by duality.

Proof

1.	$\vdash \Box \Diamond \Box p \rightarrow \Diamond \Box p$	Expansion
2.	$\vdash \Box p \rightarrow \bigcirc \Box p$	Expansion
3.	$\vdash \Diamond \Box p \rightarrow \Diamond \bigcirc \Box p$	2, Generalization
4.	$\vdash \Diamond \Box p \rightarrow \bigcirc \Diamond \Box p$	3, Commutativity
5.	$\vdash \Diamond \Box p \rightarrow \Box \Diamond \Box p$	4, Induction
6.	$\vdash \Box \Diamond \Box p \leftrightarrow \Diamond \Box p$	1, 5, Prop

■

14.3 Soundness and Completeness of $\mathscr{L}$ *

Soundness

Theorem 14.12 (Soundness of $\mathscr{L}$) *Let A be a formula of LTL. If $\vdash_{\mathscr{L}} A$ then $\models A$.*

Proof We need to show that each axiom is a valid LTL formula and that the two rules of inference preserve validity. By definition, valid formulas of propositional logic are valid, and the soundness of *MP* was shown in Theorem 3.37. The soundness of Axioms 1 and 5 was shown in Theorems 13.15 and 13.25, respectively. We leave the soundness of Axioms 2 and 3 as an exercise and show the soundness of the induction axiom and the generalization rule.

Axiom 4: $\vdash \Box(A \rightarrow \bigcirc A) \rightarrow (A \rightarrow \Box A)$.

If the formula is not valid, there exists an interpretation σ such that:

$$\sigma \models \Box(A \rightarrow \bigcirc A) \wedge A \wedge \neg \Box A.$$

Since $\sigma \models A$ and $\sigma \models \neg \Box A$ there exists a smallest value $i > 0$ such that $\sigma_i \models \neg A$ and $\sigma_j \models A$ for $0 \leq j < i$. In particular, $\sigma_{i-1} \models A$. But we also have that $\sigma \models \Box(A \rightarrow \bigcirc A)$, so by definition of the $\Box$ operator, $\sigma_{i-1} \models A \rightarrow \bigcirc A$. By *MP* we have $\sigma_{i-1} \models \bigcirc A$ and thus $\sigma_i \models A$, contradicting $\sigma_i \models \neg A$.

Generalization: If $\models A$, then $\models \Box A$.

We need to show that for all interpretations σ, $\sigma \models \Box A$. This means that for all $i \geq 0$, it is true that $\sigma_i \models A$. But $\models A$ implies that for *all* interpretation σ', $\sigma' \models A$, in particular, this must hold for $\sigma' = \sigma_i$.   ∎

Completeness

Theorem 14.13 (Completeness of $\mathscr{L}$) *Let A be a formula of LTL. If $\models A$ then $\vdash_{\mathscr{L}} A$.*

Proof If A is valid, the construction of a semantic tableau for $\neg A$ will fail, either because it closes or because all the MSCCs are non-fulfilling and were deleted. We show by induction that for every node in the tableau, the disjunction of the negations of the formulas labeling the node is provable in $\mathscr{L}$. Since the formula labeling the root is $\neg A$, it follows that $\vdash \neg \neg A$, from which $\vdash A$ follows by propositional logic.

The base case of the leaves and the inductive steps for the rules for α- and β-formulas follow by propositional reasoning together with the expansion axiom.

Suppose that the rule for an X-formula is used:

$$\bigcirc A_1, \ldots, \bigcirc A_n, \ B_1, \ldots, B_k$$
$$\downarrow$$
$$A_1, \ldots, A_n$$

where we assume that negations are pushed inwards as justified by the linearity axiom. By the inductive hypothesis, $\vdash \neg A_1 \vee \cdots \vee \neg A_n$. The following deduction proves the formula associated with the parent node:

1. $\vdash \neg A_1 \vee \cdots \vee \neg A_n$ Inductive hypothesis
2. $\vdash \Box(\neg A_1 \vee \cdots \vee \neg A_n)$ 1, Generalization
3. $\vdash \bigcirc(\neg A_1 \vee \cdots \vee \neg A_n)$ 2, Expansion
4. $\vdash \bigcirc\neg A_1 \vee \cdots \vee \bigcirc\neg A_n$ 3, Distribution
5. $\vdash \neg \bigcirc A_1 \vee \cdots \vee \neg \bigcirc A_n$ 4, Linearity
6. $\vdash \neg \bigcirc A_1 \vee \cdots \vee \neg \bigcirc A_n \vee \neg B_1 \vee \cdots \vee \neg B_k$ 5, Prop

There remains the case of a node that is part of a non-fulfilling MSCC. We demonstrate the technique on a specific example, proving $\vdash \Box p \rightarrow \bigcirc\Box p$ by constructing a semantic tableau for the negation of the formula.

$\neg\,(\Box p \rightarrow \bigcirc\Box p)$

$\downarrow$

$\Box p,\ \bigcirc\Diamond\neg\, p$

$\downarrow$

$l_s\ \boxed{p,\ \bigcirc\Box p,\ \bigcirc\Diamond\neg\, p}$

$\downarrow$

$\Box p,\ \Diamond\neg\, p$

$\downarrow$

$l_\beta\quad p,\ \bigcirc\Box p,\ \Diamond\neg\, p$

$\swarrow\qquad\qquad\searrow$

$p,\ \bigcirc\Box p,\ \neg\, p\qquad\qquad$ (To node l_s)

$\times$

The crucial part of the proof is to define the *invariant* of the loop, that is, a formula A such that $\vdash A \rightarrow \bigcirc A$. The invariant will be the conjunction of the formulas A_i, where $\bigcirc A_i$ are the next formulas in the states of the SCC, as these represent what must be true from one state to the next. In the example, for invariant is $\Box p \wedge \Diamond\neg\, p$. We proceed to prove that this formula is inductive.

1. $\vdash (\Box p \wedge \Diamond\neg\, p) \rightarrow (p \wedge \bigcirc\Box p) \wedge (\neg\, p \vee \bigcirc\Diamond\neg\, p)$ Expansion
2. $\vdash (\Box p \wedge \Diamond\neg\, p) \rightarrow (p \wedge \bigcirc\Box p \wedge \bigcirc\Diamond\neg\, p)$ 1, Prop
3. $\vdash (\Box p \wedge \Diamond\neg\, p) \rightarrow (\bigcirc\Box p \wedge \bigcirc\Diamond\neg\, p)$ 2, Prop
4. $\vdash (\Box p \wedge \Diamond\neg\, p) \rightarrow \bigcirc(\Box p \wedge \Diamond\neg\, p)$ 3, Distribution
5. $\vdash (\Box p \wedge \Diamond\neg\, p) \rightarrow \Box(\Box p \wedge \Diamond\neg\, p)$ 4, Induction

The leaf on the left of the tableau has a complementary pair of literals, so $\vdash \neg\, p \vee \neg\bigcirc\Box p \vee \neg\neg\, p$ is an axiom. We use this formula together with formula (5) to prove the formula associated with l_β.

6. $\vdash \neg p \vee \neg \bigcirc \square p \vee \neg \neg p$ Axiom 0
7. $\vdash (p \wedge \bigcirc \square p) \rightarrow \neg \neg p$ 6, Prop
8. $\vdash \square p \rightarrow \neg \neg p$ 7, Contraction
9. $\vdash (\square p \wedge \Diamond \neg p) \rightarrow \neg \neg p$ 8, Prop
10. $\vdash \square(\square p \wedge \Diamond \neg p) \rightarrow \square \neg \neg p$ 9, Generalization
11. $\vdash (\square p \wedge \Diamond \neg p) \rightarrow \square \neg \neg p$ 5, 10, Prop
12. $\vdash (p \wedge \bigcirc \square p \wedge \Diamond \neg p) \rightarrow \square \neg \neg p$ 11, Expansion
13. $\vdash (p \wedge \bigcirc \square p \wedge \Diamond \neg p) \rightarrow \neg \Diamond \neg p$ 12, Duality
14. $\vdash \neg p \vee \neg \bigcirc \square p \vee \neg \Diamond \neg p$ 13, Prop

Line 14 is the disjunction of the complements of the formulas at node l_β. ∎

The method used in the proof will almost certainly not yield the shortest possible proof of a formula, but it is an algorithmic procedure for discovering a proof of a valid LTL formula.

14.4 Axioms for the Binary Temporal Operators *

Section 13.6 presented several binary temporal operators, any one of which can be chosen as a basic operator and the others defined from it. If we choose $\mathcal{U}$ as the basic operator, a complete axiom system is obtained by adding the following two axioms to the axioms of Definition 14.1:

    Axiom 6 **Expansion of** $\mathcal{U}$ $\vdash A \mathcal{U} B \leftrightarrow (B \vee (A \wedge \bigcirc(A \mathcal{U} B)))$.
    Axiom 7 **Eventuality** $\vdash A \mathcal{U} B \rightarrow \Diamond B$.

$\mathcal{U}$ is similar to $\Diamond$: Axiom 6 requires that either B is true today or A is true today and $A \mathcal{U} B$ will be true tomorrow. Axiom 7 requires that B eventually be true.

14.5 Summary

The deductive system $\mathcal{L}$ assumes that propositional reasoning can be informally applied. There are five axioms: the distributive and expansion axioms are straightforward, while the duality axiom for $\bigcirc$ is essential to capture the linearity of interpretations of LTL. The central axiom of $\mathcal{L}$ is the induction axiom: since interpretations in LTL are infinite paths, proofs of non-trivial formulas usually require induction. The rules of inference are the familiar *modus ponens* and generalization using $\square$. As usual, the proof of soundness is straightforward. Proving completeness is based on the existence of a non-fulfilling MSCC in a tableau. The formulas labeling the nodes of the MSCC can be used to construct a formula that can be proved by induction.

14.6 Further Reading

The deductive system $\mathcal{L}$ and the proof of its soundness and completeness is based on Ben-Ari et al. (1983), although that paper used a different system of temporal logic. The definitive reference for the specification and verification of concurrent programs using temporal logic is Manna and Pnueli (1992, 1995). The third volume was never completed, but a partial draft is available (Manna and Pnueli, 1996). Axioms for the various binary temporal operators are given in Kröger and Merz (2008, Chap. 3).

14.7 Exercises

14.1 Prove $\vdash \Box(p \wedge q) \rightarrow (\Box p \wedge \Box q)$ (Theorem 14.3).

14.2 Prove $\vdash (\Box p \vee \Box q) \rightarrow \Box(p \vee q)$ (Theorem 14.5) and show that the converse is not valid.

14.3 Prove the future formulas in Theorem 14.10.

14.4 Prove that Axioms 2 and 3 are valid.

14.5 Prove $\vdash \Diamond \Box \Diamond p \leftrightarrow \Box \Diamond p$ (Theorem 14.11) and $\vdash \Diamond \Box p \rightarrow \Box \Diamond p$ (Theorem 14.8).

14.6 Prove $\vdash \Box(\Box \Diamond p \rightarrow \Diamond q) \leftrightarrow (\Box \Diamond q \vee \Diamond \Box \neg p)$.

14.7 Fill in the details of the proof of $\vdash \Box((p \vee \Box q) \wedge (\Box p \vee q)) \leftrightarrow (\Box p \vee \Box q)$.

References

M. Ben-Ari, Z. Manna, and A. Pnueli. The temporal logic of branching time. *Acta Informatica*, 20:207–226, 1983.
F. Kröger and S. Merz. *Temporal Logic and State Systems*. Springer, 2008.
Z. Manna and A. Pnueli. *The Temporal Logic of Reactive and Concurrent Systems. Vol. I: Specification*. Springer, New York, NY, 1992.
Z. Manna and A. Pnueli. *The Temporal Logic of Reactive and Concurrent Systems. Vol. II: Safety*. Springer, New York, NY, 1995.
Z. Manna and A. Pnueli. Temporal verification of reactive systems: Progress. Draft available at http://www.cs.stanford.edu/~zm/tvors3.html, 1996.

Chapter 15
Verification of Sequential Programs

A computer program is not very different from a logical formula. It consists of a sequence of symbols constructed according to formal syntactical rules and it has a meaning which is assigned by an interpretation of the elements of the language. In programming, the symbols are called *statements* or *commands* and the intended interpretation is the execution of the program on a computer. The syntax of programming languages is specified using formal systems such as BNF, but the semantics is usually informally specified.

In this chapter, we describe a formal semantics for a simple programming language, as well as a deductive system for proving that a program is correct. Unlike our usual approach, we first define the deductive system and only later define the formal semantics. The reason is that the deductive system is useful for proving programs, but the formal semantics is primarily intended for proving the soundness and completeness of the deductive system.

The chapter is concerned with sequential programs. A different, more complex, logical formalism is needed to verify concurrent programs and this is discussed separately in Chap. 16.

Our programs will be expressed using a fragment of the syntax of popular languages like Java and C. A program is a *statement* S, where statements are defined recursively using the concepts of *variables* and *expressions*:

Assignment statement	variable = expression ;
Compound statement	{ statement1 statement2 ... }
Alternative statement	if (expression) statement1 else statement2
Loop statement	while (expression) statement

We assume that the informal semantics of programs written in this syntax is familiar. In particular, the concept of the *location counter* (sometimes called the *instruction pointer*) is fundamental: During the execution of a program, the location counter stores the address of the next instruction to be executed by the processor.

In our examples the values of the variables will be integers.

M. Ben-Ari, *Mathematical Logic for Computer Science*,
DOI 10.1007/978-1-4471-4129-7_15, © Springer-Verlag London 2012

15.1 Correctness Formulas

A statement in a programming language can be considered to be a function that transforms the state of a computation. If the variables (x, y) have the values $(8, 7)$ in a state, then the result of executing the statement $x = 2*y+1$ is the state in which $(x, y) = (15, 7)$ and the location counter is incremented.

Definition 15.1 Let S be a program with n variables $(x1, \ldots, xn)$. A state s of S consists of an $n + 1$-tuple of values $(lc, x_1, \ldots, x_n)$, where lc is the value of the location counter and x_i is the value of the variable xi. ∎

The variables of a program will be written in typewriter font x, while the corresponding value of the variable will be written in italic font x. Since a state is always associated with a specific location, the location counter will be implicit and the state will be an n-tuple of the values of the variables.

In order to reason about programs within first-order logic, predicates are used to specify sets of states.

Definition 15.2 Let U be the set of all n-tuples of values over some domain(s), and let $U' \subseteq U$ be a relation over U. The n-ary predicate $P_{U'}$ is the *characteristic predicate* of U' if it is interpreted over the domain U by the relation U'. That is, $v(P_{U'}(x_1, \ldots, x_n)) = T$ iff $(x_1, \ldots, x_n) \in U'$. ∎

We can write $\{(x_1, \ldots, x_n) \mid (x_1, \ldots, x_n) \in U'\}$ as $\{(x_1, \ldots, x_n) \mid P_{U'}\}$.

Example 15.3 Let U be the set of 2-tuples over $\mathscr{Z}$ and let $U' \subseteq U$ be the 2-tuples described in the following table:

⋯	$(-2, -3)$	$(-2, -2)$	$(-2, -1)$	$(-2, 0)$	$(-2, 1)$	$(-2, 2)$	$(-2, 3)$
⋯	$(-1, -3)$	$(-1, -2)$	$(-1, -1)$	$(-1, 0)$	$(-1, 1)$	$(-1, 2)$	$(-1, 3)$
⋯	$(0, -3)$	$(0, -2)$	$(0, -1)$	$(0, 0)$	$(0, 1)$	$(0, 2)$	$(0, 3)$
⋯	$(1, -3)$	$(1, -2)$	$(1, -1)$	$(1, 0)$	$(1, 1)$	$(1, 2)$	$(1, 3)$
⋯	$(2, -3)$	$(2, -2)$	$(2, -1)$	$(2, 0)$	$(2, 1)$	$(2, 2)$	$(2, 3)$

Two characteristic predicates of U' are $(x_1 = x_1) \wedge (x_2 \leq 3)$ and $x_2 \leq 3$. The set can be written as $\{(x_1, x_2) \mid x_2 \leq 3\}$. ∎

The semantics of a programming language is given by specifying how each statement in the language transforms one state into another.

Example 15.4 Let S be the statement $x = 2*y+1$. If started in an arbitrary state (x, y), the statement terminates in the state (x', y') where $x' = 2y' + 1$. Another way of expressing this is to say that S transforms the set of states $\{(x, y) \mid true\}$ into the set $\{(x, y) \mid x = 2y + 1\}$.

The statement S also transforms the set of states $\{(x, y) \mid y \leq 3\}$ into the set $\{(x, y) \mid (x \leq 7) \wedge (y \leq 3)\}$, because if $y \leq 3$ then $2y + 1 \leq 7$. ∎

The concept of transforming a set of states can be extended from an assignment statement to the statement representing the entire program. This is then used to define correctness.

Definition 15.5 A *correctness formula* is a triple $\{p\}\, S\, \{q\}$, where S is a program, and p and q are formulas called the *precondition* and *postcondition*, respectively. S is *partially correct with respect to p and q*, $\models \{p\}\, S\, \{q\}$, iff:

> If S is started in a state where p is true and *if* the computation of S terminates, then it terminates in a state where q is true. ∎

Correctness formulas were first defined in Hoare (1969). The term is taken from Apt et al. (2009); the formulas are also called *inductive expressions*, *inductive assertions* and *Hoare triples*.

Example 15.6 $\models \{y \le 3\}\, x\ =\ 2*y+1\ \{(x \le 7) \wedge (y \le 3)\}.$ ∎

Example 15.7 For any S, p and q:

$$\models \{false\}\, S\, \{q\}, \qquad \models \{p\}\, S\, \{true\},$$

since *false* is not true in any state and *true* is true in all states.   ∎

15.2 Deductive System $\mathscr{HL}$

The deductive system $\mathscr{HL}$ (*Hoare Logic*) is sound and *relatively complete* for proving partial correctness. By relatively complete, we mean that the formulas expressing properties of the domain will not be formally proven. Instead, we will simply take all true formulas in the domain as axioms. For example, $(x \ge y) \rightarrow (x+1 \ge y + 1)$ is true in arithmetic and will be used as an axiom. This is reasonable since we wish to concentrate on the verification that a program S is correct without the complication of verifying arithmetic formulas that are well known.

Definition 15.8 (Deductive system $\mathscr{HL}$)

Domain axioms

> Every true formula over the domain(s) of the program variables.

Assignment axiom

$$\vdash \{p(x)\{x \leftarrow t\}\}\, x\ =\ t\ \{p(x)\}.$$

Composition rule

$$\frac{\vdash \{p\}\, S1\, \{q\} \qquad \vdash \{q\}\, S2\, \{r\}}{\vdash \{p\}\, S1\ \ S2\, \{r\}}.$$

Alternative rule

$$\frac{\vdash \{p \wedge B\}\ S1\ \{q\} \qquad \vdash \{p \wedge \neg B\}\ S2\ \{q\}}{\vdash \{p\}\ \texttt{if}\ \ \texttt{(B)}\ \ \texttt{S1}\ \ \texttt{else}\ \ \texttt{S2}\ \{q\}}.$$

Loop rule

$$\frac{\vdash \{p \wedge B\}\ S\ \{p\}}{\vdash \{p\}\ \texttt{while}\ \ \texttt{(B)}\ \ \texttt{S}\ \{p \wedge \neg B\}}.$$

Consequence rule

$$\frac{\vdash p_1 \to p \qquad \vdash \{p\}\ S\ \{q\} \qquad \vdash q \to q_1}{\vdash \{p_1\}\ S\ \{q_1\}}.$$

∎

The consequence rule says that we can always strengthen the precondition or weaken the postcondition.

Example 15.9 From Example 15.6, we know that:

$$\models \{y \le 3\}\ \mathtt{x}\ =\ \mathtt{2 * y + 1}\ \{(x \le 7) \wedge (y \le 3)\}.$$

Clearly:

$$\models \{y \le 1\}\ \mathtt{x}\ =\ \mathtt{2 * y + 1}\ \{(x \le 10) \wedge (y \le 3)\}.$$

The states satisfying $y \le 1$ are a subset of those satisfying $y \le 3$, so a computation started in a state where, say, $y = 0 \le 1$ satisfies $y \le 3$. Similarly, the states satisfying $x \le 10$ are a superset of those satisfying $x \le 7$; we know that the computation results in a value of x such that $x \le 7$ and that value is also less than or equal to 10. ∎

Since $\vdash p \to p$ and $\vdash q \to q$, we can strengthen the precondition without weakening the postcondition or conversely.

The assignment axiom may seem strange at first, but it can be understood by reasoning from the conclusion to the premise. Consider:

$$\vdash \{?\}\ \mathtt{x}\ =\ \mathtt{t}\ \{p(x)\}.$$

After executing the assignment statement, we want $p(x)$ to be true when the value assigned to x is the value of the expression t. If the formula that results from performing the substitution $p(x)\{x \leftarrow t\}$ is true, then when x is actually assigned the value of t, $p(x)$ will be true.

The composition rule and the alternative rule are straightforward.

The formula p in the loop rule is called an *invariant*: it describes the behavior of a single execution of the statement S in the body of the while-statement. To prove:

$$\vdash \{p_0\}\ \texttt{while}\ \ \texttt{(B)}\ \ \texttt{S}\ \{q_0\},$$

we find a formula p and prove that it is an invariant: $\vdash \{p \wedge B\}\ S\ \{p\}$.

By the loop rule:

$$\vdash \{p\} \texttt{while (B) S} \{p \wedge \neg B\}.$$

If we can prove $p_0 \rightarrow p$ and $(p \wedge \neg B) \rightarrow q_0$, then the consequence rule can be used to deduce the correctness formula. We do not know how many times the `while`-loop will be executed, but we know that $p \wedge \neg B$ holds when it does terminate.

To prove the correctness of a program, one has to find appropriate invariants. The weakest possible formula *true* is an invariant of *any* loop since $\vdash \{true \wedge B\} \texttt{S} \{true\}$ holds for any B and S. Of course, this formula is too weak, because it is unlikely that we will be able to prove $(true \wedge \neg B) \rightarrow q_0$. On the other hand, if the formula is too strong, it will not be an invariant.

Example 15.10 $x = 5$ is too strong to be an invariant of the `while`-statement:

```
while (x > 0) x = x - 1;
```

because $x = 5 \wedge x > 0$ clearly does *not* imply that $x = 5$ after executing the statement `x = x - 1`. The weaker formula $x \geq 0$ is also an invariant: $x \geq 0 \wedge x > 0$ implies $x \geq 0$ after executing the loop body. By the loop rule, if the loop terminates then $x \geq 0 \wedge \neg (x > 0)$. This can be simplified to $x = 0$ by reasoning within the domain and using the consequence rule. ∎

15.3 Program Verification

Let us use $\mathcal{HL}$ to proving the partial correctness of the following program P:

```
{true}
x = 0;
{x = 0}
y = b;
{x = 0 ∧ y = b}
while (y != 0)
    {x = (b − y) · a}
    {
        x = x + a;
        y = y - 1;
    }
{x = a · b}
```

Be careful to distinguish between braces {} used in the syntax of the program from those used in the correctness formulas.

We have *annotated* P with formulas between the statements. Given:

$$\{p_1\}\texttt{S1}\{p_2\}\texttt{S2}\cdots\{p_n\}\texttt{Sn}\{p_{n+1}\},$$

if we can prove $\{p_i\} \texttt{Si} \{p_{i+1}\}$ for all i, then we can conclude:

$$\{p_1\}\, \texttt{S1}\ \cdots\ \texttt{Sn}\, \{p_{n+1}\}$$

by repeated application of the composition rule. See Apt et al. (2009, Sect. 3.4) for a proof that $\mathscr{HL}$ with annotations is equivalent to $\mathscr{HL}$ without them.

Theorem 15.11 $\vdash \{true\}\ \texttt{P}\ \{x = a \cdot b\}$.

Proof From the assignment axiom we have $\{0 = 0\}\ \texttt{x=0}\ \{x = 0\}$, and from the consequence rule with premise $true \to (0 = 0)$, we have $\{true\}\ \texttt{x=0}\ \{x = 0\}$. The proof of $\{x = 0\}\ \texttt{y=b}\ \{(x = 0) \land (y = b)\}$ is similar.

Let us now show that $x = (b - y) \cdot a$ is an invariant of the loop. Executing the loop body will substitute $x + a$ for x and $y - 1$ for y. Since the assignments have no variable in common, we can do them simultaneously. Therefore:

$$
\begin{aligned}
(x = (b - y) \cdot a)\{x \leftarrow x + a, y \leftarrow y - 1\} &\equiv x + a = (b - (y - 1)) \cdot a \\
&\equiv x = (b - y + 1) \cdot a - a \\
&\equiv x = (b - y) \cdot a + a - a \\
&\equiv x = (b - y) \cdot a.
\end{aligned}
$$

By the consequence rule, we can strengthen the precondition:

$$\{(x = (b - y) \cdot a) \land y \neq 0\}\ \texttt{x=x+a; y=y-1;}\ \{x = (b - y) \cdot a\},$$

and then use the Loop Rule to deduce:

```
{x = (b − y) · a}
while (y != 0)
    {
        x=x+a;
        y=y-1;
    }
{(x = (b − y) · a) ∧ ¬(y ≠ 0)}
```

Since $\neg(y \neq 0) \equiv (y = 0)$, we obtain the required postcondition:

$$(x = (b - y) \cdot a) \land (y = 0) \equiv (x = b \cdot a) \equiv (x = a \cdot b).$$

∎

15.3.1 Total Correctness *

Definition 15.12 A program $\texttt{S}$ is *totally correct with respect to p and q* iff:

If $\texttt{S}$ is started in a state where p is true, *then* the computation of $\texttt{S}$ terminates and it terminates in a state where q is true. ∎

The program in Sect. 15.3 is partial correct but not totally correct: if the initial value of b is negative, the program will not terminate. The precondition needs to be strengthened to $b \geq 0$ for the program to be totally correct.

Clearly, the only construct in a program that can lead to non-termination is a loop statement, because the number of iterations of a while-statement need not be bounded. Total correctness is proved by showing that the body of the loop always decreases some value and that that value is bounded from below. In the above program, the value of the variable y decreases by one during each execution of the loop body. Furthermore, it is easy to see that $y \geq 0$ can be added to the invariant of the loop and that y is bounded from below by 0. Therefore, if the precondition is $b \geq 0$, then $b \geq 0 \rightarrow y \geq 0$ and the program terminates when $y = 0$.

$\mathcal{HL}$ can be extended to a deductive system for total correctness; see Apt et al. (2009, Sect. 3.3).

15.4 Program Synthesis

Correctness formulas may also be used in the *synthesis* of programs: the construction of a program directly from a formal specification. The emphasis is on finding invariants of loops, because the other aspects of proving a program (aside from deductions within the domain) are purely mechanical. Invariants are hypothesized as modifications of the postcondition and the program is constructed to maintain the truth of the invariant. We demonstrate the method by developing two different programs for finding the integer square root of a non-negative integer $x = \lfloor \sqrt{a} \rfloor$; expressed as a correctness formula using integers, this is:

$$\{0 \leq a\} \ S \ \{0 \leq x^2 \leq a < (x+1)^2\}.$$

15.4.1 Solution 1

A loop is used to calculate values of the variable x until the postcondition holds. Suppose we let the first part of the postcondition be the invariant and try to establish the second part upon termination of the loop. This gives the following program outline, where E1(x,a), E2(x,a) and B(x,a) represent expressions that must be determined:

```
{0 ≤ a}
x = E1(x,a);
while (B(x,a))
       {0 ≤ x² ≤ a}
       x = E2(x,a);
{0 ≤ x² ≤ a < (x+1)²}.
```

Let p denote the formula $0 \leq x^2 \leq a$ that is the first subformula of the postcondition and then see what expressions will make p an invariant:

- The precondition is $0 \leq a$, so p will be true at the beginning of the loop if the first statement is x=0.
- By the loop rule, when the while-statement terminates, the formula $p \wedge \neg B(x, a)$ is true. If this formula implies the postcondition:

$$(0 \leq x^2 \leq a) \wedge \neg B(x, a) \rightarrow 0 \leq x^2 \leq a < (x + 1)^2,$$

the postcondition follows by the consequence rule. Clearly, $\neg B(x, a)$ should be $a < (x + 1)^2$, so we choose B(x,a) to be (x+1)*(x+1)<=a.
- Given this Boolean expression, if the loop body always increases the value of x, then the loop will terminate. The simplest way to do this is x=x+1.

Here is the resulting program:

$$\{0 \leq a\}$$
```
x = 0;
while  ((x+1)*(x+1)  <=  a)
```
$$\{0 \leq x^2 \leq a\}$$
```
    x = x + 1;
```
$$\{0 \leq x^2 \leq a < (x + 1)^2\}.$$

What remains to do is to check that p is, in fact, an invariant of the loop: $\{p \wedge B\}\,\mathrm{S}\,\{p\}$. Written out in full, this is:

$$\{0 \leq x^2 \leq a \wedge (x + 1)^2 \leq a\}\,\texttt{x=x+1}\,\{0 \leq x^2 \leq a\}.$$

The assignment axiom for x=x+1 is:

$$\{0 \leq (x + 1)^2 \leq a\}\,\texttt{x=x+1}\,\{0 \leq x^2 \leq a\}.$$

The invariant follows from the consequence rule if the formula:

$$(0 \leq x^2 \leq a \wedge (x + 1)^2 \leq a) \rightarrow (0 \leq (x + 1)^2 \leq a)$$

is provable. But this is a true formula of arithmetic so it is a domain axiom.

15.4.2 Solution 2

Incrementing the variable x is not a very efficient way of computing the integer square root. With some more work, we can find a better solution. Let us introduce a new variable y to bound x from above; if we maintain $x < y$ while increasing the value of x or decreasing the value of y, we should be able to close in on a value that makes the postcondition true. Our invariant will contain the formula:

$$0 \le x^2 \le a < y^2.$$

Looking at the postcondition, we see that y is overestimated by $a + 1$, so a candidate for the invariant p is:

$$(0 \le x^2 \le a < y^2) \wedge (x < y \le a + 1).$$

Before trying to establish p as an invariant, let us check that we can find an initialization statement and a Boolean expression that will make p true initially and the postcondition true when the loop terminates.

- The statement y=a+1 makes p true at the beginning of the loop.
- If the loop terminates when $\neg B$ is $y = x + 1$, then:

$$p \wedge \neg B \rightarrow 0 \le x^2 \le a < (x + 1)^2.$$

The outline of the program is:

```
{0 ≤ a}
x = 0;
y = a+1;
while (y != x+1)
        {(0 ≤ x² ≤ a < y²) ∧ (x < y ≤ a + 1)}
    E(x,y,a);
{0 ≤ x² ≤ a < (x + 1)²}.
```

Before continuing with the synthesis, let us try an example.

Example 15.13 Suppose that $a = 14$. Initially, $x = 0$ and $y = 15$. The loop should terminate when $x = 3$ and $y = x + 1 = 4$ so that $0 \le 9 \le 14 < 16$. We need to increase x or decrease y while maintaining the invariant $0 \le x^2 \le a < y^2$. Let us take the midpoint $\lfloor (x + y)/2 \rfloor = \lfloor (0 + 15)/2 \rfloor = 7$ and assign it to either x or y, as appropriate, to narrow the range. In this case, $a = 14 < 49 = 7 \cdot 7$, so assigning 7 to y will maintain the invariant. On the next iteration, $\lfloor (x + y)/2 \rfloor = \lfloor (0 + 7)/2 \rfloor = 3$ and $3 \cdot 3 = 9 < 14 = a$, so assigning 3 to x will maintain the invariant. After two more iterations during which y receives the values 5 and then 4, the loop terminates.   ∎

Here is an outline for the annotated loop *body*; the annotations are derived from the invariant $\{p \wedge B\}$ S1 $\{p\}$ that must be proved and as well as from additional formulas that follow from the assignment axiom.

```
{p ∧ (y ≠ x + 1)}
z = (x+y) / 2;
{p ∧ (y ≠ x + 1) ∧ (⌊z = (x + y)/2⌋)}
if (Cond(x,y,z))
    {p{x ← z}}
    x = z;
else
    {p{y ← z}}
    y = z;
{p}
```

z is a new variable and $\texttt{Cond(x,y,z)}$ is a Boolean expression chosen so that:

$$(p \wedge (y \neq x + 1) \wedge (z = \lfloor(x + y)/2\rfloor) \wedge Cond(x, y, z)) \quad \rightarrow \quad p\{x \leftarrow z\},$$
$$(p \wedge (y \neq x + 1) \wedge (z = \lfloor(x + y)/2\rfloor) \wedge \neg Cond(x, y, z)) \quad \rightarrow \quad p\{y \leftarrow z\}.$$

Let us write out the first subformula of p on both sides of the equations:

$$(0 \leq x^2 \leq a < y^2) \wedge Cond(x, y, z) \quad \rightarrow \quad (0 \leq z^2 \leq a < y^2),$$
$$(0 \leq x^2 \leq a < y^2) \wedge \neg Cond(x, y, z) \quad \rightarrow \quad (0 \leq x^2 \leq a < z^2).$$

These formulas will be true if $\texttt{Cond(x,y,z)}$ is chosen to be $\texttt{z*z <= a}$.

We have to establish the second subformulas of $p\{x \leftarrow z\}$ and $p\{y \leftarrow z\}$, which are $z < y \leq a + 1$ and $x < z \leq a + 1$. Using the second subformulas of p, they follow from arithmetical reasoning:

$$(x < y \leq a + 1) \wedge \qquad\qquad z = \lfloor(x + y)/2\rfloor \quad \rightarrow \quad (z < y \leq a + 1),$$
$$(x < y \leq a + 1) \wedge (y \neq x + 1) \wedge z = \lfloor(x + y)/2\rfloor \quad \rightarrow \quad (x < z \leq a + 1).$$

Here is the final program:

```
{0 ≤ a}
x = 0;
y = a+1;
while (y != x+1)
    {0 ≤ x² ≤ a < y² ∧ x < y ≤ a + 1}
    {
    z = (x+y) / 2;
    if (z*z <= a)
        x = z;
    else
        y = z;
    }
{0 ≤ x² ≤ a < (x + 1)²}.
```

15.5 Formal Semantics of Programs *

A statement transforms a *set* of initial states where the precondition holds into a *set* of final states where the postcondition holds. In this section, the semantics of a program is defined in terms the weakest precondition that causes the postcondition to hold when a statement terminates. In the next section, we show how the formal semantics can be used to prove the soundness and relative completeness of the deductive system $\mathcal{HL}$.

15.5.1 Weakest Preconditions

Let us start with an example.

Example 15.14 Consider the assignment statement x=2*y+1. A correctness formula for this statement is:

$$\{y \le 3\}\, \texttt{x=2*y+1}\, \{(x \le 7) \wedge (y \le 3)\},$$

but $y \le 3$ is not the only precondition that will make the postcondition true. Another one is $y = 1 \vee y = 3$:

$$\{y = 1 \vee y = 3\}\, \texttt{x}\ =\ \texttt{2*y+1}\, \{(x \le 7) \wedge (y \le 3)\}.$$

The precondition $y = 1 \vee y = 3$ is 'less interesting' than $y \le 3$ because it does not characterize *all* the states from which the computation can reach a state satisfying the postcondition.   ∎

We wish to choose the least restrictive precondition so that as many states as possible can be initial states in the computation.

Definition 15.15 A formula A is *weaker than* formula B if $B \rightarrow A$. Given a set of formulas $\{A_1, A_2, \ldots\}$, A_i is the *weakest* formula in the set if $A_j \rightarrow A_i$ for all j. ∎

Example 15.16 $y \le 3$ is weaker than $y = 1 \vee y = 3$ because $(y = 1 \vee y = 3) \rightarrow (y \le 3)$. Similarly, $y = 1 \vee y = 3$ is weaker than $y = 1$, and (by transitivity) $y \le 3$ is also weaker than $y = 1$. This is demonstrated by the following diagram:

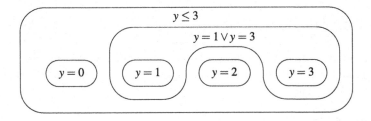

which shows that the weaker the formula, the most states it characterizes.   ∎

The consequence rule is based upon the principle that you can always strengthen an antecedent and weaken a consequent; for example, if $p \rightarrow q$, then $(p \wedge r) \rightarrow q$ and $p \rightarrow (q \vee r)$. The terminology is somewhat difficult to get used to because we are used to thinking about states rather than predicates. Just remember that the *weaker* the predicate, the *more* states satisfy it.

Definition 15.17 Given a program S and a formula q, $wp(\text{S}, q)$, the *weakest precondition of S and q*, is the weakest formula p such that $\models \{p\} \text{S} \{q\}$. ∎

E.W. Dijkstra called this the weakest *liberal* precondition *wlp*, and reserved *wp* for preconditions that ensure total correctness. Since we only discuss partial correctness, we omit the distinction for conciseness.

Lemma 15.18 $\models \{p\} \text{S} \{q\}$ *if and only if* $\models p \rightarrow wp(\text{S}, q)$.

Proof Immediate from the definition of weakest. ∎

Example 15.19 $wp(\text{x=2}*\text{y+1}, x \leq 7 \wedge y \leq 3) = y \leq 3$. Check that $y \leq 3$ really is the weakest precondition by showing that for any weaker formula p', $\not\models \{p'\} \text{x=2}*\text{y+1} \{x \leq 7 \wedge y \leq 3\}$. ∎

The weakest precondition p depends upon both the program and the postcondition. If the postcondition in the example is changed to $x \leq 9$ the weakest precondition becomes $y \leq 4$. Similarly, if S is changed to $\text{x} = \text{y+6}$ without changing the postcondition, the weakest precondition becomes $y \leq 1$.

wp is a called a *predicate transformer* because it defines a transformation of a postcondition predicate into a precondition predicate.

15.5.2 Semantics of a Fragment of a Programming Language

The following definitions formalize the semantics of the fragment of the programming language used in this chapter.

Definition 15.20 $wp(\text{x=t}, p(x)) = p(x)\{x \leftarrow t\}$. ∎

Example 15.21 $wp(\text{y=y-1}, y \geq 0) = (y - 1 \geq 0) \equiv y \geq 1$. ∎

For a compound statement, the weakest precondition obtained from the second statement and postcondition of the compound statement defines the postcondition for the first statement.

Definition 15.22 $wp(\text{S1 S2}, q) = wp(\text{S1}, wp(\text{S2}, q))$. ∎

The following diagram illustrates the definition:

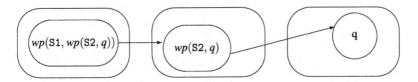

The precondition $wp(\texttt{S2}, q)$ characterizes the largest set of states such that executing S2 leads to a state in which q is true. If executing S1 leads to one of these states, then S1 S2 will lead to a state whose postcondition is q.

Example 15.23

$$
\begin{aligned}
wp(\texttt{x=x+1; y=y+2}, x < y) &= wp(\texttt{x=x+1}, wp(\texttt{y=y+2}, x < y)) \\
&\equiv wp(\texttt{x=x+1}, x < y + 2) \\
&\equiv x + 1 < y + 2 \\
&\equiv x < y + 1.
\end{aligned}
$$

∎

Example 15.24

$$
\begin{aligned}
wp(\texttt{x=x+a; y=y-1}, x &= (b - y) \cdot a) \\
&= wp(\texttt{x=x+a}, wp(\texttt{y=y-1}, x = (b - y) \cdot a)) \\
&\equiv wp(\texttt{x=x+a}, x = (b - y + 1) \cdot a) \\
&\equiv x + a = (b - y + 1) \cdot a \\
&\equiv x = (b - y) \cdot a.
\end{aligned}
$$

Given the precondition $x = (b - y) \cdot a$, the statement x=x+a; y=y-1, considered as a predicate transformer, does nothing! This is not really surprising because the formula is an invariant. Of course, the statement does transform the state of the computation by changing the values of the variables, but it does so in such a way that the formula remains true.   ∎

Definition 15.25 A predicate I is an *invariant* of S iff $wp(\texttt{S}, I) = I$.   ∎

Definition 15.26

$$wp(\texttt{if (B) S1 else S2}, q) = (B \wedge wp(\texttt{S1}, q)) \vee (\neg B \wedge wp(\texttt{S2}, q)).$$

∎

The definition is straightforward because the predicate B partitions the set of states into two disjoint subsets, and the preconditions are then determined by the actions of each Si on its subset.

From the propositional equivalence:

$$(p \rightarrow q) \land (\neg p \rightarrow r) \equiv (p \land q) \lor (\neg p \land r),$$

it can be seen that an alternate definition is:

$wp(\texttt{if (B) S1 else S2}, q) = (B \rightarrow wp(\texttt{S1}, q)) \land (\neg B \rightarrow wp(\texttt{S2}, q))$.

Example 15.27

$wp(\texttt{if (y=0) x=0; else x=y+1}, x = y)$

$\quad = \quad (y = 0 \rightarrow wp(\texttt{x=0}, x = y)) \land (y \neq 0 \rightarrow wp(\texttt{x=y+1}, x = y))$

$\quad \equiv \quad ((y = 0) \rightarrow (0 = y)) \land ((y \neq 0) \rightarrow (y + 1 = y))$

$\quad \equiv \quad true \land ((y \neq 0) \rightarrow false)$

$\quad \equiv \quad \neg (y \neq 0)$

$\quad \equiv \quad y = 0.$

∎

Definition 15.28

$wp(\texttt{while (B) S}, q) = (\neg B \land q) \lor (B \land wp(\texttt{S; while (B) S}, q)).$

∎

The execution of a `while`-statement can proceed in one of two ways.

- The statement can terminate immediately because the Boolean expression evaluates to false, in which case the state does not change so the precondition is the same as the postcondition.
- The expression can evaluate to true and cause S, the body of the loop, to be executed. Upon termination of the body, the `while`-statement again attempts to establish the postcondition.

Because of the recursion in the definition of the weakest precondition for a `while`-statement, we cannot constructively compute it; nevertheless, an attempt to do so is informative.

Example 15.29 Let W be an abbreviation for `while (x>0) x=x-1`.

$wp(\texttt{W}, x = 0)$

$\quad = \quad [\neg (x > 0) \land (x = 0)] \lor [(x > 0) \land wp(\texttt{x=x-1; W}, x = 0)]$

$\quad \equiv \quad (x = 0) \lor [(x > 0) \land wp(\texttt{x=x-1}, wp(\texttt{W}, x = 0))]$

$\quad \equiv \quad (x = 0) \lor [(x > 0) \land wp(\texttt{W}, x = 0)\{x \leftarrow x - 1\}].$

We have to perform the substitution $\{x \leftarrow x - 1\}$ on $wp(\texttt{W}, x = 0)$. But we have just computed a value for $wp(\texttt{W}, x = 0)$. Performing the substitution and simplifying gives:

$wp(\mathtt{W}, x = 0)$

$\equiv \quad (x = 0) \vee [(x > 0) \wedge$

$wp(\mathtt{W}, x = 0)\{x \leftarrow x - 1\}]$

$\equiv \quad (x = 0) \vee [(x > 0) \wedge$

$((x = 0) \vee [(x > 0) \wedge wp(\mathtt{W}, x = 0)\{x \leftarrow x - 1\}])\{x \leftarrow x - 1\}]$

$\equiv \quad (x = 0) \vee [(x - 1 > 0) \wedge$

$((x - 1 = 0) \vee [(x - 1 > 0) \wedge wp(\mathtt{W}, x = 0)\{x \leftarrow x - 1\}\{x \leftarrow x - 1\}])]$

$\equiv \quad (x = 0) \vee [(x > 1) \wedge$

$((x = 1) \vee [(x > 1) \wedge wp(\mathtt{W}, x = 0)\{x \leftarrow x - 1\}\{x \leftarrow x - 1\}])]$

$\equiv \quad (x = 0) \vee (x = 1) \vee [(x > 1) \wedge$

$wp(\mathtt{W}, x = 0)\{x \leftarrow x - 1\}\{x \leftarrow x - 1\}].$

Continuing the computation, we arrive at the following formula:

$$wp(\mathtt{W}, x = 0) \quad \equiv \quad (x = 0) \vee (x = 1) \vee (x = 2) \vee \cdots$$
$$\equiv \quad x \geq 0.$$

■

The theory of fixpoints can be used to formally justify the infinite substitution but that is beyond the scope of this book.

15.5.3   Theorems on Weakest Preconditions

Weakest preconditions distribute over conjunction.

Theorem 15.30 (Distributivity) $\models wp(\mathtt{S}, p) \wedge wp(\mathtt{S}, q) \leftrightarrow wp(\mathtt{S}, p \wedge q)$.

Proof Let s be an arbitrary state in which $wp(\mathtt{S}, p) \wedge wp(\mathtt{S}, q)$ is true. Then both $wp(\mathtt{S}, p)$ and $wp(\mathtt{S}, q)$ are true in s. Executing $\mathtt{S}$ starting in state s leads to a state s' such that p and q are both true in s'. By propositional logic, $p \wedge q$ is true in s'. Since s was arbitrary, we have proved that:

$$\{s \mid \models wp(\mathtt{S}, p) \wedge wp(\mathtt{S}, q)\} \subseteq \{s \mid \models wp(\mathtt{S}, p \wedge q)\},$$

which is the same as:

$$\models wp(\mathtt{S}, p) \wedge wp(\mathtt{S}, q) \rightarrow wp(\mathtt{S}, p \wedge q).$$

The converse is left as an exercise.   ■

Corollary 15.31 (Excluded miracle) $\models wp(S, p) \land wp(S, \neg p) \leftrightarrow wp(S, \textit{false})$.

According to the definition of partial correctness, *any* postcondition (including *false*) is vacuously true if the program does not terminate. It follows that the weakest precondition must include all states for which the program does not terminate. The following diagram shows how $wp(S, \textit{false})$ is the intersection (conjunction) of the weakest preconditions $wp(S, p)$ and $wp(S, \neg p)$:

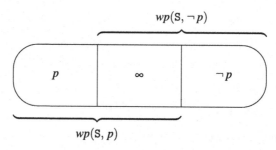

The diagram also furnishes an informal proof of the following theorem.

Theorem 15.32 (Duality) $\models \neg wp(S, \neg p) \to wp(S, p)$.

Theorem 15.33 (Monotonicity) *If* $\models p \to q$ *then* $\models wp(S, p) \to wp(S, q)$.

Proof

1.	$\models wp(S, p) \land wp(S, \neg q) \to wp(S, p \land \neg q)$	Theorem 15.30
2.	$\models p \to q$	Assumption
3.	$\models \neg(p \land \neg q)$	2, PC
4.	$\models wp(S, p) \land wp(S, \neg q) \to wp(S, \textit{false})$	1,3
5.	$\models wp(S, \textit{false}) \to wp(S, q) \land wp(S, \neg q)$	Corollary 15.31
6.	$\models wp(S, \textit{false}) \to wp(S, q)$	5, PC
7.	$\models wp(S, p) \land wp(S, \neg q) \to wp(S, q)$	4, 6, PC
8.	$\models wp(S, p) \to \neg wp(S, \neg q) \lor wp(S, q)$	7, PC
9.	$\models wp(S, p) \to wp(S, q)$	8, Theorem 15.32, PC

∎

The theorem shows that a weaker formula satisfies more states:

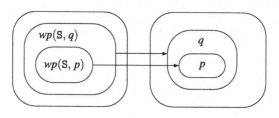

Example 15.34 Let us demonstrate the theorem where p is $x < y - 2$ and q is $x < y$ so that $\models p \to q$. We leave it to the reader to calculate:

$$wp(\mathtt{x=x+1;\ y=y+2;}, x < y - 2) \;\; = \;\; x < y - 1$$
$$wp(\mathtt{x=x+1;\ y=y+2;}, x < y) \;\;\;\;\;\;\; = \;\; x < y + 1.$$

Clearly $\models x < y - 1 \to x < y + 1$.   ∎

15.6 Soundness and Completeness of $\mathcal{HL}$ *

We start with definitions and lemmas which will be used in the proofs.

The programming language is extended with two statements $\mathtt{skip}$ and $\mathtt{abort}$ whose semantics are defined as follows.

Definition 15.35 $wp(\mathtt{skip},\, p) = p$ and $wp(\mathtt{abort},\, p) = \mathit{false}$.   ∎

In other words, $\mathtt{skip}$ does nothing and $\mathtt{abort}$ doesn't terminate.

Definition 15.36 Let W be an abbreviation for $\mathtt{while\ (B)\ S}$.

$$\mathtt{W}^0 \;\;\; = \;\; \mathtt{if\ (B)\ abort;\ else\ skip}$$
$$\mathtt{W}^{k+1} \; = \;\; \mathtt{if\ (B)\ S;W}^k\ \mathtt{else\ skip}$$

                                                                                        ∎

The inductive definition will be used to prove that an execution of W is equivalent to $\mathtt{W}^k$ for some k.

Lemma 15.37 $wp(\mathtt{W}^0,\, p) \equiv \neg B \wedge (\neg B \to p)$.

Proof

$$\begin{aligned}
&wp(\mathtt{W}^0,\, p) &&\equiv\\
&wp(\mathtt{if\ (B)\ abort;\ else\ skip},\, p) &&\equiv\\
&(B \to wp(\mathtt{abort},\, p)) \wedge (\neg B \to wp(\mathtt{skip},\, p)) &&\equiv\\
&(B \to \mathit{false}) \wedge (\neg B \to p) &&\equiv\\
&(\neg B \vee \mathit{false}) \wedge (\neg B \to p) &&\equiv\\
&\neg B \wedge (\neg B \to p).
\end{aligned}$$

                                                                                        ∎

Lemma 15.38 $\bigvee_{k=0}^{\infty} wp(\mathtt{W}^k, p) \rightarrow wp(\mathtt{W}, p)$.

Proof We show by induction that for each k, $wp(\mathtt{W}^k, p) \rightarrow wp(\mathtt{W}, p)$.

For $k = 0$:

1.	$wp(\mathtt{W}^0, p) \rightarrow \neg B \wedge (\neg B \rightarrow p)$	Lemma 15.37
2.	$wp(\mathtt{W}^0, p) \rightarrow \neg B \wedge p$	1, PC
3.	$wp(\mathtt{W}^0, p) \rightarrow (\neg B \wedge p) \vee (B \wedge wp(\mathtt{S};\mathtt{W}, p))$	2, PC
4.	$wp(\mathtt{W}^0, p) \rightarrow wp(\mathtt{W}, p)$	3, Def. 15.28

For $k > 0$:

1.	$wp(\mathtt{W}^{k+1}, p) = wp(\mathtt{if\ (B)\ S;W}^k\ \mathtt{else\ skip}, p)$	Def. 15.36
2.	$wp(\mathtt{W}^{k+1}, p) \equiv (B \rightarrow wp(\mathtt{S};\mathtt{W}^k, p)) \wedge$	Def. 15.26
	$\quad (\neg B \rightarrow wp(\mathtt{skip}, p))$	
3.	$wp(\mathtt{W}^{k+1}, p) \equiv (B \rightarrow wp(\mathtt{S}, wp(\mathtt{W}^k, p))) \wedge$	Def. 15.22
	$\quad (\neg B \rightarrow wp(\mathtt{skip}, p))$	
4.	$wp(\mathtt{W}^{k+1}, p) \equiv (B \rightarrow wp(\mathtt{S}, wp(\mathtt{W}^k, p))) \wedge (\neg B \rightarrow p)$	Def. 15.35
5.	$wp(\mathtt{W}^{k+1}, p) \rightarrow (B \rightarrow wp(\mathtt{S}, wp(\mathtt{W}, p))) \wedge (\neg B \rightarrow p)$	Ind. hyp.
6.	$wp(\mathtt{W}^{k+1}, p) \rightarrow (B \rightarrow wp(\mathtt{S};\mathtt{W}, p)) \wedge (\neg B \rightarrow p)$	Def. 15.22
7.	$wp(\mathtt{W}^{k+1}, p) \rightarrow wp(\mathtt{W}, p)$	Def. 15.28

∎

As k increases, more and more states are included in $\bigvee_{i=0}^{k} wp(\mathtt{W}^i, p)$:

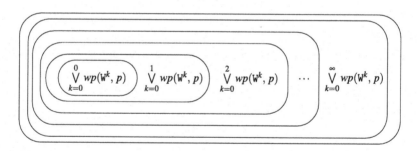

Theorem 15.39 (Soundness of $\mathcal{HL}$) *If* $\vdash_{HL} \{p\}\ \mathtt{S}\ \{q\}$ *then* $\models \{p\}\ \mathtt{S}\ \{q\}$.

Proof The proof is by induction on the length of the $\mathcal{HL}$ proof. By assumption, the domain axioms are true, and the use of the consequence rule can be justified by the soundness of *MP* in first-order logic.

By Lemma 15.18, $\models \{p\}\ \mathtt{S}\ \{q\}$ iff $\models p \rightarrow wp(\mathtt{S}, q)$, so it is sufficient to prove $\models p \rightarrow wp(\mathtt{S}, q)$. The soundness of the assignment axioms is immediate by Definition 15.20.

Suppose that the composition rule is used. By the inductive hypothesis, we can assume that $\models p \rightarrow wp(\mathtt{S1}, q)$ and $\models q \rightarrow wp(\mathtt{S2}, r)$. From the second assumption and monotonicity (Theorem 15.33),

$$\models wp(\texttt{S1}, q) \rightarrow wp(\texttt{S1}, wp(\texttt{S2}, r)).$$

By the consequence rule and the first assumption, $\models p \rightarrow wp(\texttt{S1}, wp(\texttt{S2}, r))$, which is $\models p \rightarrow wp(\texttt{S1};\texttt{S2}, r)$ by the definition of wp for a compound statement.

We leave the proof of the soundness of the alternative rule as an exercise.

For the loop rule, by structural induction we assume that:

$$\models (p \land B) \rightarrow wp(\texttt{S}, p)$$

and show:

$$\models p \rightarrow wp(\texttt{W}, p \land \neg B).$$

We will prove by numerical induction that for all k:

$$\models p \rightarrow wp(\texttt{W}^k, p \land \neg B).$$

For $k = 0$, the proof of

$$\models wp(\texttt{W}^0, p \land \neg B) = wp(\texttt{W}, p \land \neg B)$$

is the same as the proof of the base case in Lemma 15.38. The inductive step is proved as follows:

1.	$\models p \rightarrow (\neg B \rightarrow (p \land \neg B))$	PC
2.	$\models p \rightarrow (\neg B \rightarrow wp(\texttt{skip}, p \land \neg B))$	Def. 15.35
3.	$\models (p \land B) \rightarrow wp(\texttt{S}, p)$	Structural ind. hyp.
4.	$\models p \rightarrow wp(\texttt{W}^k, p \land \neg B)$	Numerical ind. hyp.
5.	$\models (p \land B) \rightarrow wp(\texttt{S}, wp(\texttt{W}^k, p \land \neg B))$	3, 4, Monotonicity
6.	$\models (p \land B) \rightarrow wp(\texttt{S};\texttt{W}^k, p \land \neg B)$	5, Composition
7.	$\models p \rightarrow (B \rightarrow wp(\texttt{S};\texttt{W}^k, p \land \neg B))$	6, PC
8.	$\models p \rightarrow wp(\texttt{if (B) S};\texttt{W}^k \texttt{ else skip}, p \land \neg B)$	2, 7, Def. 15.26
9.	$\models p \rightarrow wp(\texttt{W}^{k+1}, p \land \neg B)$	Def. 15.36

By infinite disjunction:

$$\models p \rightarrow \bigvee_{k=0}^{\infty} wp(\texttt{W}^k, p \land \neg B),$$

and:

$$\models p \rightarrow wp(\texttt{W}, p \land \neg B)$$

follows by Lemma 15.38.   ∎

Theorem 15.40 (Completeness of $\mathcal{HL}$)   *If* $\models \{p\}\, S\, \{q\}$, *then* $\vdash_{HL} \{p\}\, S\, \{q\}$.

Proof We have to show that if $\models p \to wp(S, q)$, then $\vdash_{HL} \{p\}\, S\, \{q\}$. The proof is by structural induction on S. Note that $p \to wp(S, q)$ is just a formula of the domain, so $\vdash p \to wp(S, q)$ follows by the domain axioms.

Case 1: Assignment statement x=t.

$$\vdash \{q\{x \leftarrow t\}\}\, \texttt{x=t}\, \{q\}$$

is an axiom, so:

$$\vdash \{wp(\texttt{x=t}, q)\}\, \texttt{x=t}\, \{q\}$$

by Definition 15.20. By assumption, $\vdash p \to wp(\texttt{x=t}, q)$, so by the consequence rule $\vdash \{p\}\, \texttt{x=t}\, \{q\}$.

Case 2: Composition S1 S2.

By assumption:

$$\models p \to wp(\texttt{S1 S2}, q)$$

which is equivalent to:

$$\models p \to wp(\texttt{S1}, wp(\texttt{S2}, q))$$

by Definition 15.22, so by the inductive hypothesis:

$$\vdash \{p\}\, \texttt{S1}\, \{wp(\texttt{S2}, q)\}.$$

Obviously:

$$\models wp(\texttt{S2}, q) \to wp(\texttt{S2}, q),$$

so again by the inductive hypothesis (with $wp(\texttt{S2}, q)$ as p):

$$\vdash \{wp(\texttt{S2}, q)\}\, \texttt{S2}\, \{q\}.$$

An application of the composition rule gives $\vdash \{p\}\, \texttt{S1 S2}\, \{q\}$.

Case 3: if-statement. Exercise.

Case 4: while-statement, W = while (B) S.

1.	$\models wp(\texttt{W}, q) \wedge B \to wp(\texttt{S;W}, q)$	Def. 15.28
2.	$\models wp(\texttt{W}, q) \wedge B \to wp(\texttt{S}, wp(\texttt{W}, q))$	Def. 15.22
3.	$\vdash \{wp(\texttt{W}, q) \wedge B\}\, \texttt{S}\, \{wp(\texttt{W}, q)\}$	Inductive hypothesis
4.	$\vdash \{wp(\texttt{W}, q)\}\, \texttt{W}\, \{wp(\texttt{W}, q) \wedge \neg B\}$	Loop rule
5.	$\vdash (wp(\texttt{W}, q) \wedge \neg B) \to q$	Def. 15.28, Domain axiom
6.	$\vdash \{wp(\texttt{W}, q)\}\, \texttt{W}\, \{q\}$	4, 5, Consequence rule
7.	$\vdash p \to wp(\texttt{W}, q)$	Assumption, domain axiom
8.	$\vdash \{p\}\, \texttt{W}\, \{q\}$	Consequence rule

∎

15.7 Summary

Computer programs are similar to logical formulas in that they are formally defined by syntax and semantics. Given a program and two correctness formulas—the precondition and the postcondition—we aim to verify the program by proving: if the input to the program satisfies the precondition, then the output of the program will satisfy the postcondition. Ideally, we should perform program synthesis: start with the pre- and postconditions and derive the program from these logical formulas.

The deductive system Hoare Logic $\mathscr{HL}$ is sound and relatively complete for verifying sequential programs in a programming language that contains assignment statements and the control structures `if` and `while`.

15.8 Further Reading

Gries (1981) is the classic textbook on the verification of sequential programs; it emphasizes program synthesis. Manna (1974) includes a chapter on program verification, including the verification of programs written as flowcharts (the formalism originally used by Robert W. Floyd). The theory of program verification can be found in Apt et al. (2009), which also treats deductive verification of concurrent programs.

SPARK is a software system that supports the verification of programs; an open-source version can be obtained from http://libre.adacore.com/.

15.9 Exercises

15.1 What is $wp(S, true)$ for any statement S?

15.2 Let S1 be x=x+y and S2 be y=x*y. What is $wp(S1\ S2, x < y)$?

15.3 Prove $\models wp(S, p \wedge q) \rightarrow wp(S, p) \wedge wp(S, q)$, (the converse direction of Theorem 15.30).

15.4 Prove that

$$wp(\text{if } (B) \ \{ \ S1\ S3 \ \} \text{ else } \{ \ S2\ S3 \ \}, q) =$$
$$wp(\{\text{if } (B) \ S1 \text{ else } S2\} \ S3, q).$$

15.5 * Suppose that $wp(S, q)$ is defined as the weakest formula p that ensures *total* correctness of S, that is, if S is started in a state in which p is true, then it *will* terminate in a state in which q is true. Show that under this definition $\models \neg wp(S, \neg q) \equiv wp(S, q)$ and $\models wp(S, p) \vee wp(S, q) \equiv wp(S, p \vee q)$.

15.6 Complete the proofs of the soundness and completeness of $\mathcal{HL}$ for the alternative rule (Theorems 15.39 and 15.40).

15.7 Prove the partial correctness of the following program.

```
{a ≥ 0}
x = 0; y = 1;
while (y <= a)
   {
      x = x + 1;
      y = y + 2*x + 1;
   }
{0 ≤ x² ≤ a < (x + 1)²}
```

15.8 Prove the partial correctness of the following program.

```
{a > 0 ∧ b > 0}
x = a; y = b;
while (x != y)
   if (x > y)
      x = x-y;
   else
      y = y-x;
{x = gcd(a, b)}
```

15.9 Prove the partial correctness of the following program.

```
{a > 0 ∧ b > 0}
x = a; y = b;
while (x != y)
   {
      while (x > y) x = x-y;
      while (y > x) y = y-x;
   }
{x = gcd(a, b)}
```

15.10 Prove the partial correctness of the following program.

```
{a ≥ 0 ∧ b ≥ 0}
x = a; y = b; z = 1;
while (y != 0)
   if (y % 2 == 1) { /* y is odd */
      y = y - 1;
      z = x*z;
   }
   else {
      x = x*x;
      y = y / 2;
   }
{z = aᵇ}
```

15.11 Prove the partial correctness of the following program.

```
{a ≥ 2}
y = 2; x = a; z = true;
while (y < x)
   if (x % y == 0)
      z = false;
      break;
   }
   else
      y = y + 1;
{z ≡ (a is prime)}
```

References

K.R. Apt, F.S. de Boer, and E.-R. Olderog. *Verification of Sequential and Concurrent Programs (Third Edition)*. Springer, London, 2009.

D. Gries. *The Science of Programming*. Springer, New York, NY, 1981.

C.A.R. Hoare. An axiomatic basis for computer programming. *Communications of the ACM*, 12(10): 576–580, 583, 1969.

Z. Manna. *Mathematical Theory of Computation*. McGraw-Hill, New York, NY, 1974. Reprinted by Dover, 2003.

Chapter 16
Verification of Concurrent Programs

Verification is routinely used when developing computer hardware and concurrent programs. A sequential program can always be tested and retested, but the nondeterministic nature of hardware and concurrent programs limits the effectiveness of testing as a method to demonstrate that the system is correct. Slight variations in timing, perhaps caused by congestion on a network, mean that two executions of the same program might give different results. Even if a bug is found by testing and then fixed, we have no way of knowing if the next test runs correctly because we fixed the bug or because the execution followed a different *scenario*, one in which the bug cannot occur.

We start this chapter by showing how temporal logic can be used to verify the correctness of a concurrent program deductively. Deductive verification has proved to be difficult to apply in practice; in many cases, an alternate approach called model checking is used. Model checking examines the reachable states in a program looking for a state where the correctness property does not hold. If it searches all reachable states without finding an error, the correctness property holds. While model checking is easier in practice than deductive verification, it is difficult to implement efficiently. We will show how binary decision diagrams (Chap. 5) and SAT solvers (Chap. 6) can be used to implement model checkers. The chapter concludes with a short overview of CTL, a branching-time temporal logic that is an alternative to the linear-time temporal logic that we studied so far. Traditionally, CTL has found wide application in the verification of (synchronous) hardware systems, while LTL was used for (asynchronous) software systems.

This chapter is a survey only, demonstrating the various concepts and techniques by examples. For details of the theory and practice of the verification of concurrent programs, see the list of references at the end of the chapter.

M. Ben-Ari, *Mathematical Logic for Computer Science*,
DOI 10.1007/978-1-4471-4129-7_16, © Springer-Verlag London 2012

16.1 Definition of Concurrent Programs

Our concurrent programs will be composed of the same statements used in the sequential programs of Chap. 15. A concurrent program is a set of sequential programs together with a set of global variables.

Definition 16.1 A concurrent program is a set of *processes* {p1, p2, ..., pn}, where each process is a program as defined in Definition 15.1. The variables declared in each process are its *local variables*; a local variable can be read and written only by the process where it is declared. In addition, there may be *global variables* that can be read and written by all of the processes. ■

Processes are also known as *threads*; in some contexts, the two terms have different meanings but the difference is not relevant here.

Example 16.2 The following concurrent program consists of two processes p and q, each of which is a sequential program with two assignment statements (and an additional label end). There is one global variable n initialized to 0 and no local variables.

	int n = 0
Process p	Process q
1: n = n + 1	1: n = n + 1
2: n = n + 1	2: n = n + 1
end:	end:

■

The state of a concurrent programs consists of the values of its variables (both local and global), together with the location counters of its processes.

Definition 16.3 Let S be a program with processes {p1,p2,...,pn} and let the *statements* of process i be labeled by $L^i = (L_1^i, L_2^i, \ldots, L_{k_i}^i)$. Let $(v1, v2, \ldots, vm)$ be the variables of S (both global and local). A *state s* of a computation of S is an $m + n$-tuple:

$$(v_1, v_2, \ldots, v_m, l^1, l^2, \ldots, l^n),$$

where v_j is the value of the jth variable in the state and $l^i \in L^i$ is the value in the location counter of the ith process. ■

Example 16.4 For the program of Example 16.2, there are $5 \times 3 \times 3 = 45$ different states, because the variable n can have the values 0, 1, 2, 3, 4 and there are three labels for each process. These seems like quite a large number of states for such a simple program, but many of the states (for example, (0, *end*, *end*)) will never occur in any computation. ■

Interleaving

A computation of a concurrent program is obtained by *asynchronous interleaving of atomic instructions*.

Definition 16.5 A *computation* of a concurrent program S is a sequence of states. In the initial state s_0, v_j contains the initial value of the variable vj and l^i is set to the initial statement l_1^i of the *i*th process. A *transition* from state s to state s' is done by selecting a process i and executing the statement labeled l^i. The components of s' are the same as those of s except:

- If the statement at l^i is an assignment statement v=e, then v', the value of the variable v in s', is the value obtained by evaluating the expression e given the values of the variables in s.
- $l^{i'}$, the value of the *i*th location counter in s', is set using the rules for control structures.

The computation is said to be obtained by *interleaving* statements from the processes of the program.   ∎

Example 16.6 Although there are 45 possible states for the program of Example 16.2, only a few of them will actually occur in any computation. Here are two computations, where each triple is (n, l^p, l^q):

$$(0, 1, 1) \rightarrow (1, 2, 1) \rightarrow (2, end, 1) \rightarrow (3, end, 2) \rightarrow (4, end, end),$$
$$(0, 1, 1) \rightarrow (1, 2, 1) \rightarrow (2, 2, 2) \rightarrow (3, end, 2) \rightarrow (4, end, end).$$

In the first computation, process p executes its statements to termination and only then does process q execute its statements. In the second computation, the interleaving is obtained by alternating execution of statements from the two processes. The result—the final value of n—is the same in both cases.   ∎

Atomic Operations

In the definition of a computation, *statements* are interleaved, that is, each statement is executed to completion before the execution of another statement (from the same process or another process) is started. We say that the statements are *atomic operations*. It is important to define the atomic operations of a system before you can reason about it. Consider a system where an assignment statement is not executed atomically; instead, each separate access to memory is an atomic operation and they can be interleaved. We demonstrate the effect of the specification of atomic operations by comparing the computations of the following two programs.

In the first program, an assignment statement is an atomic operation:

int n = 0	
Process p	Process q
1: n = n + 1	1: n = n + 1
end:	end:

In the second program, local variables are used to simulate a computer that evaluates expressions in a register; the value of n is loaded into the register and then stored back into memory when the expression has been evaluated:

int n = 0	
Process p int temp = 0	Process q int temp = 0
1: temp = n 2: temp = temp + 1 3: n = temp end:	1: temp = n 2: temp = temp + 1 3: n = temp end:

Clearly, the final value of n in the first program will be 2. For the second program, if all the statements of p are executed before the statements of q, the same result will be obtained. However, consider the following computation of the second program obtained by interleaving one statement at a time from each process, where the 5-tuple is $(n, temp^p, temp^q, l^p, l^q)$:

$$(0, 0, 0, 1, 1) \rightarrow (0, 0, 0, 2, 1) \rightarrow (0, 0, 0, 2, 2) \rightarrow (0, 1, 0, 3, 2) \rightarrow (0, 1, 1, 3, 3) \rightarrow$$
$$(1, 1, 1, end, 3) \rightarrow (1, 1, 1, end, end).$$

The result of this computation—n has the value 1—is not the same as the result of the previous computation. Unlike a sequential program which has only one computation, a concurrent program has many computations and they may have different results, not all of which may be correct. Consider the correctness property expressed in LTL as $\Diamond(n = 2)$, eventually the value of the variable n is 2. The formula is true for some computations but not for all computations, so the correctness property does not hold for the program.

16.2 Formalization of Correctness

We will use Peterson's algorithm for solving the critical section problem for two processes as the running example throughout this chapter.

Definition 16.7 The *critical section problem for two processes* is to design an algorithm that for synchronizing two concurrent processes according to the following specification:

Each process consists of a *critical section* and a *non-critical section*. A process may stay indefinitely in its non-critical section, or—at any time—it may request to enter its critical section. A process that has entered its critical section will eventually leave it.

The solution must satisfy the following two correctness properties:

- **Mutual exclusion:** It is forbidden for the two processes to be in their critical sections simultaneously.
- **Liveness:** If a process attempts to enter its critical section, it will eventually succeed. ∎

The problem is difficult to solve. In a classic paper (Dijkstra, 1968), Dijkstra went through a series of four *attempts* at solving the problem, each one of which contained a different type of error, before arriving at a solution called Dekker's algorithm (see Ben-Ari (2006)). Here, we choose to work with Peterson's algorithm, which is much more concise than Dekker's.

Peterson's Algorithm

Here is Peterson's algorithm (Peterson, 1981):

| boolean wantp = false, wantq = false |
| int turn = 1 |

Process p	Process q

```while (true) {```
 ``` non-critical-section```
 ``` wantp = true```
 ``` turn = 1```
 ``` wait until```
 ```   (!wantq or turn == 2)```
 ``` critical-section```
 ``` wantp = false```
 ```}```
	```while (true) {```
	 ``` non-critical-section```
	 ``` wantq = true```
	 ``` turn = 2```
	 ``` wait until```
	 ``` (!wantp or turn == 1)```
	 ``` critical-section```
	 ``` wantq = false```
	 ```}```

The statement:

```
wait until (!wantq or turn == 2)
```

is a more intuitive way of writing:

```
while (!(!wantq or turn == 2)) /* do nothing */
```

The intuitive explanation of Peterson's algorithm is as follows. The variables wantp and wantq are set to true by the processes to indicate that they are trying to enter their critical sections and reset to false when they leave their critical sections. A trying-process waits until the other process is neither trying to enter its critical section nor is it in its critical section (!wantq or !wantp). Since the algorithm is symmetric, the variable turn is used to break ties when both processes are trying to enter their critical sections. A tie is broken in favor of the first process which set `turn`. Suppose that process p set `turn` to 1 and then process q set `turn` to 2. The expression `turn==2` will be true and allow process p to enter its critical section.

16.2.1 An Abbreviated Algorithm

Before proceeding to specifying and proving the correctness of Peterson's algorithm, we simplify it to reduce the number of states and transitions:

`boolean wantp = false, wantq = false`
`int turn = 1`

Process p	Process q

```
while (true) {               while (true) {
  tryp: wantp = true; turn = 1   tryq: wantq = true; turn = 2
  waitp: wait until            waitq: wait until
      (!wantq or turn == 2)         (!wantp or turn == 1)
  csp: wantp = false          csq: wantq = false
}                            }
```

First, we omit the critical and non-critical section! This may seem strange because the whole point of the algorithm is the execute a critical section, but we are not at all interested in the contents of the critical section. It is simply a no-operation that we are assured must terminate. A process will be considered to be 'in' its critical section when its location counter is at the statement `wantp=false` or `wantq=false`. A process will be considered to be in its non-critical section when its location counter is at the statement `wantp=true` or `wantq=true`.

Second, the two assignments before the `wait` are written on one line and executed as one atomic operation. It follows that we are allowing *fewer* computations than in the original algorithm. We leave it as an exercise to show the correctness of the algorithm without this simplification.

Correctness Properties

The following two LTL formulas express the correctness of Peterson's algorithm for the critical section problem:

Mutual exclusion: $\Box \neg (csp \wedge csq),$

Liveness: $\Box(tryp \rightarrow \Diamond csp) \wedge \Box(tryq \rightarrow \Diamond csq).$

In these formulas, the labels of the statements of the algorithm are used as atomic propositions meaning that the location counter of the corresponding process is at that label. For example, in the state:

$$(true, false, 2, csp, tryq),$$

wantp is true, *wantq* is false, the value of the variable `turn` is 2 and the processes are at `csp` and `tryq`, respectively.

Mutual exclusion forbids (always false) a computation from including a state where both processes are in their critical section, while liveness requires that (always) if a computation includes a state where a process is trying to enter its critical section then (eventually) the computation will include a state where the process is in its critical section.

16.3 Deductive Verification of Concurrent Programs

Invariants

A safety property can be verified using the induction rule (Sect. 14.2):

$$\frac{\vdash A \rightarrow \bigcirc A}{\vdash A \rightarrow \Box A}.$$

Assume that A is true in a state and prove that it holds in the next state; if A is also true in the initial state, then $\Box A$ is true. In other words, we have to show that A is an *invariant* (cf. Sect. 15.2).

If the formula that is supposed to be an invariant is an implication $A \rightarrow B$, the effort needed to prove the inductive step can often be significantly reduced. By the inductive hypothesis, $A \rightarrow B$ is assumed to be true and there are only two ways for a true implication to become false. Either A and B are both true and B 'suddenly' becomes false while A remains true, or A and B are both false and A 'suddenly' becomes true while B remains false. By 'suddenly' we mean that a single transition changes the truth value of a formula.

Lemma 16.8

(a) $\vdash \Box((\mathit{turn} = 1) \vee (\mathit{turn} = 2))$.
(b) $\vdash \Box(\mathit{wantp} \leftrightarrow (\mathit{waitp} \vee \mathit{csp}))$.
(c) $\vdash \Box(\mathit{wantq} \leftrightarrow (\mathit{waitq} \vee \mathit{csq}))$.

Proof The proof of (a) is trivial since `turn` is initialized to 1 and is only assigned the values 1 and 2. We prove the forward direction of (b) and leave the other direction of (b) as an exercise. Since the program is symmetric in p and q, the same proof holds for (c).

The formula $\mathit{wantp} \rightarrow (\mathit{waitp} \vee \mathit{csp})$ is true initially since `wantp` is initialized to false and an implication is true if its antecedent is true regardless of the truth of its consequent (although here the initial location counter of process p is set to `tryp` so the consequent is also false).

Suppose that the formula is true. It can be falsified if both the antecedent and consequent are true and the consequent suddenly becomes false, which can only occur when the transition `csp`→`tryp` is taken. However, the assignment to `wantp` at `csp` falsifies the antecedent, so the formula remains true.

The formula could also be falsified if both the antecedent and consequent are false and the antecedent suddenly becomes true. That can only occur when the transition tryp→waitp that assigns true to wantp is taken. However, the location counter is changed so that *waitp* becomes true, so the consequent *waitp* ∨ *csp* becomes true and the formula remains true.   ∎

The proof has been given in great detail, but you will soon learn that invariants where the value of a variable is coordinated with the value of the location counter are easily proved. By the properties of material implication, the truth of an invariant is preserved by any transition such as waitp→csp that cannot make the antecedent true nor the consequent false. Similarly, no transition of process q can affect the truth value of the formula.

Mutual Exclusion

To prove that the mutual exclusion property holds for Peterson's algorithm, we need to prove that ¬ (*csp* ∧ *csq*) is an invariant. Unfortunately, we cannot prove that directly; instead, we show that two other formulas are invariant and then deduce the mutual exclusion property from them.

Lemma 16.9 *The following formulas are invariant in Peterson's algorithm:*

$$(waitp \land csq) \ \rightarrow \ (wantq \land turn = 1),$$
$$(csp \land waitq) \ \rightarrow \ (wantp \land turn = 2).$$

Theorem 16.10 *In Peterson's algorithm,* ¬ (*csp* ∧ *csq*) *is an invariant.*

Proof The formula is true initially.

The definition of a computation of a concurrent program is by interleaving, where only one statement from one process is executed at a time. Therefore, either process q was already in its critical section when process p entered its critical section, or p was in its critical section when q entered. By the symmetry of the algorithm, it suffices to consider the first possibility.

To falsify the formula ¬ (*csp* ∧ *csq*), the computation must execute the transition waitp→csp while *waitp* ∧ *csq* is true. By Lemma 16.9, this implies that *wantq* ∧ *turn* = 1 is true. We have the following chain of logical equivalences:

$$wantq \land turn = 1 \qquad \equiv$$
$$\neg \neg (wantq \land turn = 1) \qquad \equiv$$
$$\neg (\neg wantq \lor \neg (turn = 1)) \ \equiv$$
$$\neg (\neg wantq \lor (turn = 2)).$$

The last equivalence used the invariant in Lemma 16.8(a).

However, the transition waitp→csp is enabled only if ¬ *wantq* ∨ *turn* = 2 is true, but we have just shown that it is false. It follows that ¬ (*csp* ∧ *csq*) can never become true.   ∎

Proof of Lemma 16.9 By symmetry it suffices to prove the first formula.

Clearly, the formula is true initially since the location counters are initialized to tryp and tryq.

Suppose that the antecedent of $(waitp \land csq) \rightarrow (wantq \land turn = 1)$ becomes true because the transition tryp→waitp is taken in a state where *csq* is true. By Lemma 16.8(c), *wantq* is true and the transition assigns 1 to *turn*, so the consequent remains or becomes true.

Suppose now that the antecedent of $(waitp \land csq) \rightarrow (wantq \land turn = 1)$ becomes true because the transition waitq→csq is taken in a state where *waitp* is true. By Lemma 16.8(c), *wantq* is true, so we have to show that $turn = 1$. But, by Lemma 16.8(b), *waitp* implies that *wantp* is true; therefore, the only way that the transition waitq→csq could have been taken is if $turn = 1$, so the consequent remains or becomes true.

It remains to check the possibility that the consequent becomes false while the antecedent remains or becomes true. But the only transitions that change the value of the consequent are tryq→waitq and csq→tryq, both of which falsify *csq* in the antecedent.   ∎

Progress

The axiom system $\mathcal{HL}$ for proving correctness of sequential programs provides the semantics of the execution of statements in a program (Definition 15.5). It defines, for example, the effect of an assignment statement—in the new state, the value of the assigned variable is the value of the expression—but it does not actually require that the assignment statement will ever be executed. In order to prove the liveness of a program like Peterson's algorithm, we need to add progress axioms for each type of statement.

In this section, we assume that the interleaving is *fair* (Definition 16.20). For a detailed discussion of this concept see Ben-Ari (2006, Sect. 2.7).

Definition 16.11 Here are the *progress axioms* for each statement:

	Statement	Progress axioms
li: li+1:	v = expression;	$\vdash l_i \rightarrow \Diamond l_{i+1}$

li:
lt:
 	
lf:
	if (B)
	    S1;
	else
	    S2;
		$\vdash l_i \rightarrow \Diamond(l_t \lor l_f)$
		$\vdash (l_i \land \Box B) \rightarrow \Diamond l_t$
		
		$\vdash (l_i \land \Box \neg B) \rightarrow \Diamond l_f$

li: lt: lf:	while (B) S1;	$\vdash l_i \rightarrow \Diamond(l_t \lor l_f)$ $\vdash (l_i \land \Box B) \rightarrow \Diamond l_t$ $\vdash (l_i \land \Box \neg B) \rightarrow \Diamond l_f$

∎

An assignment statement will be unconditionally executed eventually. However, for control statements with alternatives (if- and while-statement), all we can say for sure is that it will eventually be executed and one of the two alternatives taken $\vdash l_i \rightarrow \Diamond(l_t \lor l_f)$, but without more information we cannot know *which* branch will be taken. $\vdash (l_i \land B) \rightarrow \Diamond l_t$ is not acceptable as an axiom because by the time that this transition is taken, another process could have modified a global variable falsifying B. Only if B is held true or false indefinitely can we prove which branch will be taken.

For Peterson's algorithm, we do not assume progress at the statements tryp and tryq; this models the specification that a process need not leave its non-critical section.

Liveness

We can now prove the liveness of Peterson's algorithm. By symmetry, it is sufficient to prove liveness for one process; for process p, the correctness formula is $waitp \rightarrow \Diamond csp$. To prove the formula, we assume that it is not true ($waitp \land \Box \neg csp$) and deduce a contradiction.

Lemma 16.12 $\vdash waitp \land \Box \neg csp \rightarrow \Box \Diamond (wantq \land last \neq 2)$.

Proof Recall that the statement at waitp:

```
waitp: wait until (!wantq or turn == 2)
```

is an abbreviation for the while-statement:

```
while (!(!wantq or turn == 2)) /* do nothing */
```

By the progress axiom:

$$\vdash waitp \land \Box \neg B \rightarrow \Diamond csp,$$

where B is the expression in the while-loop. By propositional reasoning and duality, we have:

$$\vdash waitp \land \Box \neg csp \rightarrow \Diamond B,$$

which is:

$$\vdash waitp \land \Box \neg csp \rightarrow \Diamond (wantq \land turn \neq 2).$$

By generalization:

$$\vdash \Box(waitp \land \Box \neg csp) \rightarrow \Box \Diamond (wantq \land turn \neq 2),$$

and we leave it as an exercise to show that:

$$\vdash waitp \land \Box \neg csp \rightarrow \Box(waitp \land \Box \neg csp).$$

■

Lemma 16.13 $\vdash \Diamond \Box \neg\, wantq \,\vee\, \Diamond (turn = 2)$.

Proof If $\Diamond (turn = 2)$, the formula is true, so we ask what can happen if it is not true. This is done by cases on the location counter of process q. If the location counter is at `tryq` and the computation never leaves there (because it is simulating a non-critical section), then $\Box \neg\, wantq$ (Lemma 16.8(c)). If the computation leaves `tryq`, then by the progress axiom, eventually the assignment statement `turn=2` must be executed. If the location counter is at `csq`, by progress it reaches `tryq` and we have just shown what happens in that case. Finally, if the computation is at `waitq` and $turn = 2$ is never true, then $turn = 1$ is always true (Lemma 16.8(a)) and by the progress axiom, the computation proceeds to `csq` and we have already shown what happens in that case.   ∎

Lemma 16.14 $\vdash waitp \,\wedge\, \Box \neg\, csp \,\wedge\, \Diamond (turn = 2) \;\rightarrow\; \Diamond \Box (turn = 2)$.

Proof The only way that $turn = 2$ could be falsified is for process p to execute the assignment at `tryp`, assigning 1 to `turn`, but $waitp \,\wedge\, \Box \neg\, csp$ in the antecedent of the formula implies $\Box waitp$.   ∎

Theorem 16.15 $\vdash waitp \rightarrow \Diamond csp$.

Proof Assume to the contrary that $\vdash waitp \,\wedge\, \Box \neg\, csp$. By Lemmas 16.13 and 16.14, we conclude that $\vdash \Diamond \Box \neg\, wantq \,\vee\, \Diamond \Box (turn = 2)$. But:

$$\vdash \Diamond \Box A \,\vee\, \Diamond \Box B \rightarrow \Diamond \Box (A \vee B)$$

is a theorem of LTL, so:

$$\vdash \Diamond \Box \neg\, wantq \,\vee\, \Diamond \Box (turn = 2) \;\rightarrow\; \Diamond \Box (\neg\, wantq \,\vee\, (turn = 2)).$$

Therefore, we have:

$$\vdash waitp \,\wedge\, \Box \neg\, csp \;\rightarrow\; \Diamond \Box (\neg\, wantq \,\vee\, (turn = 2)),$$

which contradicts Lemma 16.12.   ∎

16.4 Programs as Automata

There is a different approach to the verification of the correctness of a program: generate all possible computations and check that the correctness property holds for each of them. Of course, this is possible only if there are a finite number of states so that each computation is finite or finitely presented. For the program for integer square root, we could prove its correctness this way for any specific value of a, but we could not prove it in general for all values of a. However, many concurrent algorithms have a finite number of states: the synchronization achieved by Peterson's algorithm needs only three variables with two values each and two processes with three possible values for their location counters. The critical and non-critical sections might contain sophisticated mathematical computations, but to prove the correctness of the synchronization we do not need to know these details.

This approach to verification is called *model checking*. A concurrent system is represented by an abstract finite model that ignores details of the computation; then, the correctness of this model is verified. A second reason for the terminology is technical: a correctness property is expressed as a formula (usually in temporal logic) and we wish to show that the program is a model of the formula, that is, an interpretation in which the formula is true.

The remainder of this chapter provides an overview of model checking. We will continue to use Peterson's algorithm as the running example.

16.4.1 Modeling Concurrent Programs as Automata

Concurrent programs can be *modeled* as finite automata. The abbreviated version of Peterson's algorithm (Sect. 16.2.1) can be represented as a pair of finite automata, one for each process (Fig. 16.1).

Each value of the location counter is a state of one of the automata, while each transition is labeled with the Boolean condition that enables it to be taken or with the assignment statements that change the values of the variables.

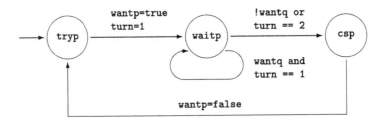

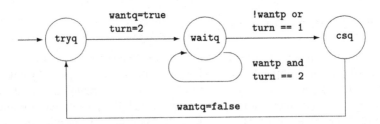

Fig. 16.1 Finite automata for Peterson's algorithm

The automata for the individual processes do not define the entire concurrent program. We must combine these automata into one automaton. This is done by constructing an automaton that is the *asynchronous product* of the automata for each process. The states are defined as the Cartesian product of the states of the automata for the individual processes. There is a transition corresponding to each transition of the individual automata. Because concurrent computation is defined by interleaving of atomic operations, a transition represents the execution of one atomic operation by one process.

The following diagram shows the beginning of the construction of the product automaton for Peterson's algorithm:

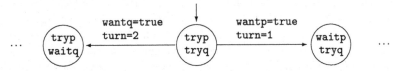

The initial state is one in which both processes are at their `try` state. From this initial state, a transition may be taken from either the automaton for process p or the one for process q; these lead to the states (`waitp,tryq`) and (`tryp,waitq`), respectively.

16.4.2 The State Space

The concept of a state of the computation of a concurrent program was given in Definition 16.3. For Peterson's algorithm, the number of possible states is finite. There are two location counters each of which can have one of three values. The two Boolean variables obviously have two possible values each, while the variable `turn` can take only two values by Lemma 16.8(a). Therefore, there are $3 \times 3 \times 2 \times 2 \times 2 = 72$ possible states in the algorithm.

Clearly, not all these states will occur in any computation. By Lemma 16.8(b–c), the values of `wantp` and `wantq` are fully determined by the location counters of the programs. For example, in no state is the location counter of process p at `tryp` and the value of `wantp` true. Therefore, the number of states is at most $3 \cdot 3 \cdot 2 = 18$, since only the variable `turn` can have different values for the same pair of values of the location counters.

Definition 16.16 The *reachable states* of a concurrent program are the states that can actually occur in a computation. The *state space* of the program is a directed graph: each reachable state is a node and there is an edge from state s_1 to state s_2 if some transition of the program which is enabled in s_1 moves the state of the computation to s_2.   ∎

The state space can be generated algorithmically by traversing the product automaton. The initial state of the state space is the initial state of the automaton together with the initial values of the variables. For each node already constructed, consider each transition of the automaton from this state in turn and create new nodes in the state space; if the new node already exists, the edge will point to the existing node.

Be careful to distinguish between the *automaton* which is the program and the *state space* which describes the computation. In practice, the automaton is usually rather small, but the state space can be extremely large because each variable multiplies the number of possible states by the range of its values.

In Peterson's algorithm, the initial value of turn is 1, so the initial state in the state space is (tryp,tryq,1). For conciseness, we do not explicitly write the values of wantp and wantq that can be determined from the location counters. There are two transitions from this state, so we create two new nodes (waitp,tryq,1) and (tryp,waitq,2). Continuing this way, we obtain the state space shown in Fig. 16.2. The left arrow out of each state points to the state obtained by taking a transition from process p, while the right arrow points to the state obtained by taking a transition from process q. Note that taking the p transition in state 4 results in a state that is the same as state 1 so we don't create a new state; instead, the left edge from 4 points to state 1.

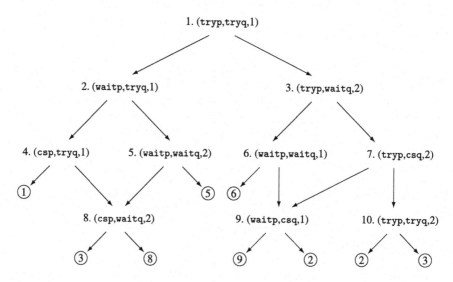

Fig. 16.2 State space for Peterson's algorithm

16.5   Model Checking of Invariance Properties

We now consider the second meaning of the term *model*: Is the state space a model of a correctness property? Consider the correctness property for mutual exclusion in Peterson's algorithm $A = \Box\neg(csp \wedge csq)$. Since the state space in Fig. 16.2 represents all the reachable states and all the transitions between them, any interpretation for A must be an infinite path in this directed graph. A quick inspection of the graph shows that all of the ten reachable states satisfy the formula $\neg(csp \wedge csq)$; therefore, for any interpretation (that is, for any path constructed from these states), $\Box\neg(csp \wedge csq)$ is true.

We have proved that the mutual exclusion property holds for Peterson's algorithm and have done so purely mechanically. Once we have written the program and the correctness property, there are algorithms to perform the rest of the proof: compile the program to a set of automata, construct the product automaton, generate the state space and check the truth of the formula expressing the correctness property at each state.

In this section we show how to verify invariance properties; Sect. 16.6 describes the extension of the algorithms to verify liveness properties.

16.5.1 Algorithms for Searching the State Space

Algorithms for searching a directed graph are described in any textbook on data structures. There are two approaches: *breadth-first search (BFS)*, where all the children of a node are visited before searching deeper in the graph, and *depth-first search (DFS)*, where as soon as a node is visited, the search continues with its children.

Searching the state space for Peterson's algorithm (Fig. 16.2) proceeds as follows, where the numbers in parentheses indicate nodes that have already been visited, so the search backtracks to try another child or backtracks to a parent when all children have been searched:

Breadth-first: 1, 2, 3, 4, 5, 6, 7, (1), 8, (8), (5), (6), 9, (9), 10, (3), (8), (9), (2), (2), (3).

Depth-first: 1, 2, 4, (1), 8, 3, 6, (6), 9, (9), (2), 7, (9), 10, (2), (3), (8), 5, (5).

Normally, DFS is preferred because the algorithm need only store a stack of the nodes visited from the root to the current node. In BFS, the algorithm has to store an indication of which child has been visited for all nodes at the current depth, so much more memory is required. BFS is preferred if you believe that there is a state relatively close to the root of the graph that does not satisfy the correctness property. In that case, DFS is likely to search deep within the graph without finding such a state.

The state space generates infinite paths, so they can be finitely represented only as directed graphs, not trees. This means that nodes will be revisited and the algorithm must avoid commencing a new search from these nodes. For example, in the DFS of Peterson's algorithm, node 2 is a child of node 9, but we obviously don't want to search again the subgraph rooted at node 2. The node 2 is not on the stack of the DFS (which is 1, 3, 6, 9), so an additional data structure must be maintained to store the set of all the nodes that have been visited. When a new node is generated, it is checked to see if it has been visited before; if so, the search skips the node and moves on to the next one. The most appropriate data structure is a hash table because of its efficiency. The memory available to store the hash table and the quality of the hashing function significantly affect the practicality of model checking.

16.5.2 On-the-Fly Searching

Here is an attempt to solve the critical section problem:

boolean wantp = false, wantq = false	
Process p	Process q
```while (true) {``` ``` waitp: wait until !wantq``` ``` tryp: wantp = true``` ``` csp: wantp = false``` ```}```	```while (true) {``` ``` waitq: wait until !wantp``` ``` tryq: wantq = true``` ``` csq: wantq = false``` ```}```

This is Dijkstra's *Second Attempt*; see Ben-Ari (2006, Sect. 3.6).

The state space for this algorithm is shown in Fig. 16.3, where we have explicitly written the values of the variables wantp and wantq although they can be inferred from the location counters. Clearly, $\neg (csp \wedge csq)$ does not hold in state 10 and there are (many) computations starting in the initial state that include this state. Therefore, $\Box \neg (csp \wedge csq)$ does not hold so this algorithm is not a solution to the critical section problem.

A DFS of the state space would proceed as follows:

$$1, 2, 4, (1), 7, 3, 5, (7), 8, 10.$$

The search terminates at state 10 because the formula $\neg (csp \wedge csq)$ is falsified. However, by generating the entire state space, we have wasted time and memory because the DFS finds the error without visiting all the states. Here state 6 is not visited.

This is certainly a trivial example, but in the verification of a real program, the search is likely to find an error without visiting millions of states. Of course, if the program is correct, the search will have to visit all the nodes of the state space, but (unfortunately) we tend to write many incorrect programs before we write a correct program. Therefore, it makes sense to optimize the generation of the state space and the search of the space so that errors can be found more efficiently.

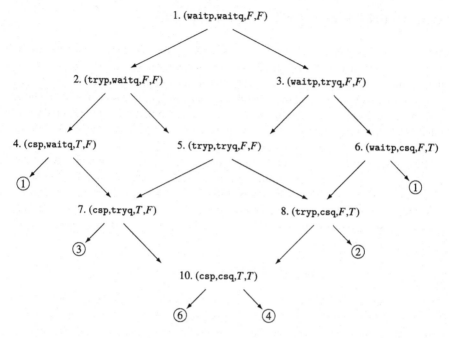

Fig. 16.3 State space for the Second Attempt

An efficient algorithm for model checking is to generate the state space incrementally and to check the correctness property *on-the-fly*:

```
while (true) {
    generate a new state;
    if (there are no more states) break;
    evaluate the correctness property in the new state;
    if (the correctness property fails to hold) break;
}
```

Since each new state is checked immediately after it is generated, the algorithm terminates as soon as an error is detected. Furthermore, the states on the DFS stack define a computation from the initial state that is in error:

$$1, 2, 4, 7, 3, 5, 8, 10.$$

This example shows that computations found by DFS are very often not the shortest ones with a given property. Clearly, 1, 2, 5, 7, 10 and 1, 3, 6, 8, 10 are shorter paths to the error state, and the first one will be found by a breadth-first search. Nevertheless, DFS is usually preferred because it needs much less memory.

16.6 Model Checking of Liveness Properties

Safety properties that are defined by the values of a state are easy to check because they can be evaluated locally. Given a correctness property like $\Box\neg(csp \wedge csq)$, the formula $\neg(csp \wedge csq)$ can be evaluated in an individual state. Since all the states generated by a search are by definition reachable, once a state is found where $\neg(csp \wedge csq)$ does not hold, it is easy to construct a path that is an interpretation that falsifies $\Box\neg(csp \wedge csq)$. Liveness properties, however, are more difficult to prove because no single state can falsify $\Box\Diamond csp$.

Before showing how to check liveness properties, we need to express the model checking algorithm in a slightly different form. Recall that a correctness property like $A = \Box\neg(csp \wedge csq)$ holds iff it is true in *all* computations. Therefore, the property does *not* hold iff there *exists* a computation is which A is *false*. Using negation, we have: the correctness property does *not* hold iff there *exists* a computation is which $\neg A$ is *true*, where:

$$\neg A \equiv \neg\Box\neg(csp \wedge csq) \equiv \Diamond(csp \wedge csq).$$

The model checking algorithm 'succeeds' if it finds a computation where $\neg A$ is true; it succeeds by finding a counterexample proving that the program is incorrect. Model checking can be understood as a 'bet' between you and the model checker: the model checker wins and you lose if it can find a model for the negation of the correctness property.

The liveness property of Peterson's algorithm is expressed by the correctness formula $\Box(waitp \rightarrow \Diamond csp)$, but let us start with the simpler property $A = waitp \rightarrow \Diamond csp$. Its negation is:

$$\neg(waitp \rightarrow \Diamond csp) \equiv waitp \wedge \neg\Diamond csp \equiv waitp \wedge \Box\neg csp.$$

A computation $\pi = s_0, s_1, \ldots$ satisfies $\neg A$ if *waitp* is true in its initial state s_0 and $\neg csp$ holds in all states s_i, $i \geq 0$. Therefore, to show that an interpretation satisfies $\neg A$, the negation of the correctness property, and thus falsifies A, the correctness property itself, we have to produce an entire computation and not just a state. Based upon the discussion in Sect. 13.5.5, the computation will be defined by a maximal strongly connected component (MSCC). For example, if the state space contained a subgraph of the following form:

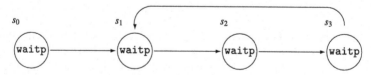

then this subgraph would define a computation that satisfies $waitp \wedge \Box\neg csp$ and thus falsifies the liveness property $waitp \rightarrow \Diamond csp$.

For the full liveness property, the negation is:

$$\neg\,\Box(waitp \rightarrow \Diamond csp) \equiv \Diamond(waitp \wedge \neg\,\Diamond csp) \equiv \Diamond(waitp \wedge \Box\neg\,csp).$$

This would be satisfied by a computation defined by the following subgraph:

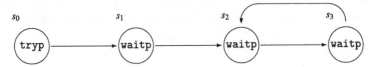

In the computation $\pi = s_0, s_1, s_2, s_3, s_2, s_3, \ldots,$ *tryp* is true in state s_0, so $\pi_0 \not\models$ *waitp* $\wedge \Box\neg\,csp$, but $\pi_1 \models$ *waitp* $\wedge \Box\neg\,csp$, so $\pi \models \Diamond(waitp \wedge \Box\neg\,csp)$.

The states on the stack of a depth first search form a path. If the construction ever tries to generate a state that already exists higher up on the stack, the transition to this node defines a finitely-presented infinite computation like the ones shown above. What we need is a way of checking if such a path is a model of the negation of the correctness property. If so, it falsifies the property and the path is a counterexample to the correctness of the program. Of course, we could generate the entire state space and then check each distinct path to see if it model, but it is more efficient if the checking can be done on-the-fly as we did for safety properties. The key is to transform an LTL formula into an automaton whose computations can be generated at the same time as those of the program.

16.7 Expressing an LTL Formula as an Automaton

An LTL formula can be algorithmically transformed into an automaton that accepts an input if and only if the input represents a computation that satisfies the LTL formula. The automaton is a *nondeterministic Büchi automaton (NBA)*, which is the same as a *nondeterministic finite automaton (NFA)* except that it reads an infinite string as its input and its definition of acceptance is changed accordingly. An NFA accepts an input string iff the state reached when the reading of the (finite) input is completed is an accepting state. Since the input to an NBA is infinite, the definition of acceptance is modified to:

Definition 16.17 A nondeterministic Büchi automaton accepts an infinite input string iff the computation that reads the string is in an accepting state *infinitely often*.

To demonstrate NBA's, we construct one NBA corresponding to the LTL formula $\Box A \equiv \Box(waitp \rightarrow \Diamond csp)$ that expresses the liveness property of Peterson's algorithm, followed by an NBA corresponding to the negation of the formula. The second NBA will be used in the following section to show that the liveness property holds.

Example 16.18 The formula A can be transformed using the inductive decomposition of $\lozenge$:

$$waitp \rightarrow \lozenge csp \equiv \neg waitp \vee (csp \vee \bigcirc \lozenge csp) \equiv (\neg waitp \vee csp) \vee \bigcirc \lozenge csp.$$

$\square A$ is true as long as $\neg waitp \vee csp$ holds, but if $\neg waitp \vee csp$ ever becomes false, then tomorrow $\lozenge csp$ must be true. The NBA constructed from this analysis is:

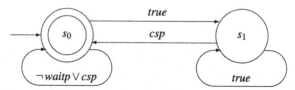

Since state s_0 is an accepting state, if the computation *never* executes the statement at `tryp` to get to `waitp`, the automaton is always in an accepting state and the formula holds. Otherwise (expressed as *true*), if the computation chooses to execute `tryp` and gets to `waitp`, $\neg waitp \vee csp$ becomes false (state s_1). The only way to (re-)enter the accepting state s_0 is if eventually the transition to s_0 is taken because csp true, as required by $\lozenge csp$. If not (expressed as *true*), the computation is not accepted since s_1 is not an accepting state. The accepting computations of this NBA are precisely those in which the process decides not to enter its critical section or those in which every such attempt is eventually followed by a return of the computation to the accepting state s_0.   ∎

Example 16.19 Let us now consider the NBA for:

$$\neg \square A \equiv \neg \square (waitp \rightarrow \lozenge csp) \equiv \lozenge (waitp \wedge \square \neg csp),$$

the negation of the liveness formula. The intuitive meaning of the formula is that the computation can do anything (expressed as *true*), but it may nondeterministically decide to enter a state where $waitp$ is true and csp is and remains false from then on. Such a computation falsifies the liveness property. The corresponding NBA is:

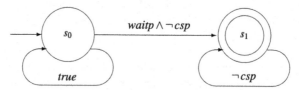

In state s_1, if csp ever becomes true, there is no transition from the state; as with NFA, an automaton that cannot continue with its computation is considered to have rejected its input.   ∎

16.8 Model Checking Using the Synchronous Automaton

On-the-fly model checking for an invariance property (Sect. 16.5.2) simply evaluates the correctness property as each new state is generated:

```
while (true) {
    generate a new state;
    if (there are no more states) break;
    evaluate the correctness property in the new state;
    if (the correctness property fails to hold) break;
}
```

When checking a liveness property (or a safety property expressed in LTL as $\Box A$), every step of the program automaton—the asynchronous product automaton of the processes—is immediately followed by a step of the NBA corresponding to the LTL formula expressing the negation of the correctness property. The product of the asynchronous automaton and the NBA is called a *synchronous automaton* since the steps of the two automata are synchronized. The model checking algorithm becomes:

```
while (true) {
    generate a new state of the program automaton;
    if (there are no more states) break;
    generate a new state of the NBA;
    if (the correctness property fails to hold) break;
}
```

How does the algorithm decide if the correctness property fails to hold? The intuitive meaning of the NBA for the negation of the correctness property is that it should *never* accept an input string. For example, in Peterson's algorithm, $\Diamond(waitp \land \Box \neg csp)$ should never hold in any computation. Therefore, if the NBA corresponding to the formula *accepts* a computation, the search should terminate because it defines a counterexample, a model for the negation of the correctness property of the program.

Acceptance by the NBA is checked on-the-fly: whenever a future formula is encountered in a state, a *nested depth-first search* is initiated. If a state is generated that already exists on the stack, it is easy to extract an interpretation that falsifies the formula. For the liveness of Peterson's algorithm, the correctness property is $\Box(waitp \to \Diamond csp)$ and its negation is $\Diamond(waitp \land \Box \neg csp)$. In any state where *waitp* holds, a nested DFS is commenced and continued as long as $\neg csp$ holds. If the search reaches a state on the stack, a model for the negation of the correctness property has been found and the model checker wins the bet. The details of a nested DFS are beyond the scope of this book and the reader is referred to Baier and Katoen (2008, Sect. 4.4) and Holzmann (2004, Chap. 8).

Let us trace the model checking algorithm for the liveness of Peterson's algorithm. The state space is shown again in Fig. 16.4. Starting from the initial state 1, state 2 is reached and $\Diamond(waitp \land \Box \neg csp)$ will be true, provided that we can find a reachable MSCC where $\neg csp$ holds in all its states. A nested DFS is initiated.

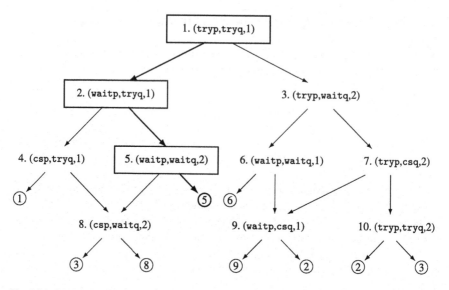

Fig. 16.4 Model checking the liveness of Peterson's algorithm

Clearly, states 4 and 8 cannot be part of the MSCC since $\neg csp$ is false in those states. However, the computation can continue:

$$1, 2, 5, 5, 5, \ldots,$$

and the state 5 with its self-loop forms an MSCC such that $\neg csp$ is false in all its states!

This is strange because it is a counterexample to the liveness of Peterson's algorithm which we have already proved deductively. The problem is that this computation is not fair.

Definition 16.20 A computation is *(weakly) fair* if a transition that is always enabled is eventually executed in the computation.   ∎

The statement:

```
wait until (!wantq or turn == 2)
```

is always enabled because $turn = 2$, but it is never taken. Therefore, we reject this counterexample.

Continuing the DFS, we encounter two more states 6 and 9 where *waitp* is true. We leave it as an exercise to show that the nested DFS will find computations in which $\neg csp$ holds in all states, but that these computations are also unfair. Therefore, the liveness holds for Peterson's algorithm.

16.9 Branching-Time Temporal Logic *

In linear temporal logic, there is an implicit universal quantification over the computations—the paths in the state space. The formula expressing the liveness of Peterson's algorithm $\square(waitp \rightarrow \lozenge csp)$ must be true for *all* computations. In *branching-time temporal logic*, universal and existential quantifiers are used as explicit prefixes to the temporal operators. In this section, we give an overview of the most widely used branching-time logic called *Computational Tree Logic (CTL)*.

16.9.1 The Syntax and Semantics of CTL

The word *tree* in the name of CTL emphasizes that rather than choosing a single path as an interpretation (see Definition 13.28 for LTL), a formula is interpreted as true or false in a state that is the root of tree of possible computations. Figure 16.5 shows the state space of Peterson's algorithm unrolled into a tree. Four levels of the tree are shown with the labels of the states of the lowest level abbreviated to save space.

Here are the temporal operators in CTL with their intended meaning:

- $s \models \forall \square A$: A is true in *all* states of *all* paths rooted at s.
- $s \models \forall \lozenge A$: A is true in *some* state of *all* paths rooted at s.
- $s \models \forall \bigcirc A$: A is true in *all* the children of s.
- $s \models \exists \square A$: A is true in *all* states of *some* path rooted at s.
- $s \models \exists \lozenge A$: A is true in *some* state of *some* path rooted at s.
- $s \models \exists \bigcirc A$: A is true in *some* child of s.

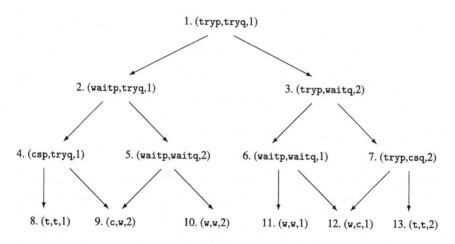

Fig. 16.5 The state space of Peterson's algorithm as a tree

We have made two changes to simplify the presentation: As in LTL, the formal definition of CTL is based on the binary operator $\mathcal{U}$ (Sect. 13.6), but we limit the discussion to the unary operators. The syntax we use is based on the LTL syntax and is different from CTL syntax which uses capital letters: AG, AF, AX, EG, EF, EX for the operators in the list above and AU, EU for the binary operators.

Example 16.21 Let s_i be the state labeled by i in Fig. 16.5. It is easy to check that $\exists\bigcirc(turn = 1)$ is true in s_1 and $\forall\bigcirc(turn = 2)$ is true in s_5 just by examining the next states. The formula $\exists\Box waitp$ is true is s_5 and represents the unfair computation where process p is never given a chance to execute. Similarly, $\forall\Diamond(turn = 1)$ is *not* true in s_5 by considering its negation and using duality:

$$\neg\forall\Diamond(turn = 1) \;\equiv\; \exists\Box\neg(turn = 1) \;\equiv\; \exists\Box(turn = 2).$$

The unfair computation is a computation whose states all satisfy $turn = 2$. Finally, the operator $\forall\Box$ can be used to express the correctness properties of Peterson's algorithm:

$$\forall\Box\neg(csp \wedge csq), \qquad \forall\Box(waitp \rightarrow \forall\Diamond csp).$$

∎

16.9.2 Model Checking in CTL

Model checking in CTL is based upon the following decomposition of the temporal operators:

$$\forall\Box A \;\equiv\; A \wedge \forall\bigcirc\forall\Box A,$$
$$\forall\Diamond A \;\equiv\; A \vee \forall\bigcirc\forall\Diamond A,$$
$$\exists\Box A \;\equiv\; A \wedge \exists\bigcirc\exists\Box A,$$
$$\exists\Diamond A \;\equiv\; A \vee \exists\bigcirc\exists\Diamond A.$$

The model checking algorithm is rather different from that of LTL. The truth of a formula is checked *bottom-up* from its subformulas.

Example 16.22 We want to show that the formula $\forall\Diamond csp$ expressing the liveness of Peterson's algorithm is true in the interpretation shown in Fig. 16.5. By its decomposition $A \vee \forall\bigcirc\forall\Diamond A$, it is clearly true in the states s_4 and s_8 where csp is true (these states are marked with thick borders in Fig. 16.6). Let $S_0 = \{s_4, s_8\}$ be the set of states that we know satisfy $\forall\Diamond A$. By the decomposition, let us create S_1 as the union of S_0 and all states for which $\forall\bigcirc\forall\Diamond A$ holds, that is, all states from which a single transition leads to a state in S_0. The set of predecessors of s_4 is $\{s_2\}$ and the set of predecessors of s_8 is $\{s_4, s_5\}$. So $S_1 = S_0 \cup \{s_2\} \cup \{s_4, s_5\} = \{s_2, s_4, s_5, s_8\}$, where the added states are marked with dashed borders. Continuing with the predecessors of S_1, we obtain $S_2 = \{s_1, s_2, s_4, s_5, s_8, s_9, s_{10}\}$ (where the added states are marked

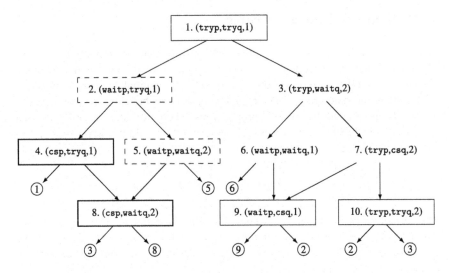

Fig. 16.6 CTL model checking of Peterson's algorithm

with thin borders). Two more steps of the algorithm will add the remaining states to S_3 and then S_4, proving that $\forall \diamond csp$ holds in all states. ∎

Example 16.23 Consider now the formula $\exists \square waitp$. In this case, the algorithm works *top-down* by removing states where it does *not* hold. Initially, S_0, the set of states where the formula is true, is tentatively assumed to be the set of all states. By the decomposition:

$$\exists \square wantp \equiv wantp \wedge \exists \bigcirc \exists \square wantp,$$

wantp must be true in a state for $\exists \square waitp$ to be true; therefore, remove from S_0 all states where *wantp* does not hold. The states that remain are $S_1 = \{s_2, s_5, s_6, s_9\}$. Additionally, $\exists \bigcirc \exists \square wantp$ must be true in a state for $\exists \square waitp$ to be true. Repeatedly, remove from the set any state that does not have *some* successor ($\exists \bigcirc$) already in the set. This causes no change to S_1.

Check that from all of the states in S_1, there *exists* an infinite path in *all* of whose states *waitp* is true. ∎

16.10 Symbolic Model Checking *

In *symbolic model checking*, the states and transitions are not represented explicitly; instead, they are encoded as formulas in propositional logic. Model checking algorithms use efficient representations like BDDs to manipulate these formulas.

A state in the state space of Peterson's algorithm can be represented as a propositional formula using five atomic propositions. There are three locations in each process, so two bits for each process can represent these values $\{p_0, p_1, q_0, q_1\}$. Let us encode the locations as follows:

tryp	$p_0 \wedge p_1$	tryq	$q_0 \wedge q_1$
waitp	$\neg p_0 \wedge p_1$	waitq	$\neg q_0 \wedge q_1$
csp	$p_0 \wedge \neg p_1$	csq	$q_0 \wedge \neg q_1$

The variable turn can take two values so one bit is sufficient. The atomic proposition t will encode *turn*: true for *turn* $= 1$ and false for *turn* $= 2$. As usual, we don't bother to represent the variables wantp and wantq since their values can be deduced from the location counters.

The initial state of the state space is encoded by the formula:

$$p_0 \wedge p_1 \wedge q_0 \wedge q_1 \wedge t,$$

and, for example, the state $s_8 = (csp, waitq, 2)$ of Fig. 16.2 is encoded by:

$$p_0 \wedge \neg p_1 \wedge \neg q_0 \wedge q_1 \wedge \neg t.$$

To encode the transitions, we need another set of atomic propositions: the original set will encode the state *before* the transition and the new set (denoted by primes) will encode the state *after* the transition. The encoding of the transition from $s_5 = (waitp, waitq, 2)$ to s_8 is given by the formula:

$$(\neg p_0 \wedge p_1 \wedge \neg q_0 \wedge q_1 \wedge \neg t) \wedge (p_0' \wedge \neg p_1' \wedge \neg q_0' \wedge q_1' \wedge \neg t').$$

There are two ways of proceeding from here. One is to encode the formulas using BDDs. CTL model checking, described in the previous chapter, works on sets of states. A set of states is represented by the disjunction of the formulas representing each state. The algorithms on BDDs can be used to compute the formulas corresponding to new sets of states: union, predecessor, and so on.

The other approach to symbolic model checking is called *bounded model checking*. Recall that a formula in temporal logic has the finite model property (Corollary 13.67): if a formula is satisfiable then it is satisfied in a finitely-presented model. For an LTL formula, we showed that a model consists of MSCCs that are reachable from the initial state. In fact, by unwinding the MSCCs, we can always find a model that consists of a single cycle reachable from the initial state (cf. Sect. 16.6):

In bounded model checking, a maximum size k for the model is guessed. The behavior of the program and the negation of a correctness property are expressed as a propositional formula obtained by encoding each state that can appear at distance i from the initial state $0 \le i \le k$. This formula is the input to a SAT solver (Chap. 6); if a satisfying interpretation is found, then there is a computation that satisfies the negation of the correctness property is true and the program is not correct.

16.11 Summary

The computation of a concurrent program can be defined as the interleaving of the atomic operations of its processes, where each process is a sequential program. Since a concurrent program must be correct for every possible computation, it is not possible to verify or debug programs by testing.

Correctness properties of concurrent programs can be expressed in linear temporal logic. There are two types of properties: safety properties that require that something bad never happens and liveness properties that require that something good eventually happen. A safety property is proved by showing inductively that it is an invariant. Proving a liveness property is more difficult and requires that the progress of a program be specified.

Model checking is an alternative to deductive systems for verifying the correctness of concurrent programs. A model checker verifies that a concurrent program is correct with respect to a correctness formula by searching the entire state space of the program for a counterexample: a state or path that violates correctness. The advantage of model checking is that once the program and the correctness property have been written, model checking is purely algorithmic and no intervention is required. Algorithms and data structures have been developed that enable a model checker to verify very large state spaces.

Model checking with correctness properties specified in LTL is done by explicitly generating the state space. If the correctness property is a safety property expressed as an assertion or an invariant, the correctness can be checked on-the-fly at each state as it is generated. Liveness properties require the use of nested search whenever a state is reached that could be part of a path that is a counterexample. LTL formulas are translated into Büchi automata so that the path in the computation can be synchronized with a path specified by the correctness formula.

Model checking can also be based upon the branching-time logic CTL. Here, computations are encoded in binary decision diagrams and the algorithms for BDDs are used to efficiently search for counterexamples. SAT solvers have also been used in model checkers in place of BDDs.

16.12 Further Reading

For an introduction to concurrent programming, we recommend (of course) Ben-Ari (2006), which contains deductive proofs of algorithms as well as verifications using the SPIN model checker. Magee and Kramer (1999) is an introductory textbook that takes a different approach using transition systems to model programs.

The deductive verification of concurrent programs is the subject of Manna and Pnueli (1992, 1995): the first volume presents LTL and the second volume defines rules for verifying safety properties. The third volume on the verification of liveness properties was never completed, but a partial draft is available (Manna and Pnueli, 1996). Deductive verification is also the subject of the textbook by Apt et al. (2009).

Textbooks on model checking are Baier and Katoen (2008), and Clarke et al. (2000).

The SPIN model checker is particular easy to use as described in Ben-Ari (2008). Holzmann (2004) describes SPIN in detail: both practical aspects of using it and the important details of how the algorithms are implemented.

Bounded model checking with SAT solvers is presented in Biere et al. (2009, Chap. 14).

16.13 Exercises

16.1 Show that Peterson's algorithm remains correct if the assignments in `wantp = true; turn = 1` and in `wantq = true; turn = 2` are not executed as one atomic operation, but rather as two operations. Show that if the order of the separate assignments is reversed, the algorithm is not correct.

16.2 Complete the proof the invariants of Peterson's algorithm (Lemma 16.8).

16.3 Complete the proof of Lemma 16.12 by proving:

$$\vdash waitp \wedge \Box \neg csp \rightarrow \Box(waitp \wedge \Box \neg csp).$$

16.4 Complete the analysis of liveness in Peterson's algorithm (Sect. 16.8) and show that computations in which $\neg csp$ holds in all states are unfair.

16.5 Generate the state space for *Third Attempt* (Ben-Ari, 2006, Sect. 3.7):

boolean wantp = false, wantq = false	
Process p	Process q
`while (true) {`	`while (true) {`
`  tryp:  wantp = true`	`  tryq:  wantq = true`
`  waitp: wait until !wantq`	`  waitq: wait until !wantp`
`  csp:   wantp = false`	`  csq:   wantq = false`
`}`	`}`

Is the algorithm correct?

16.6 * Show that the CTL operators are not independent:

$$\models \exists \Diamond p \leftrightarrow \neg \forall \Box \neg p, \qquad \models \forall \Diamond p \leftrightarrow \neg \exists \Box \neg p.$$

16.7 * A CTL formula is said to be *equivalent* to an LTL formula if the LTL formula is obtained by erasing the quantifiers from the CTL formula and the formulas are true of the same programs. Use the following automaton to show that the CTL formula $\forall \Diamond \forall \Box p$ and the LTL formula $\Diamond \Box p$ are not equivalent.

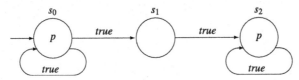

References

K.R. Apt, F.S. de Boer, and E.-R. Olderog. *Verification of Sequential and Concurrent Programs (Third Edition)*. Springer, London, 2009.

C. Baier and J.-P. Katoen. *Principles of Model Checking*. MIT Press, 2008.

M. Ben-Ari. *Principles of Concurrent and Distributed Programming (Second Edition)*. Addison-Wesley, Harlow, UK, 2006.

M. Ben-Ari. *Principles of the Spin Model Checker*. Springer, London, 2008.

A. Biere, M. Heule, H. Van Maaren, and T. Walsh, editors. *Handbook of Satisfiability*, volume 185 of *Frontiers in Artificial Intelligence and Applications*. IOS Press, 2009.

E.M. Clarke, O. Grumberg, and D.A. Peled. *Model Checking*. MIT Press, Cambridge, MA, 2000.

E.W. Dijkstra. Cooperating sequential processes. In F. Genuys, editor, *Programming Languages*. Academic Press, New York, NY, 1968.

G.J. Holzmann. *The Spin Model Checker: Primer and Reference Manual*. Addison-Wesley, Boston, MA, 2004.

J. Magee and J. Kramer. *Concurrency: State Models & Java Programs*. John Wiley, Chichester, 1999.

Z. Manna and A. Pnueli. *The Temporal Logic of Reactive and Concurrent Systems. Vol. I: Specification*. Springer, New York, NY, 1992.

Z. Manna and A. Pnueli. *The Temporal Logic of Reactive and Concurrent Systems. Vol. II: Safety*. Springer, New York, NY, 1995.

Z. Manna and A. Pnueli. Temporal verification of reactive systems: Progress. Draft available at http://www.cs.stanford.edu/~zm/tvors3.html, 1996.

G.L. Peterson. Myths about the mutual exclusion problem. *Information Processing Letters*, 12(3):115–116, 1981.

Appendix
Set Theory

Our presentation of mathematical logic is based upon an informal use of set theory, whose definitions and theorems are summarized here. For an elementary, but detailed, development of set theory, see Velleman (2006).

A.1 Finite and Infinite Sets

The concept of an *element* is undefined, but informally the concept is clear: an element is any identifiable object like a number, color or node of a graph. Sets are built from elements.

Definition A.1 A *set* is composed of *elements*. $a \in S$ denotes that a is an element of set S and $a \notin S$ denotes that a is *not* an element of S. The set with no elements is the *empty set*, denoted $\emptyset$. Capital letters like S, T and U are used for sets. ∎

There are two ways to define a set: (a) We can explicitly write the elements comprising the set. If a set is large and if it is clearly understood what its elements are, an ellipsis '...' is used to indicate the elements not explicitly listed. (b) A set may be defined by *set comprehension*, where the set is specified to be composed of all elements that satisfy a condition. In either case, braces are used to contain the elements of the set.

Example A.2

- The set of colors of a traffic light is {*red, yellow, green*}.
- The set of atomic elements is {*hydrogen, helium, lithium,* ...}.
- $\mathscr{Z}$, the set of *integers*, is $\{\ldots, -2, -1, 0, 1, 2, \ldots\}$.
- $\mathscr{N}$, the set of *natural numbers*, is $\{0, 1, 2, \ldots\}$. $\mathscr{N}$ can also be defined by set comprehension: $\mathscr{N} = \{n \mid n \in \mathscr{Z} \text{ and } n \geq 0\}$. Read this as: $\mathscr{N}$ is the set of all n such that n is an integer and $n \geq 0$.
- $\mathscr{E}$, the set of even natural numbers, is $\{n \mid n \in \mathscr{N} \text{ and } n \bmod 2 = 0\}$.

M. Ben-Ari, *Mathematical Logic for Computer Science*,
DOI 10.1007/978-1-4471-4129-7, © Springer-Verlag London 2012

- $\mathscr{P}$, the set of prime numbers, is:

$$\{n \mid n \in \mathscr{N} \text{ and } n \geq 2 \text{ and } (n \bmod m = 0 \text{ implies } m = 1 \text{ or } m = n)\}.$$

∎

There is no meaning to the order of the elements in a set or to repetition of elements: $\{3, 2, 1, 1, 2, 3\} = \{1, 2, 3\} = \{3, 1, 2\}$. A set containing a single element (a *singleton set*) and the element itself are not the same: $5 \in \{5\}$.

A.2 Set Operators

Set Inclusion

Definition A.3 Let S and T be sets. S is a *subset* of T, denoted $S \subseteq T$, iff every element of S is an element of T, that is, $x \in S \rightarrow x \in T$. S is a *proper subset* of T, denoted $S \subset T$, iff $S \subseteq T$ and $S \neq T$. ∎

Example A.4 $\mathscr{N} \subset \mathscr{Z}$, $\mathscr{E} \subset \mathscr{N}$, $\{red, green\} \subset \{red, yellow, green\}$. ∎

Theorem A.5 $\emptyset \subseteq T$.

The intuition behind $\emptyset \subseteq T$ is as follows. To prove $S \subseteq T$, we have to show that $x \in S \rightarrow x \in T$ holds for all $x \in S$. But there are no elements in $\emptyset$, so the statement is vacuously true.

The relationships among sets can be shown graphically by the use of *Venn diagrams*. These are closed curves drawn in the plane and labeled with the name of a set. A point is in the set if it is within the interior of the curve. In the following diagram, since every point within S is within T, S is a subset of T.

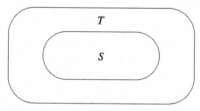

Theorem A.6 *The subset property is transitive*:

$$\text{If } S \subseteq T \text{ and } T \subseteq U \text{ then } S \subseteq U.$$
$$\text{If } S \subset T \text{ and } T \subseteq U \text{ then } S \subset U.$$
$$\text{If } S \subseteq T \text{ and } T \subset U \text{ then } S \subset U.$$
$$\text{If } S \subset T \text{ and } T \subset U \text{ then } S \subset U.$$

The relationship between equality of sets and set inclusion is given by the following theorem.

Theorem A.7 $S = T$ *iff* $S \subseteq T$ *and* $T \subseteq S$.

Union, Intersection, Difference

Definition A.8

- $S \cup T$, the *union* of S and T, is the set consisting of those elements which are elements of *either S or T* (or both).
- $S \cap T$, the *intersection* of S and T, is the set consisting of those elements which are elements of *both S and T*. If $S \cap T = \emptyset$ then S and T are *disjoint*.
- $S - T$, the *difference* of S and T, is the set of elements of S that are not elements of T.
- Let S be understood as a universal set; then $\bar{T}$, the *complement* of T, is $S - T$. ∎

The following Venn diagram illustrates these concepts.

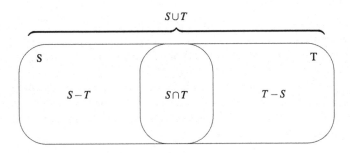

Example A.9 Here are some examples of operations on sets:

$$
\begin{aligned}
\{red, yellow\} \cup \{red, green\} &= \{red, yellow, green\}, \\
\{red, yellow\} \cap \{red, green\} &= \{red\}, \\
\{red, yellow\} - \{red, green\} &= \{yellow\}, \\
\mathscr{P} \cap \mathscr{E} &= \{2\}, \\
\mathscr{P} \cap \mathscr{N} &= \mathscr{P}, \\
\mathscr{P} \cup \mathscr{N} &= \mathscr{N}.
\end{aligned}
$$

∎

The operators $\cup$ and $\cap$ are commutative, associative and distributive.

Theorem A.10

$$
\begin{aligned}
S \cup T &= T \cup S, \\
S \cap T &= T \cap S, \\
(S \cup T) \cup U &= S \cup (T \cup U), \\
(S \cap T) \cap U &= S \cap (T \cap U), \\
S \cup (T \cap U) &= (S \cup T) \cap (S \cup U), \\
S \cap (T \cup U) &= (S \cap T) \cup (S \cap U).
\end{aligned}
$$

The following theorem states some simple properties of the set operators.

Theorem A.11

$$T = (T - S) \cup (S \cap T).$$
If $S \subseteq T$ then: $S \cap T = S$, $S \cup T = T$, $S - T = \emptyset$.
If S and T are disjoint then $S - T = S$.
$S \cup \emptyset = S$, $S \cap \emptyset = \emptyset$, $S - \emptyset = S$.

A.3 Sequences

Definition A.12 Let $\mathscr{S}$ be a set.

- A *finite sequence* f on $\mathscr{S}$ is a function from $\{0, \ldots, n-1\}$ to $\mathscr{S}$. The length of the sequence is n.
- An *infinite sequence* f on $\mathscr{S}$ is a mapping from $\mathscr{N}$ to $\mathscr{S}$. ∎

Example A.13 Let $\mathscr{S}$ be the set of three colors $\{red, yellow, green\}$. Suppose that you see a green light but don't manage to cross the road before it changes. The sequence of colors that you will see before you cross the road is the sequence f on $\{0, 1, 2, 3\}$ defined by:

$$f_0 = green, \qquad f_1 = yellow, \qquad f_2 = red, \qquad f_3 = green.$$

The infinite sequence of colors that the light shows (assuming that it is never turned off or malfunctions) is:

$$f_0 = green, \qquad f_1 = yellow, \qquad f_2 = red, \qquad \ldots,$$

where the ellipsis ... indicates that we know how to continue constructing the sequence. Alternatively, we could formally define the sequence as:

$$f_i = green \quad \text{if } i \bmod 3 = 0,$$
$$f_i = yellow \quad \text{if } i \bmod 3 = 1,$$
$$f_i = red \quad \text{if } i \bmod 3 = 2.$$

∎

In place of functional notation, one usually lists the elements of a sequence within parentheses ( ) to differentiate a sequence from a set which is written within braces $\{\}$:

Definition A.14 Let f be a sequence on $\mathscr{S}$. The sequence is denoted:

$$(s_0, s_1, s_2, \ldots)$$

where $s_i = f(i)$. ∎

Definition A.15 A finite sequence of length n is an *n-tuple*. The following terms are also used: a 2-tuple is a *pair*, a 3-tuple is a *triple* and a 4-tuple is a *quadruple*. ■

Example A.16 Examples of sequences:

- A 1-tuple: (*red*).
- A pair: $(5, 25)$.
- A triple: (*red, yellow, green*).
- A different triple: (*red, green, yellow*).
- A triple with repeated elements: (*red, green, green*).
- An infinite sequence: $(1, 2, 2, 3, 3, 3, 4, 4, 4, 4, \ldots)$. ■

Definition A.17 Let S and T be sets. $S \times T$, their *Cartesian product* , is the set of all pairs (s, t) such that $s \in S$ and $t \in T$.

Let $S_1, \ldots, S_n$ be sets. $S_1 \times \cdots \times S_n$,, their *Cartesian product*, is the set of n-tuples $(s_1, \ldots, s_n)$, such that $s_i \in S_i$. If all the sets S_i are the same set S, the notation S^n is used for $S \times \cdots \times S$. ■

Example A.18

- $\mathcal{N} \times \mathcal{N} = \mathcal{N}^2$ is the set of all pairs of natural numbers. This can be used to represent discrete coordinates in the plane.
- $\mathcal{N} \times \{red, yellow, green\}$ is the set of all pairs whose first element is a number and whose second is a color. This could be used to represent the color of a traffic light at different points of time. ■

A.4 Relations and Functions

Two central concepts in mathematics are that of relation (3 is less that 5) and function (the square of 5 is 25). Formally, a relation is a subset of a Cartesian product of sets and a function is a relation with a special property.

Relations

Definition A.19 An *n-ary relation* $\mathcal{R}$ is a subset of $S_1 \times \cdots \times S_n$. $\mathcal{R}$ is said to be a relation *on* $S_1 \times \cdots \times S_n$. A 1-ary (unary) relation is simply a subset. ■

Example A.20 Here are some relations over $\mathcal{N}^k$ for various $k \geq 1$:

- The set of prime numbers $\mathcal{P}$ is a relation on $\mathcal{N}^1$.
- $\mathcal{S2} = \{(n_1, n_2) \mid n_2 = n_1^2\}$ is a relation on $\mathcal{N}^2$; it is the set of pairs of numbers and their squares: $(4, 16) \in \mathcal{S}$, $(7, 49) \in \mathcal{S}$.
- The following relation on $\mathcal{N}^2$:

$$\mathcal{R} = \{(n, m) \mid n \bmod k = 0 \text{ and } m \bmod k = 0 \text{ implies } k = 1\}$$

is the set of relatively prime numbers. Examples are: $(4, 9) \in \mathscr{R}$, $(15, 28) \in \mathscr{R}$, $(7, 13) \in \mathscr{R}$.

- Pythagorean triples $\{(x, y, z) \mid x^2 + y^2 = z^2\}$ are a relation on $\mathscr{N}^3$. They are the values that can be the lengths of right-angled triangles. Examples are $(3, 4, 5)$ and $(6, 8, 10)$.
- Let $\mathscr{F}$ be the set of quadruples $\{(x, y, z, n) \mid n > 2 \text{ and } x^n + y^n = z^n\}$. Fermat's Last Theorem (which was recently proved) states that this relation $\mathscr{F}$ on $\mathscr{N}^4$ is the empty set $\emptyset$. ∎

Properties of Relations

Definition A.21 Let R be a binary relation on S^2.

- R is *reflexive* iff $R(x, x)$ for all $x \in S$.
- R is *symmetric* iff $R(x_1, x_2)$ implies $R(x_2, x_1)$.
- R is *transitive* iff $R(x_1, x_2)$ and $R(x_2, x_3)$ imply $R(x_1, x_3)$.

R^*, the *reflexive transitive closure of* $\mathscr{R}$, is defined as follows:

- If $R(x_1, x_2)$ then $R^*(x_1, x_2)$.
- $R^*(x_i, x_i)$ for all $x_i \in S$.
- $R^*(x_1, x_2)$ and $R^*(x_2, x_3)$ imply $R^*(x_1, x_3)$. ∎

Example A.22 Let $\mathscr{C}$ be the relation on the set of ordered pairs of strings (s_1, s_2) such that $s_1 = s_2, s_1 = c \cdot s_2$, or $s_1 = s_2 \cdot c$, for some c in the underlying character set. Then $\mathscr{C}^*$ is the substring relation between strings. Let us check the three properties:

- For each of the three conditions defining $\mathscr{C}$, $\mathscr{C}(s_1, s_2)$ implies that s_1 is a substring of s_2.
- $\mathscr{C}^*$ is reflexive because every string is a substring of itself.
- 'Substring of' is a transitive relation. For example, suppose that the following relations hold: abc is a substring of xxabcyy and xxabcyy is a substring of aaxxabcyycc; then the transitive relation also holds: abc is a substring of aaxxabcyycc. ∎

Functions

Consider the relation $\mathscr{SQ} = \{(n_1, n_2) \mid n_2 = n_1^2\}$ on $\mathscr{N}^2$. It has the special property that for any n_1, there is a most one element n_2 such that $\mathscr{S}(n_1, n_2)$. In fact, there is exactly one such n_2 for each n_1.

Definition A.23 Let $\mathscr{F}$ be a relation on $S_1 \times \cdots \times S_n$. $\mathscr{F}$ is a *function* iff for every $n-1$-tuple $(x_1, \ldots, x_{n-1}) \in S_1 \times \cdots \times S_{n-1}$, there is at most one $x_n \in S_n$, such that $\mathscr{F}(x_1, \ldots, x_n)$. The notation $x_n = \mathscr{F}(x_1, \ldots, x_{n-1})$ is used.

- The *domain* of $\mathscr{F}$ is the set of all $(x_1, \ldots, x_{n-1}) \in S_1 \times \cdots \times S_{n-1}$ for which (exactly one) $x_n = \mathscr{F}(x_1, \ldots, x_{n-1})$ exists.

- The *range* of $\mathscr{F}$ is the set of all $x_n \in S_n$ such that $x_n = \mathscr{F}(x_1, \ldots, x_{n-1})$ for at least one $(x_1, \ldots, x_{n-1})$.
- $\mathscr{F}$ is *total* if the domain of $\mathscr{F}$ is (all of) $S_1 \times \cdots \times S_{n-1}$; otherwise, $\mathscr{F}$ is *partial*.
- $\mathscr{F}$ is *injective* or *one-to-one* iff $(x_1, \ldots, x_{n-1}) \neq (y_1, \ldots, y_{n-1})$ implies that

$$\mathscr{F}(x_1, \ldots, x_{n-1}) \neq \mathscr{F}(y_1, \ldots, y_{n-1}).$$

- $\mathscr{F}$ is *surjective* or *onto* iff its range is (all of) S_n.
- $\mathscr{F}$ is *bijective* (*one-to-one and onto*) iff it is injective and surjective. ∎

Example A.24 $\mathscr{S2} = \{(n_1, n_2) \mid n_2 = n_1^2\}$ is a total function on $\mathscr{N}^2$. Its domain is all of $\mathscr{N}$, but its range is only the subset of $\mathscr{N}$ consisting of all squares. Therefore $\mathscr{S}q$ is not surjective and thus not bijective. The function is injective, because given an element in its range, there is exactly one square root in $\mathscr{N}$, symbolically, $x \neq y \to x^2 \neq y^2$, or equivalently, $x^2 = y^2 \to x = y$. If the domain were taken to be $\mathscr{Z}$, the set of integers, the function would no longer be injective, because $n \neq -n$ but $(n)^2 = (-n)^2$. ∎

A.5 Cardinality

Definition A.25 The *cardinality* of a set is the number of elements in the set. The cardinality of a S is *finite* iff there is an integer n such that the number of elements in S is the same that the number of elements in the set $\{1, 2, \ldots, n\}$. Otherwise the cardinality is *infinite*. An infinite set S is *countable* if its cardinality is the same as the cardinality of $\mathscr{N}$. Otherwise the set is *uncountable*. ∎

To show that the cardinality of a set S is finite, we can *count* the elements. Formally, we define a bijective function from the finite set $\{1, \ldots, n\}$ to S. To show that an infinite set is countable, we do exactly the same thing, defining a bijective function from (all of) $\mathscr{N}$ to S. Clearly, we can't define the function by listing all of its elements, but we can give an expression for the function.

Example A.26 $\mathscr{E}$, the set of even natural numbers, is countable. Define $f(i) = 2i$ for each $i \in \mathscr{N}$:

$$0 \mapsto 0, \quad 1 \mapsto 2, \quad 2 \mapsto 4, \quad 3 \mapsto 6, \quad \ldots.$$

We leave it to the reader to show that f is bijective. ∎

We immediately see that non-finite arithmetic can be quite non-intuitive. The set of even natural numbers is a *proper* subset of the set of natural numbers, because, for example, $3 \in \mathscr{N}$ but $3 \notin \mathscr{E}$. However, the cardinality of $\mathscr{E}$ (the number of elements in $\mathscr{E}$) is the same as the cardinality of $\mathscr{N}$ (the number of elements in $\mathscr{N}$)! It takes just a bit of work to show that $\mathscr{Z}$, the set of integers, is countable, as is the set of rational numbers $\mathscr{Q}$.

Georg Cantor first proved the following theorem:

Theorem A.27 *The set of real numbers $\mathscr{R}$ is uncountable.*

Proof Suppose to the contrary that there is a bijective function $f : \mathscr{N} \mapsto \mathscr{R}$, so that it makes sense to talk about r_i, the ith real number. Each real number can be represented as an infinite decimal number:

$$r_i = d_i^1 d_i^2 d_i^3 d_i^4 d_i^5 \cdots .$$

Consider now the real number r defined by:

$$r = e_1 e_2 e_3 e_4 e_5 \cdots ,$$

where $e_i = (d_i^i + 1) \bmod 10$. That is, the first digit of r is different from the first digit of r_1, the second digit of r is different from the second digit of r_2, and so on. It follows that $r \neq r_i$ for all $i \in \mathscr{N}$, contradicting the assumption that f was surjective. ∎

This method of proof, called the *diagonalization argument* for obvious reasons, is frequently used in computer science to construct an entity that cannot be a member of a certain countable set.

Powersets

Definition A.28 The *powerset* of a set S, denoted 2^S, is the set of all subsets of S. ∎

Example A.29 Here is the powerset of the finite set $S = \{red, yellow, green\}$:

$$
\begin{aligned}
&\{ \\
&\{red, yellow, green\}, \\
&\{red, yellow\}, \{red, green\}, \{yellow, green\}, \\
&\{red\}, \{yellow\}, \{green\}, \\
&\emptyset \\
&\}.
\end{aligned}
$$

The cardinality of S is 3, while the cardinality of the powerset is $8 = 2^3$. ∎

This is true for any finite set:

Theorem A.30 *Let S be a finite set of cardinality n; then the cardinality of its powerset is 2^n.*

A.6 Proving Properties of Sets

To show that two sets are equal, use Theorem A.7 and show that each set is a subset of the other. To show that a set S is a subset of another set T, choose an *arbitrary* element $x \in S$ and show $x \in T$. This is also the way to prove a property $R(x)$ of a set S by showing that $S \subseteq \{x \mid R(x)\}$.

Example A.31 Let S be the set of prime numbers greater than 2. We prove that every element of S is odd. Let n be an *arbitrary* element of S. If n is greater than 2 and even, then $n = 2k$ for some $k > 1$. Therefore, n has two factors other than 1 and itself, so it cannot be a prime number. Since n was an arbitrary element of S, all elements of S are odd.   ∎

Induction

Let S be an arbitrary set, let $s = (s_0, s_1, s_2, \ldots)$ be a (finite or infinite) sequence of elements of $\mathscr{S}$ and let R be any unary relation on S, that is, $R \subseteq S$. Suppose that we want to prove that $s_i \in R$ for all $i \geq 0$. The can be done using the *rule of induction*, which is a two-step proof method:

- Prove that $s_0 \in R$; this is the *base case*.
- Assume $s_i \in R$ for an *arbitrary* element s_i, and prove $s_{i+1} \in R$. This is the *inductive step* and the assumption is the *inductive hypothesis*.

The rule of induction enables us to conclude that the set of elements appearing in the sequence s is a subset of R.

Example A.32 Let s be the sequence of non-zero even numbers in $\mathscr{N}$:

$$s = (2, 4, 6, 8, \ldots),$$

and let R be the subset of elements of $\mathscr{N}$ that are the sum of two odd numbers, that is, $r \in R$ if and only if there exist odd numbers r_1 and r_2 such that $r = r_1 + r_2$. We wish to prove that s, consider as a set of elements of $\mathscr{N}$, is a subset of R:

$$\{2, 4, 6, 8, \ldots\} \subseteq R.$$

Base case: The base case is trivial because $2 = 1 + 1$.
Inductive step: Let $2i$ be the ith non-zero even number. By the *inductive hypothesis*, $2i$ is the sum of two odd numbers $2i = (2j + 1) + (2k + 1)$. Consider now, $2(i + 1)$, the $i + 1$st element of S and compute as follows:

$$
\begin{aligned}
2(i + 1) &= 2i + 2 \\
&= (2j + 1) + (2k + 1) + 2 \\
&= (2j + 1) + (2k + 3) \\
&= (2j + 1) + (2(k + 1) + 1).
\end{aligned}
$$

The computation is just arithmetic except for the second line which uses the inductive hypothesis. We have shown that $2(i + 1)$ is the sum of two odd numbers $2j + 1$ and $2(k + 1) + 1$. Therefore, by the rule of induction, we can conclude that $\{2, 4, 6, 8, \ldots\} \subseteq R$. ■

The method of proof by induction can be generalized to any mathematical structure which can be ordered—larger structures constructed out of smaller structures. The two-step method is the same: Prove the base case for the smallest, indivisible structures, and then prove the induction step assuming the inductive hypothesis. We will use induction extensively in the form of *structural induction*. Since formulas are built out of subformulas, to prove that a property holds for all formulas, we show that it holds for the smallest, indivisible atomic formulas and then inductively show that is holds when more complicated formulas are constructed. Similarly, structural induction is used to prove properties of trees that are built out of subtrees and eventually leaves.

References

D.J. Velleman. *How to Prove It: A Structured Approach (Second Edition)*. Cambridge University Press, 2006.

Index of Symbols

M. Ben-Ari, *Mathematical Logic for Computer Science*,
DOI 10.1007/978-1-4471-4129-7, © Springer-Verlag London 2012

Name Index

A

Apt, K., 275, 278, 279, 293, 324

B

Baier, C., 99, 110, 261, 317, 324
Ben-Ari, M., 261, 272, 301, 305, 312, 324
Bratko, I., 221
Bryant, R., 99, 103, 110

C

Cantor, G., 334
Church, A., 223
Clarke, E.M., 324
Clocksin, W.F., 221

D

Davis, M., 128
de Boer, F.S., 275, 278, 279, 293, 324
Dijkstra, E.W., 284, 301
Dreben, B., 226, 227, 229

E

Even, S., 254

F

Fitting, M., 45, 92, 153, 182, 202
Floyd, R.W., 293

G

Gödel, K., 2, 228
Goldfarb, W., 226, 227, 229
Gopalakrishnan, G.L., 128
Gries, D., 293
Grumberg, O., 324

H

Heule, M., 128, 324
Hilbert, D., 2

Hoare, C.A.R., 275
Holzmann, G.J., 317, 324
Hopcroft, J.E., 14, 128, 224
Huth, M., 71

K

Katoen, J.-P., 99, 110, 261, 317, 324
Kramer, J., 324
Kripke, S.A., 232
Kröger, F., 261, 272

L

Lewis, H., 226, 229
Lloyd, J.W., 182, 194, 202, 212, 215, 221
Logemann, G., 128
Loveland, D., 128, 202
Łukasiewicz, J., 12

M

Magee, J., 324
Malik, S., 128
Manna, Z., 47, 223, 261, 272, 293, 324
Martelli, A., 190, 202
Mellish, C.S., 221
Mendelson, E., 6, 69, 71, 164, 165, 182, 226, 228, 229
Merz, S., 261, 272
Minsky, M., 223, 224
Monk, D., 69, 182, 229
Montanari, U., 190, 202
Motwani, R., 128

N

Nadel, B.A., 128
Nerode, A., 6, 45, 92, 153

O

Olderog, E.-R., 275, 278, 279, 293, 324

M. Ben-Ari, *Mathematical Logic for Computer Science*,
DOI 10.1007/978-1-4471-4129-7, © Springer-Verlag London 2012

Subject Index

1 7 AUG 2017

Lightning Source UK Ltd.
Milton Keynes UK
UKOW01f0609040817
306671UK00004B/100/P